Springer Texts in Business and Economics

More information about this series at http://www.springer.com/series/10099

Burkhard Heer

Public Economics

The Macroeconomic Perspective

Springer

Burkhard Heer
Department of Business and Economics
University of Augsburg
Augsburg, Germany

ISSN 2192-4333 ISSN 2192-4341 (electronic)
Springer Texts in Business and Economics
ISBN 978-3-030-00987-8 ISBN 978-3-030-00989-2 (eBook)
https://doi.org/10.1007/978-3-030-00989-2

Library of Congress Control Number: 2018962259

Dedication illustration "Give Way to Blue" by Pete Kilkenny

Cover illustration: Cover design to be commissioned direclty at estudio calamar

This Springer imprint is published by the registered company Springer Nature Switzerland AG
The registered company address is: Gewerbestrasse 11, 6330 Cham, Switzerland

To Barbara, Sarah, and Carla

(Illustration "Give Way to Blue" by Pete Kilkenny with kind permission by the artist)

Preface

The writing and the quality of a comprehensive graduate textbook on "Public Economies" such as the present one relies upon many discussions, recommendations, and comments from students, colleagues, and fellow researchers. This book grew out of first- and second-year graduate-level introductory courses to public economics at the Universities of Augsburg and Bolzano. Parts of this book have also been taught as PhD courses on pension and debt policies at the Universities of Luxembourg and Leipzig. I would like to thank many graduate students from Augsburg, Bolzano, Leipzig, and Luxembourg for their valuable input into this book and for their patience during my courses while I was testing the material on them.

I would also like to thank Andreas Irmen, Alfred Maußner, Ludger Linnemann, Vito Polito, Kerstin Roeder, Christian Scharrer, Bernd Süssmuth, and Mark Trede for helpful comments and suggestions. A significant amount of computer code that was originally composed by Alfred has also found its way into my code for this book. Without Alfred's input, the book would not have been possible. Of course, all remaining errors are mine.

Part of this book was written during my stays at Fordham University, New York, the University of Luxembourg, and the Federal Reserve Bank at St. Louis. I would like to thank Paul McNelis, Andreas Irmen, Christopher Waller, and Christian Zimmermann for their hospitality. Of course, the views expressed in this book are all mine and do not necessarily reflect the official positions of the Federal Reserve Bank of St. Louis, the Federal Reserve System, or the Board of Governors.

For their assistance in the preparation of the computer program download page, the statistics, and the illustrations, I would like to thank Sijmen Duineveld, Alexander Lerf, Stefan Rohrbacher, and Benjamin Weiß. For her proofreading, I am also grateful to Anja Erdl. I also thank Martina Bihn at Springer Publishing Company for editorial suggestions and corrections and for her support and help through the editorial process.

For inspiration, I would like to thank Pete Kilkenny. His fabulous paintings of cows helped me to envision resilience and stoicism, which were valuable inputs while writing this book.

Most emphatically, I would like to express my gratitude to my wife, Barbara, and my daughters, Sarah and Carla, for their patience and the harmonious environment they provided me.

Augsburg, Germany Burkhard Heer

Contents

List of Figures

List of Programs

Chapter 2

Chapter 3

Chapter 4

Chapter 5

Chapter 6

Chapter 7

1 Introduction

1.1 Motivation

Public economics studies the problems of government expenditures and revenues. Almost all major graduate textbooks on this issue focus on an exclusively microeconomic presentation. Modern problems in public economics, however, are inherently macroeconomic in nature. Just consider what are arguably the most important problems facing modern industrialized countries: demographics, debt, and pensions. As a consequence, many textbooks on public economics are awfully shy about these pressing issues that were aggravated by the recent financial crisis.

The present book is intended to fill this gap and adopts a macroeconomist's perspective. It considers the main issues facing modern governments: (1) taxation, (2) pensions, (3) debt, (4) stabilization policies, and the (5) demographic transition. To study these questions and to provide adequate answers to both researchers and politicians, I base my textbook on three fundamental principles:

1. I use micro-founded macroeconomic models.
2. Given the inherently dynamic nature of most problems in public economics, I predominantly apply the two standard intertemporal models, the Ramsey model and the overlapping generations model, in my analysis.
3. In addition to theoretical results, I often provide computational analysis to present an estimate of the quantitative effects.

1. My approach is deeply founded in microeconomics. Let me illustrate the reasoning in this book with the help of a standard problem in public economics: labor income taxation. The standard line of argument is to start by presenting the first theorem of welfare economics: The competitive equilibrium (in the absence of externalities and public goods) is Pareto-efficient. If a labor income tax is introduced, welfare losses arise. This is demonstrated in a partial equilibrium model of the labor market, where labor supply l^s is derived from the individual household's

B. Heer, *Public Economics*, Springer Texts in Business and Economics,
https://doi.org/10.1007/978-3-030-00989-2_1

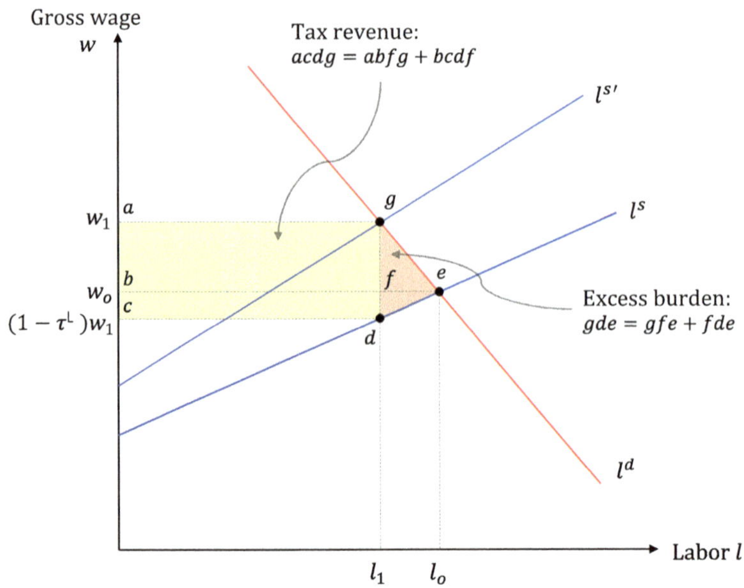

Fig. 1.1 Equilibrium in the labor market and the effect of a labor income tax

utility maximization, while labor demand l^d is derived from the firm's profit maximization. As a consequence, market equilibrium is given as the intersection of labor demand and supply, which is illustrated as point e in Fig. 1.1. Notice that both labor supply l^s and labor demand l^d are graphed as functions of the gross wage w (before taxes) and that labor income taxes are zero at point e.

A tax on wage income induces an upward shift in labor supply, as illustrated by the shift of the curve from l^s to $l^{s'}$. Taking everything else as given, the welfare effects can be evaluated with the help of the triangle deg, and a quantitative evaluation of this welfare effect can be estimated with the help of the empirical values of the labor demand and labor supply elasticities.

In most textbooks, the analysis stops here. However, for a macroeconomist, this is not the end. Because employment falls, households receive less income and reduce their demand. As goods demand falls, firms reduce labor demand, and the l^d curve also shifts inward (not illustrated). This *general equilibrium* effect is usually ignored in textbooks on public economics, even though it is quantitatively important.

2. Most of the problems in public economics are dynamic in nature, e.g., the accumulation of government debt or the financing of future pensions through present-day contributions. For this reason, a natural framework for the analysis of such public finance issues is provided by models of intertemporal household optimization, e.g., the Ramsey model and the overlapping generations (OLG) model.

The Ramsey model is useful for the analysis of most growth and business cycle problems. It considers a representative household with an infinite lifetime (or, alternatively, a finite lifetime with altruism toward the next generation). However, for some of the most pressing modern problems in government finance, the Ramsey model is not the appropriate framework. Consider two examples: (1) pension reforms and (2) government debt. (1) Due to the aging of the population, the current 'pay-as-you-go' systems in most industrialized countries are no longer affordable because a decreasing share of young people has to finance the pensions of an increasing share of old people. Therefore, either contributions to the pension system have to be raised, or pension benefits have to be cut.[1] As a consequence, income is redistributed from young to old agents or vice versa. (2) In the standard Ramsey model,[2] Ricardian equivalence holds, i.e., the financing of government expenditures does not affect the equilibrium value of savings or income. The empirical evidence, however, does not support the hypothesis of Ricardian equivalence. The redistribution of income between generations and the failure of Ricardian equivalence cannot be analyzed in the standard Ramsey model, and we will introduce the OLG model as an alternative framework that is able to model this redistribution among generations.

There are many excellent expositions of the Ramsey model and the OLG model.[3] In this book, I attempt to avoid restating what these other textbooks have done and instead focus on the features of the models that are important for the subsequent material in the book. Therefore, I leave out some unnecessary technical details; instead, I extend the usual textbook exposition to incorporate numerical evaluations of the models. In addition, I have to choose whether to consider the two models in discrete or continuous time (or both). I choose discrete time. It has the advantage that the periodic values of, for example, income, production, and government expenditures are readily available in the data. In addition, stochastic processes, e.g., for total factor productivity, are often easier to handle in discrete time than in continuous time.[4]

3. For some problems, exact theoretical results can be derived. For example, I will present the famous Chamley-Judd result whereby the optimal capital income tax rate and/or the optimal wealth tax rate is zero in steady state. However, this result per se does not help the politician who is interested in the quantitative gains and losses of a policy measure, e.g., an increase in the wealth tax. Therefore, we will use computational methods to obtain an estimate of what the quantitative effect of

---

[1] In addition, the retirement age can be raised or the social security system can be subsidized by the government with the help of tax revenues.

[2] In heterogeneous-agent extensions of the Ramsey model, in which households face credit market restrictions such as in Aiyagari (1994), Ricardian equivalence fails.

[3] For the Ramsey model, see, for example, Ljungqvist and Sargent (2012). The OLG model is studied extensively in Blanchard and Fischer (1989) and de la Croix and Michel (2002).

[4] Of course, there are also disadvantages of discrete-time models. Often, one has to make artificial assumptions about the timing of events, e.g., whether consumption takes place at the beginning or the end of the period, or if many events occur at the same point in time, in what order they take place, e.g., whether a shock is observed prior to or after the labor supply decision. In many cases, the results will be sensitive to the specification of the timing, and we will highlight this at the relevant points.

Table 1.1 Modeling choices in top journals, 2017

Model type	AER (%)	RED (%)	JME (%)	Average (%)
General equilibrium	42	48	51	47
Ramsey model	35	46	51	45
OLG model	8	30	2	15
Discrete time	35	65	49	52
Computational model	38	57	51	50

Notes: The entries in this table report the prevalence of different model types in macroeconomic studies in the *American Economic Review*, *Review of Economic Dynamics*, and *Journal of Monetary Economics* in the year 2017. The average is computed relative to the number of all macroeconomic studies in these journals in 2017

a one-percentage-point increase in a wealth tax would be. In addition, the result only holds for the steady state. However, at present, we can hardly argue that economies such as Italy or Greece are in steady state. Accordingly, the result is of little help to the Italian or Greek Minister of Finance. Let us assume that the Italian Prime Minister would like to raise the wealth tax by five percentage points to pay off Italian government debt. What are the consequences? In such cases, we need to use computational methods to approximate the general equilibrium effects on aggregates such as consumption, income, wealth, and welfare.

Table 1.1 reports the use of the different model types in three prominent journals in 2017. I simply identified the articles on macroeconomic questions (totaling 117) in the *American Economic Review* (AER), the *Review of Economic Dynamics* (RED), and the *Journal of Monetary Economics* (JME) and counted the number of studies that accord with the approach in this book summarized by the points (1)–(3) above. The sample gives ample support to my approach. Among the articles on macroeconomic problems in these three top journals, 47% used a general equilibrium rather than just a partial equilibrium analysis.[5] The Ramsey model is the dominant model type with a share of 45% in the macroeconomic articles in these journals, even though the overlapping generations (OLG) framework is important, too. The models are usually formalized in discrete time for the convenience in their analysis, in particular in the presence of uncertainty. And, finally, the sample seems to support the hypothesis that a large fraction of the macroeconomic models in modern macroeconomic theory is evaluated quantitatively and employs computational methods, with an average share amounting to 50% in the sample.

1.2 Organization of the Book

The book is divided into three parts. In the first part, Chaps. 2 and 3, the two most prominent models of dynamic macroeconomic analysis are presented, the Ramsey model and the OLG model. While the Ramsey model studies a representative agent,

[5]Most of the studies which were not based on a general-equilibrium model were empirical ones.

the OLG model considers individuals who differ in age and wealth. In the latter model, Ricardian equivalence fails, and the study of debt problems in this model allows for more realistic and meaningful conclusions.

In the second part, we consider fiscal policy. In Chap. 4, I present the effects of government consumption on output and employment and compute government consumption multipliers. The role of the substitutability of private and public consumption is a key element in this chapter. In addition, we analyze countercyclical government spending policy that helps to stabilize output fluctuations over business cycles. In Chap. 5, we examine income taxation, including the study of its output and employment effects, results from the theory of optimal taxation, and the growth effects of income taxes. Given the tight fiscal budgets and high debt in many industrialized countries, we also examine empirical Laffer curves, which we estimate and compute for the US economy.

The third part considers issues of social security, demographics, and debt. Chapter 6 describes the pay-as-you-go pension system that prevails in OECD countries. The effect of the demographic transition on modern pension systems is analyzed, and optimal pensions are demonstrated to be much smaller than the present ones observed empirically. We also contrast the pay-as-you-go system with a fully funded system, in which the government invests the contributions to the pension system in the capital market. We will also take a look at the sustainability of public finances in modern industrialized countries and point out that, in some countries, a debt crisis is imminent over the next 10–20 years if no drastic reforms of public finances and pensions are seized. In Chap. 7, we consider public debt in detail. First, we will analyze Ricardian equivalence and the causes for its failure. The financing of government expenditures through either taxes or debt affects output, investment, and consumption. We derive the quantitative effects of debt financing on macroeconomic variables and study the role of government debt to alleviate the consequences of the demographic transition on the welfare of present and future generations over the next 50–100 years. For the US economy, we derive a threshold of the public debt level and study the role of debt financing to ease the transition to a more sustainable pension policy for the present generations. Our welfare analyses in Chaps. 6 and 7 will emphasize that the majority of the present voters would oppose policies that help to improve the sustainability of public finances.

The book is self-contained, and the student should be able to follow its analysis without any additional material or prior knowledge of public economics or dynamic macroeconomic theory. The only prerequisite for understanding the material is a solid background in mathematical methods, including analysis and linear algebra. In addition, if the student would like to use the computer code and adapt it for her/his purposes, knowledge of one of the two computer languages Gauss or MATLAB is needed. If the student would like to learn more about computational methods, the book and Matlab/Gauss code provided by Miranda and Fackler (2002) and Heer and Maußner (2009) are useful.[6] Gauss and MATLAB computer code as well as teaching material (slides) are available as downloads from the author's homepage 'http://www.wiwi.uni-augsburg.de/vwl/heer/pubec_buch/'.

The book is aimed at graduate students or advanced undergraduate students. It may be used for both in-class and self study. The material in the book, however, cannot easily be covered in one semester, but one can conveniently choose parts of it as a one-semester course. For example, a course on fiscal policy could be based on Chaps. 2, 4, and 5, while a course on social security and debt could be adapted from Chaps. 3, 6, and 7.

References

Aiyagari, R. S. (1994). Uninsured idiosyncratic risk and aggregate saving. *Quarterly Journal of Economics, 109*, 659–684.

Blanchard, O. J., & Fischer, S. (1989). *Lectures on macroeconomics*. Cambridge: MIT Press.

de la Croix, D., & Michel, P. (2002). *A theory of economic growth: Dynamics and policy in overlapping generations*. Cambridge: Cambridge University Press.

Fehr, H., & Kindermann, F. (2018). *Introduction to computational economics using Fortran*. Oxford: Oxford University Press.

Heer, B., & Maußner, A. (2009). *Dynamic general equilibrium modeling: Computational methods and applications* (2nd ed.). Heidelberg; Springer.

Ljungqvist, L., & Sargent, T. J. (2012). *Recursive macroeconomic theory*. Cambridge: MIT Press.

Miranda, M. J., & Fackler, P. L. (2002). *Applied computational economics and finance*. Cambridge: MIT Press.

[6]Fehr and Kindermann (2018) provide an introduction to computational economics using Fortran with an extensive treatment of OLG models and public policy.

Part I

Useful Models

2 Ramsey Model

2.1 Introduction

This chapter presents the Ramsey model. It is the benchmark model for most dynamic macroeconomic models that study growth and business cycle phenomena. We first study the deterministic Ramsey model in which the total factor productivity is certain. We contrast the effects of a once-and-for-all change with those of a temporary change in productivity on investment, output, and labor supply. In addition, we distinguish the effects of this change when it is known in advance or only observed at the beginning of the period, t, when the shock occurs. Finally, we also introduce uncertainty with respect to the technology level and discuss the real business cycle (RBC) model.

2.2 Central Planner

In the following, we consider the Ramsey model, which was initially formulated by Ramsey (1928) and later advanced by Cass (1965) and Koopmans (1965). The model is also often referred to as the Ramsey-Cass-Koopmanns model. The Ramsey model is the most fundamental neoclassical model of economic growth and dynamic macroeconomics. The original question studied by Ramsey (1928) was how much an economy should save over an infinite time horizon to maximize the lifetime utility of its agents. We will first analyze the case of inelastic labor that was studied by Ramsey (1928) before we endogenize the labor supply decision.

2.2.1 Inelastic Labor Supply

Let us assume that a central planner owns the means of production and is benevolent, meaning that he wants to maximize lifetime household utility. We also assume that

B. Heer, *Public Economics*, Springer Texts in Business and Economics,
https://doi.org/10.1007/978-3-030-00989-2_2

a household's lifetime is infinite.[1] Let the intertemporal utility function in period $t = 0$ be given by a time-separable function:

$$U = \sum_{t=0}^{\infty} \beta^t u(c_t), \quad 0 < \beta < 1. \tag{2.1}$$

Instantaneous utility is a function of per capita consumption c_t only and is discounted by the discount factor β.[2] The discount factor is below one, $\beta < 1$, meaning that lifetime utility is finite. We further assume that $u(.)$ is concave, $u' > 0$, $u'' < 0$. For expositional purposes, we will use the following constant elasticity of substitution (CES) utility function[3]:

$$u(c) = \begin{cases} \frac{c^{1-\sigma}-1}{1-\sigma} & \sigma \neq 1, \\ \ln c & \sigma = 1, \end{cases} \tag{2.2}$$

where $1/\sigma$ denotes the intertemporal elasticity of substitution (IES) of consumption.[4] Furthermore, the household inelastically supplies one unit of labor.

The number of households is equal to N_t and grows at the constant rate n:

$$N_t = (1+n)N_{t-1}. \tag{2.3}$$

[1] One way to justify this assumption is that a household with a finite lifetime also cares about the utility of its descendants and applies the same discount factor β to their (representative) lifetime utility.

[2] Take care to distinguish between the discount factor β and the discount rate $\theta > 0$ that is given by

$$\frac{1}{1+\theta} = \beta \quad \Leftrightarrow \quad \theta = \frac{1}{\beta} - 1.$$

[3] Why have we added '−1' in the nominator of the utility function in (2.2) in the case $\sigma \neq 1$? First notice that the additive constant $-1/(1-\sigma)$ does not change the solution of the utility maximization problem and, therefore, does not affect optimal consumption. Furthermore, we know from calculus that

$$\lim_{x \to 0} \left(\frac{a^x - 1}{x} \right) = \ln a.$$

Therefore, $\ln c$ is just the limit of the function $(c^{1-\sigma} - 1)/(1-\sigma)$ for $\sigma \to 1$.

In order to derive the limit formula above, notice that from the *L'Hôspital rule*—which states that if the functions $f(x)$ and $g(x)$ in the nominator and denominator have the limit equal to zero, $\lim_{x\to\infty} f(x) = 0$ and $\lim_{x\to\infty} g(x) = 0$, the value of the limit $\lim_{x\to 0} (f(x)/g(x))$, if it exists, is given by $\lim_{x\to 0} \left(f'(x)/g'(x)\right)$—implies

$$\lim_{x \to 0} \left(\frac{a^x - 1}{x} \right) = \lim_{x \to 0} \left(\frac{a^x \ln a}{1} \right) = \ln a.$$

[4] Appendix 2.1 derives the IES in a simplified two-period model.

The central planner uses labor N_t and capital K_t for production:

$$Y_t = Z_t F(K_t, N_t), \tag{2.4}$$

where Z_t denotes the level of technology. We assume that the production technology is characterized by constant returns to scale, meaning that per capita production, $y = Y/N$, is a function of the capital intensity, $k = K/N$,

$$y_t \equiv \frac{Y_t}{N_t} = Z_t f(k_t) \equiv Z_t F(K_t/N_t, 1). \tag{2.5}$$

In addition, we assume a constant elasticity of substitution, $\sigma_p = 1/(1-\rho)$, in production[5]:

$$Y_t = Z_t \left[\alpha K_t^\rho + (1-\alpha) N_t^\rho\right]^{\frac{1}{\rho}}. \tag{2.6}$$

In the following, we set $\sigma_p = 1$, meaning that (2.6) reduces to the well-known Cobb-Douglas production function[6]:

$$Y_t = Z_t K_t^\alpha N_t^{1-\alpha}. \tag{2.7}$$

The central planner owns the capital stock K_t and saves $S_t = Y_t - C_t$ for the next period. In equilibrium, his savings are equal to his investment, $I_t = S_t$, and capital accumulates according to

$$K_{t+1} = (1-\delta)K_t + I_t = (1-\delta)K_t + Y_t - C_t. \tag{2.8}$$

Capital depreciates at the rate δ. Using the definitions $c_t \equiv C_t/N_t$ and $k_t \equiv K_t/N_t$, the resource constraint in per capita terms is represented by (after dividing (2.8) by N_t and noticing that $K_{t+1}/N_t = (1+n)k_{t+1}$)

$$(1+n)k_{t+1} = (1-\delta)k_t + Z_t f(k_t) - c_t. \tag{2.9}$$

---

[5]The elasticity of substitution σ_p is defined as follows:

$$\sigma_p = \frac{\frac{d\left(\frac{K}{L}\right)}{\frac{K}{L}}}{\frac{d\left(\frac{w}{r}\right)}{\frac{w}{r}}},$$

where w and r denote the marginal products of labor and capital.

[6]You are asked to compute the dynamics for the case in which $\sigma_p = 3/4$ in Problem 2.1.

The first-order conditions of the central planner's optimization problem follow from the derivation of the Lagrangian

$$\mathscr{L} = \sum_{t=0}^{\infty} \beta^t \left[\frac{c_t^{1-\sigma} - 1}{1-\sigma} + \lambda_t \left(Z_t f(k_t) + (1-\delta)k_t - c_t - (1+n)k_{t+1} \right) \right] \tag{2.10}$$

with respect to c_t and k_{t+1}. In particular, the first-order conditions are given by:

$$\lambda_t = c_t^{-\sigma}, \tag{2.11a}$$

$$(1+n)\lambda_t = \lambda_{t+1}\beta \left[1 + Z_{t+1} f'(k_{t+1}) - \delta \right]. \tag{2.11b}$$

The planner chooses his savings such that the marginal utility from consumption today λ_t is equal to the discounted marginal utility of consumption next period $\beta\lambda_{t+1}$ times the return from investing one unit of the consumption good $1 + Z_{t+1} f'(k_{t+1}) - \delta$.

The two first-order conditions can be combined to yield the so-called Euler condition[7]:

$$\left(\frac{c_{t+1}}{c_t} \right)^{\sigma} = \frac{\beta}{1+n} \left[1 + Z_{t+1} f'(k_{t+1}) - \delta \right]. \tag{2.12}$$

Accordingly, per capita consumption c_t grows (falls) if the marginal product of capital less depreciation $Z_{t+1} f'(k_{t+1}) - \delta$ is above (below) the rate $\hat{r} = (1+n)/\beta - 1$.

In dynamic models, it is often convenient to analyze the so-called steady state, which, in the present model, is the long-run equilibrium of the economy when the technology level is constant, $Z_t \equiv 1$. The steady state is a rest point of a dynamic system. Its analysis is crucial for the study of economic dynamics. In public economics, for example, we are interested in the question of how a fiscal policy, such as a tax increase, affects the equilibrium values of output and employment in the long run. Therefore, we compare the steady states under the old and the new tax rate. In addition, we are also interested in the dynamics during the transition to the new steady state. Therefore, the new steady state needs to be locally stable.[8] Two possible ways to analyze the dynamics of economic models are as follows: (1) We locally approximate the dynamics around the steady state. This procedure is often applied in business cycle theory, in which the economy is assumed to fluctuate

[7] More generally, an *Euler equation* is the intertemporal first-order condition for a dynamic choice problem and is usually formulated as a difference of differential equation. Equation (2.12) is also referred to as the *Keynes-Ramsey rule* that describes the growth rate of consumption as a result of intertemporal utility maximization.

[8] By local stability we mean that if we perturb the initial condition slightly, then the system stays in the neighborhood of that steady state. If we use the term global stability, the system returns to the steady state even if the starting point is not very close to the steady state.

around an equilibrium growth path. (2) We consider the transition dynamics after a policy change and attempt to solve a boundary value problem of a system of difference equations, where the boundary values might be presented by the values of the state variable, such as the capital stock, in the initial and final periods. For the value in the final period, we use the new steady-state value of the state variable.

Let us define a steady state as an equilibrium in which the per capita economic variables are constant, i.e., $c_t = c_{t+1} = c$, $k_{t+1} = k_t = k$, and $Z_t = Z_{t+1} = Z$. Without loss of generality, we can set $Z \equiv 1$. The Euler equation in steady state takes the following form:

$$\frac{1+n}{\beta} - 1 + \delta = f'(k).$$

In the case of a Cobb-Douglas production function, $y = f(k) = k^{\alpha}$, the above equation implies the steady-state capital intensity

$$k = \left(\frac{\alpha}{\frac{1+n}{\beta} - 1 + \delta} \right)^{\frac{1}{1-\alpha}}.$$

To illustrate the dynamics in the Ramsey model, we will choose specific values for the parameters α, β, δ, n, and σ. This process is also know as *calibration*. In essence, there are three ways to fix the parameter values of a quantitative models. (1) One can use time-series studies to compute historical averages of the variables. Often, empirical studies from the literature that use econometric or time series methods are available. (2) One can estimate single equations of the model. For example, Attanasio and Low (2004) use (nonlinear) generalized methods of moments (GMM) to estimate the Euler equation of the Ramsey model and to yield consistent estimates of the preference parameters. (3) One can search for the (unobservable) parameter values that optimize the behavior of the model with respect to certain statistics.[9]

In our calibration, we choose a period length of one year and set $\alpha = 0.36$ and $\beta = 0.96$. As we will see subsequently, this implies a capital (labor) income share of 36% (64%) and a real annual interest rate equal to 4%. In addition, capital depreciates at $\delta = 8\%$ annually, and population is constant, $n = 0$. Finally, we set the IES to one half, $1/\sigma = 1/2$, in accordance with empirical evidence.[10] Therefore, in steady state, $k = 5.447$, $y = k^{\alpha} = 1.841$, and $c = y - (\delta + n)k = 1.405$.

To study the dynamics following a shock, we assume that the economy is in steady state in period $t = 0$ and that the technology level $Z_t = 1.0$ is constant during period $t = 0, 1, \ldots, 9$. In period $t = 10$, the technology level increases from

[9]A recommendable introduction to the methods of calibration is provided by deJong and Dave (2011).

[10]Mehra and Prescott (1985), Auerbach and Kotlikoff (1987), and Prescott (1986) review empirical studies which suggest a range for $\sigma \in [1, 2]$. Most business cycle studies either use $\sigma = 1.0$ or $\sigma = 2.0$.

1.0 to 1.1 for three periods, $Z_{10} = Z_{11} = Z_{12} = 1.1$, and declines to $Z_t = 1.0$ for $t > 12$ thereafter:

$$Z_t = \begin{cases} 1.1 & t = 10, 11, 12 \\ 1.0 & \text{else.} \end{cases} \tag{2.13}$$

We will distinguish two different scenarios. In the first scenario, the central planner only finds out about this increase in period $t = 10$ when the shock occurs. In the second scenario, the central planner learns about the three-period increase in period $t = 0$.

To compute the model, we need to assume that the economy is in steady state prior to period $t = 0$ and that it converges to the steady state in finite time after the shocks in periods $t = 10, 11, 12$. We assume that the economy is again in steady state in period $t = 101$.[11] Let us consider, in turn, the two scenarios of an expected and an unexpected change. In the first case, we have to solve a non-linear system of equations in 100 variables, $k_t, t = 1, \ldots, 100$ with $k_0 = k_{101} = k$[12]:

$$\left(\frac{Z_t k_t^\alpha + (1-\delta)k_t - (1+n)k_{t+1}}{Z_{t-1}k_{t-1}^\alpha + (1-\delta)k_{t-1} - (1+n)k_t} \right)^\sigma = \frac{\beta}{1+n} \left[1 + \alpha Z_t k_t^{\alpha-1} - \delta \right]. \tag{2.14}$$

Equation (2.14) is obtained after inserting (2.9) into (2.12). In the first scenario, we solve (2.14) for $k_t, t = 11, \ldots, 100$, assuming that $k_1 = k_2 = \ldots = k_{10} = k_{101} = k$. In the second scenario, we assume that the initial and final per capita capital stock k_0 and k_{101} are equal to the steady-state value k.

The dynamics for the capital stock in the expected (unexpected) case are graphed by the solid (broken) line in Fig. 2.1. Let us first consider the case of an unexpected shock. Prior to period $t = 10$, the economy is in steady state. In period $t = 10$, the central planner learns about the higher productivity during the next three periods $t = 10, 11, 12$. The production factors K_t and N_t in period $t = 10$ do not change because the capital stock is accumulated during the last period, $k_{10} = [Z_9 k_9^\alpha + (1-\delta)k_9 - c_9]/(1+n) = [k^\alpha + (1-\delta)k - c]/(1+n)$, where the variables c_t and k_t in period $t = 9$ are still equal to their respective steady-state values. Furthermore, labor supply is exogenous. As a consequence, output per capita increases by 10% in period $t = 10$, as illustrated in Fig. 2.2. The higher output is used for both consumption and

[11] Of course, we should check whether this number of periods is sufficient to guarantee a smooth approximation of the new steady state. If not, we should increase the number of periods. For example, I first used 40 periods and found the number to be insufficient. Use the computer code and test for different values of the number of periods.

[12] Appendix 2.2 provides an overview of how this numerical problem can be solved. The MATLAB/Gauss programs `Ch2_ramsey1.m`/`Ch2_ramsey1.g` compute the solution presented in Figs. 2.1, 2.2, 2.3, and 2.4 and can be downloaded from my homepage with all the other programs used in this book.

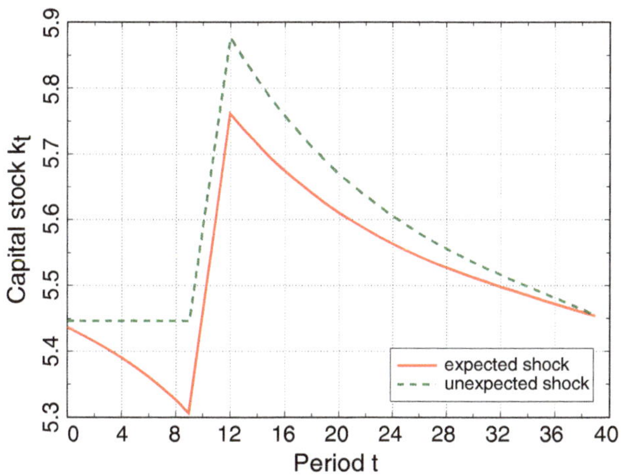

Fig. 2.1 Dynamics in the Ramsey model: capital stock k_t

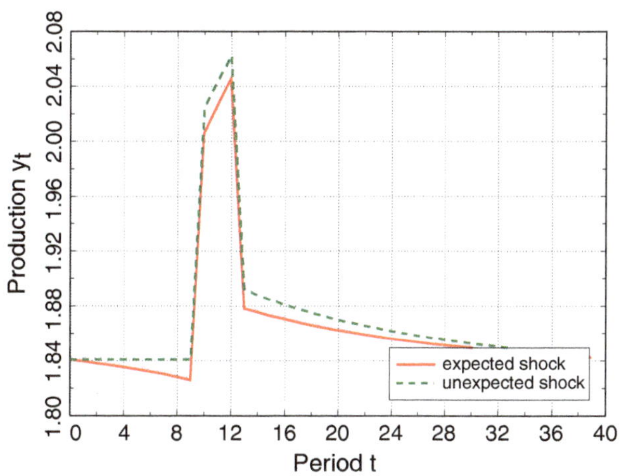

Fig. 2.2 Dynamics in the Ramsey model: production y_t

investment, and the capital stock attains its maximum at the beginning of period 12. As a consequence, output actually increases to a level that exceeds the steady-state value by more than 10% during periods $t = 11$ and $t = 12$. Subsequently, output falls because the technology level Z_t falls back to $Z_t = 1$ but remains above the steady-state level because of the higher capital stock k_t.

The effect on consumption is illustrated in Fig. 2.3. Because the change in the technology level is unexpected, consumption c_t (the broken line) is in steady state until period $t = 9$. In period $t = 10$, the central planner increases consumption. Since it is optimal to smooth consumption over time, he tries to spread out consumption over $t = 10, 11, \ldots$. This is a direct consequence of the fact that

Fig. 2.3 Dynamics in the Ramsey model: consumption c_t

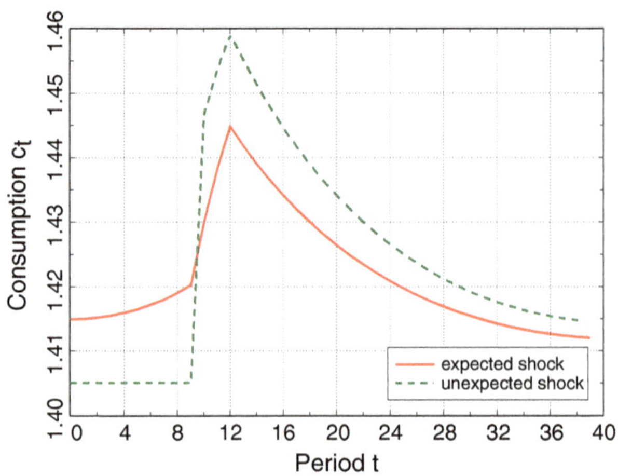

marginal utility is declining with higher consumption. Notice, in particular, that consumption first increases until period $t = 12$ and then decreases. This behavior can be explained with the help of the Euler equation (2.12). During periods $t = 10, 11$, the right-hand side (RHS) of the equation is above 1, as the marginal product of capital, $Z_{t+1}f'(k_{t+1}) = \alpha Z_{t+1}k_{t+1}^{\alpha-1}$, rises due to the increase in the technology level Z_{t+1}. In period $t = 12$, however, the RHS of (2.12) falls below 1, meaning that the growth rate of consumption $c_{t+1}/c_t - 1$ falls below zero.

In the case of an expected change, the central planner adjusts his behavior in period $t = 0$. Therefore, he tries to spread consumption more evenly over the next periods $t = 0, \ldots, 40$ and increases c_t in period $t = 0$, as depicted by the solid line in Fig. 2.3. As a consequence, investment declines, and the capital stock is reduced until period $t = 10$. Therefore, output falls until period $t = 9$ prior to the shock. In case of the expected shock, lifetime utility (2.1) increases to a larger extent than in the case of the unexpected shock.[13]

Notice another interesting observation from these two different cases. In the case of the unexpected change, consumption, output, and the capital stock are synchronized, and the correlations are very close to one. This observation also holds in so-called RBC models, in which business cycles are driven by unexpected shocks to the technology level Z_t. In the case of an expected shock, however, the strong co-movement of output and consumption is broken. Consumption c_t increases above steady-state levels during $t = 0, \ldots, 9$, while output y_t falls. Consequently, output and consumption are less perfectly correlated over time and are in better accordance with empirical observations. A recent strand of literature that reproduces

[13]The argument for this result is straightforward: The central planner could also choose to behave exactly the same in the case of an expected change as in the case of an unexpected change. Since he chooses a different policy, this must be superior, and it yields a higher value of the objective function.

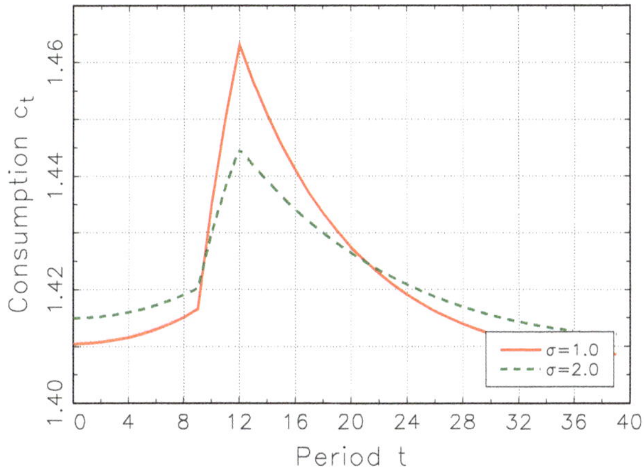

Fig. 2.4 Intertemporal consumption smoothing and the IES, $1/\sigma$

this observation considers so-called *news-driven* cycles.[14] In these models, a shock occurs in period t that signals a change $\tilde{n}$ periods ahead in a variable such as the technology level $Z_{t+\tilde{n}}$ (or other variables, for example, monetary policy).

Figure 2.4 illustrates the effect of the IES $1/\sigma \in \{1, 1/2\}$ on the household's consumption behavior (for the case of an expected shock). For a higher IES, $1/\sigma = 1$, the household decreases its consumption smoothing over time. This can already be seen from the Euler condition (2.12):

$$\left(\frac{c_{t+1}}{c_t}\right) = \left(\frac{\beta}{1+n}\left[1 + Z_{t+1} f'(k_{t+1}) - \delta\right]\right)^{\frac{1}{\sigma}}.$$

An increase in the IES from $1/\sigma = 1/2$ to $1/\sigma = 1$ increases the response of the consumption growth rate to an increase in the marginal product of capital $Z_{t+1} f'(k_{t+1})$. Accordingly, consumption declines more rapidly to the steady state after period $t = 12$. The household is more willing to intertemporally substitute consumption. Since the rise in consumption is spread out over a shorter time horizon, the household can increase consumption in the initial periods, $t = 11, 12, 13$, to a larger extent than in the case of a lower IES with $1/\sigma = 1/2$.

2.2.2 Stability Analysis and Saddle Path

In the numerical computation of the dynamics above, we assumed that the capital stock asymptotically converges to its steady-state value, $\lim_{t\to\infty} k_t = k$. In the

[14] Beaudry and Portier (2004, 2006) are two prominent articles in this literature.

following, I demonstrate the local stability of the steady state for the economy that is described by (2.14).

Formally, (2.14) is a second-order difference equation in k_t. We conveniently reformulate it into a difference equation of the first order in the variable (k_t, x_t), where we choose $x_{t+1} = k_t$:

$$\begin{pmatrix} k_{t+1} \\ x_{t+1} \end{pmatrix} = g(k_t, x_t) = \begin{pmatrix} g^1(k_t, x_t) \\ g^2(k_t, x_t) \end{pmatrix}$$
$$= \begin{pmatrix} \frac{Z_t k_t^\alpha + (1-\delta)k_t - \left(\frac{\beta}{1+n}\left[1+\alpha Z_t k_t^{\alpha-1} - \delta\right]\right)^{\frac{1}{\sigma}} \left[Z_{t-1} x_t^\alpha + (1-\delta)x_t - (1+n)k_t\right]}{1+n} \\ k_t \end{pmatrix}. \tag{2.15}$$

The local stability of the difference equation $g(k_t, x_t)$ at the steady state $k_t = x_t = k$ is determined by the absolute values of the eigenvalues of the Jacobian matrix, again evaluated at the steady state[15]:

$$J(k_t, x_t) = \begin{pmatrix} \frac{\partial g^1(k_t, x_t)}{\partial k_t} & \frac{\partial g^1(k_t, x_t)}{\partial x_t} \\ \frac{\partial g^2(k_t, x_t)}{\partial k_t} & \frac{\partial g^2(k_t, x_t)}{\partial x_t} \end{pmatrix}. \tag{2.16}$$

Evaluated at the steady state, the numerical value of the Jacobian is equal to

$$J(k, k) = \begin{pmatrix} 2.051 & -1.042 \\ 1.000 & 0 \end{pmatrix} \tag{2.17}$$

with the eigenvalues $\rho_1 = 0.924$ and $\rho_2 = 1.125$.[16] Accordingly, the system is saddle-path stable.[17]

We will use an alternative presentation of the dynamics that is more easily amenable to a graphical illustration. Therefore, we rewrite the second-order system of difference equations (2.15) in the variables (k_t, k_{t-1}) in a first-order system of

[15]You are asked to compute the Jacobian and its value in Problem 2.2.

[16]If you take the eigenvalues of the Jacobian provided in (2.17), the eigenvalues are slightly different due to rounding errors. I used the value of the Jacobian with an accuracy of 10^{-8} to compute the eigenvalues ρ_1 and ρ_2.

[17]In a two-dimensional difference equation system, the steady state is a saddle if one of the eigenvalues has an absolute value below one and the other above. The steady state is locally saddle-path stable if one of the two variables is predetermined and the other is a jump variable (not predetermined). (In addition, divergent paths must be ruled out by boundary conditions.) To learn more about the stability analysis in systems of difference equations, consult Azariadis (1993).

difference equations in the variables k_t and c_t. Straightforward rearranging of the terms in (2.15) results in the following system of difference equations:

$$\begin{pmatrix} k_{t+1} \\ c_{t+1} \end{pmatrix} = h(k_t, c_t) = \begin{pmatrix} h^1(k_t, c_t) \\ h^2(k_t, c_t) \end{pmatrix}$$

$$= \begin{pmatrix} \frac{1}{1+n}\left[Z_t k_t^\alpha + (1-\delta)k_t - c_t\right] \\ c_t \left(\frac{\beta}{1+n} \left[1 + \alpha \left(\frac{Z_t k_t^\alpha + (1-\delta)k_t - c_t}{1+n} \right)^{\alpha-1} - \delta \right] \right)^{\frac{1}{\sigma}} \end{pmatrix}. \tag{2.18}$$

Again, it can be shown that the eigenvalues of this system of difference equations are equal to $\rho_1 = 0.924$ and $\rho_2 = 1.125$ such that the system is saddle-path stable. An informal argument of this stability criterion will be presented in the following.

To characterize the stability properties of the system of difference equations, we distinguish between, on the one hand, the endogenous state or sluggish variables that are predetermined and adjust only slowly and, on the other hand, jump variables that are not predetermined. In our model, the capital stock k_t is a sluggish variable and consumption c_t is a jump variable. For a given initial value of the sluggish variable in period $t = 0$, k_0, consumption jumps to the value c_0 on the stable saddle path that provides the optimal solution.

To illustrate the dynamics and the concept of a saddle path, consider the phase diagram depicted in Fig. 2.5. The steady state is represented by point A, with $(k_t, c_t) = (k, c)$. We assume that the technology level is constant, $Z_t \equiv 1$. Two curves are drawn in the figure. At the points of the vertical line denoted by

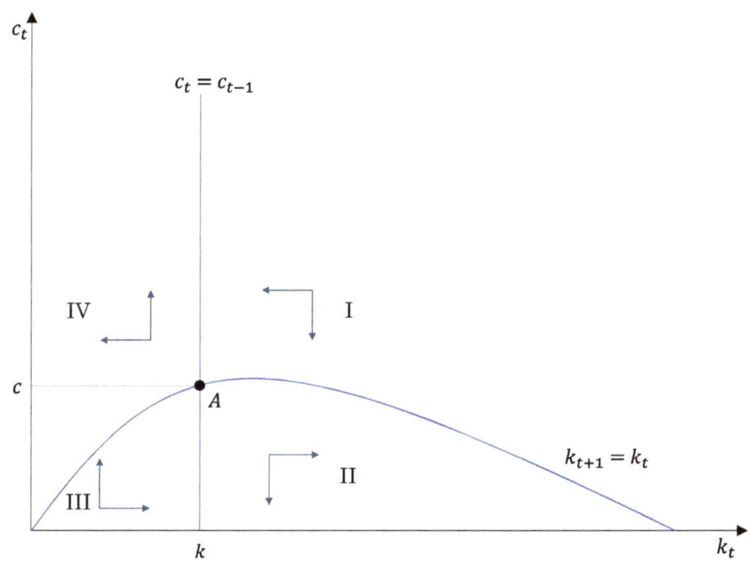

Fig. 2.5 Phase diagram of the Ramsey model

"$c_t = c_{t-1}$," consumption is constant, $c_t = c_{t-1}$.[18] From the second difference equation in (2.18) we notice that this is the case if $k_t = k$:

$$\left(\frac{c_t}{c_{t-1}}\right)^{\sigma} = \frac{\beta}{1+n}\left[1 + \alpha k_t^{\alpha-1} - \delta\right].$$

On the left-hand side (LHS) of this curve (in segments III and IV), the capital stock is below the steady-state value k, meaning that the marginal product of capital (minus depreciation) increases above the rate $(1+n)/\beta - 1$. Hence, the RHS of the above equation increases above one, and consumption must grow over time. This is indicated by the vertical arrows on the LHS of the curve. On the RHS of this curve (in segments I and II), consumption must decline over time.

The second curve, denoted by "$k_t = k_{t+1}$," describes the equilibrium values of k_t and c_t for which the capital stock is constant, $k_{t+1} = k_t$. If we solve the first difference equation in (2.18) using the condition $k_{t+1} = k_t$, we obtain

$$c_t = k_t^{\alpha} - (\delta + n)k_t.$$

The curve intersects with the abscissa at points $k_t = 0$ and $k_t = (1/(\delta+n))^{1/(1-\alpha)}$. Its maximum is to the right of the steady-state value k.[19] For all points that lie below the line, consumption is below $c_t = k_t^{\alpha} - (\delta+n)k_t$ and the capital stock increases, $k_{t+1} > k_t$. This movement is indicated by the arrows that point to the right in the two segments II and III. If c_t is chosen at a point above the line, the capital stock decreases over time, $k_{t+1} < k_t$, which is indicated by the arrows in regions I and IV that point to the left.

Next, consider Fig. 2.6. Let us assume that the initial value of the capital stock is given by $k_0 > k$. In this case, consumption can be in either region I or II. Let us first assume that consumption lies in region II, e.g., at point A. Consequently, consumption declines while the capital stock increases, and therefore, we can never reach the steady state (nor can we leave region II and move to region I). In finite time, consumption is zero or negative. Since this cannot be an optimal policy, all consumption values that lie in region II can be excluded as possible starting values in period $t = 0$.

Similarly, we can exclude all starting values c_t in region IV, if the initial capital stock is below the steady state, $k_0 < k$. Only consumption choices lying in regions I and III are left as possible choices.

In Fig. 2.6, I have also inserted two additional curves. The upper curve denoted by "$k_{t+1} = 0$" depicts all values for which the next-period capital stock is zero:

$$k_{t+1} = \frac{1}{1+n}\left[k_t^{\alpha} - (1-\delta)k_t - c_t\right] = 0.$$

[18]We used the condition "$c_t = c_{t-1}$" rather than "$c_{t+1} = c_t$" so that both functions which are graphed in Fig. 2.5 have the same argument k_t (and not k_{t+1} as in the case "$c_{t+1} = c_t$").

[19]To verify this statement, differentiate c_t with respect to k_t and solve for $k_t = \left(\frac{\alpha}{\delta+n}\right)^{1/(1-\alpha)}$. For $\beta < 1$, the value of k_t is above the steady state $k = \left(\frac{\alpha}{(1+n)/\beta-1+\delta}\right)^{1/(1-\alpha)}$.

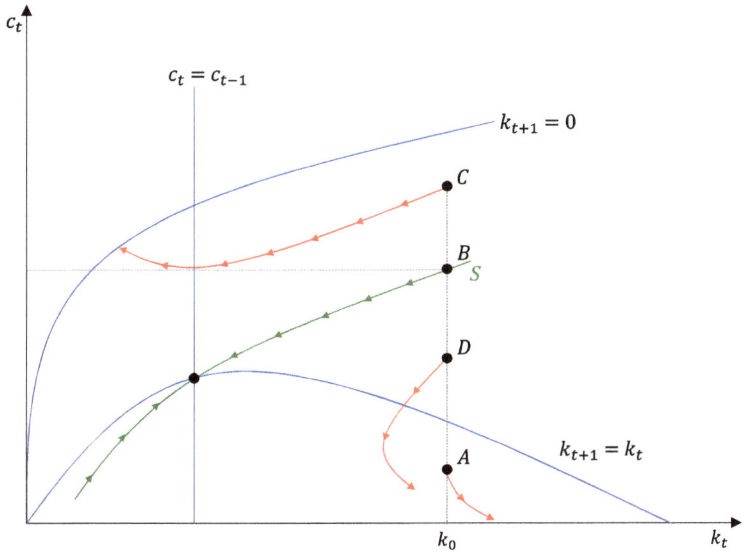

Fig. 2.6 Saddle path in the phase diagram

The second curve denoted by S is the saddle path. Given the initial value of the capital stock k_t, the points on this path converge to the steady state. In the case of k_0, for example, we have to place consumption c_0 at point B so that the economy moves to the steady state. Notice that the arrows on the saddle path S in Fig. 2.6 accord with the arrows drawn in regions I and III in Fig. 2.5. If we chose a higher value of consumption, e.g., as at point C, the trajectory would hit the $k_{t+1} = 0$ curve in finite time.[20] If the capital stock is depleted, production and, hence, consumption are zero, which cannot be optimal either. Similarly, the transition path that starts at point D will cross the curve $k_{t+1} = k_t$ in finite time, and consumption will also be zero in finite time.

For our numerical example, we are able to compute the saddle path in the (k_t, c_t)-plane for a particular starting value of the capital stock. Therefore, we take the values of k_t and c_t from Figs. 2.1 and 2.3 above for the subperiod $t = 13, \ldots, 100$ when $Z_t \equiv 1$ for all remaining periods. At the beginning of period $t = 13$, when Z_t falls back to its steady state value $Z = 1$, the capital stock is given by $k_{13} = 5.88$. For this value of the capital stock, (2.18) only converges to the steady state $(k, c) = (5.447, 1.405)$ for $c_{13} = 1.460$. Over time, (k_t, c_t) slowly approach the steady state from above as presented in Fig. 2.7.[21]

---

[20]One can show that all transition paths that start above S remain above and, similarly, that all paths that start below S remain below it. In addition, paths with the same initial capital stock k_0 but with different consumption values c_0 do not cross.

[21]To compute the saddle path in Figs. 2.7 and 2.8, I increased the number of transition periods to 100 in the program *Ch2_ramsey1.g* so that the approximation of the new steady state is smooth.

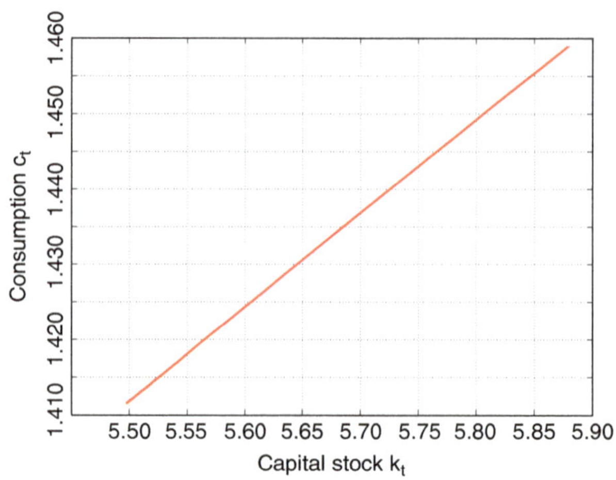

Fig. 2.7 Saddle path in the numerical example

For a unique saddle path, we need one stable (absolute value less than one) and one unstable (absolute value above one) eigenvalue of the difference equation because we have one sluggish variable and one jump variable. More generally, the number of eigenvalues with absolute values less than one must be equal to the number of sluggish variables, and the number of eigenvalues larger than one must be equal to the number of jump variables.

We will demonstrate the solution for our example.[22] Therefore, we consider a linear approximation of the policy function around the steady state (k, c).[23] For this reason, reconsider Eq. (2.18) (with $Z_t = Z_{t-1} = 1$):

$$k_{t+1} = h^1(k_t, c_t) = \frac{1}{1+n}\left[k_t^\alpha + (1-\delta)k_t - c_t\right], \tag{2.19a}$$

$$c_{t+1} = h^2(k_t, c_t) = c_t\left(\frac{\beta}{1+n}\left[1+\alpha\left(\frac{k_t^\alpha + (1-\delta)k_t - c_t}{1+n}\right)^{\alpha-1} - \delta\right]\right)^{\frac{1}{\sigma}}. \tag{2.19b}$$

A linear approximation of these two equations at point $(k_t, c_t) = (k, c)$ is provided by

$$k_{t+1} = k + \frac{\partial h^1(k,c)}{\partial k_t}(k_t - k) + \frac{\partial h^1(k,c)}{\partial c_t}(c_t - c), \tag{2.20a}$$

[22]A more general treatment is contained in Gandolfo (2009) or Azariadis (1993).

[23]The closer we are to the steady state, the better the fit of our linear approximation will be.

$$c_{t+1} = c + \frac{\partial h^2(k,c)}{\partial k_t}(k_t - k) + \frac{\partial h^2(k,c)}{\partial c_t}(c_t - c), \tag{2.20b}$$

or more formally,

$$\begin{pmatrix} k_{t+1} - k \\ c_{t+1} - c \end{pmatrix} = J \begin{pmatrix} k_t - k \\ c_t - c \end{pmatrix}. \tag{2.21}$$

The Jacobian matrix J in our particular case is equal to

$$J = \begin{pmatrix} \frac{1}{1+n}\left(\alpha k^{\alpha-1} + 1 - \delta\right) & -\frac{1}{1+n} \\ \frac{c}{1+n}\frac{\beta}{(1+n)\sigma}\alpha(\alpha-1)k^{\alpha-2}\left(\alpha k^{\alpha-1} + 1 - \delta\right) & 1 - \frac{c}{1+n}\frac{\beta}{(1+n)\sigma}\alpha(\alpha-1)k^{\alpha-2} \end{pmatrix}, \tag{2.22}$$

where we have already used the observation that, in steady state,

$$\frac{\beta}{1+n}\left[1 + \alpha\left(\frac{k^\alpha + (1-\delta)k - c}{1+n}\right)^{\alpha-1} - \delta\right] = 1.$$

For our calibration,

$$J = \begin{pmatrix} 1.0417 & -1.000 \\ -0.0100 & 1.0096 \end{pmatrix}.$$

Next, we use the Schur decomposition $J = \tilde{T}\tilde{S}\tilde{T}^{-1}$ to rewrite Jacobian matrix J, where $\tilde{T}$ is a unitary matrix,[24] and $\tilde{S}$ is an upper triangular matrix:

$$J = \tilde{T}\tilde{S}\tilde{T}^{-1}$$

with

$$\tilde{T} = \begin{pmatrix} 0.9964 & 0.08512 \\ -0.08516 & 0.9964 \end{pmatrix}, \quad \tilde{S} = \begin{pmatrix} 1.127 & -0.9900 \\ 0.0000 & 0.9242 \end{pmatrix},$$

where the eigenvalues of matrices J and $\tilde{S}$ are located on the diagonal of matrix $\tilde{S}$. Notice that one eigenvalue, $\rho_1 = 0.9242$, is below one and the other, $\rho_2 = 1.1271$, is above one.[25] The Schur decomposition is not unique, and we use Givens rotation

[24]In the case of a real matrix $\tilde{T}$, the inverse $\tilde{T}^{-1}$ of a unitary matrix is just the transpose $\tilde{T}'$.

[25]Standard software, such as MATLAB or Gauss, provides commands to compute the Schur factorization. MATLAB also provides a routine, *ordschur(.)*, that can change the order of the eigenvalues if needed.

to rearrange S such that the eigenvalues are located on the diagonal in ascending order[26]:

$$T = \begin{pmatrix} 0.9932 & -0.1167 \\ 0.1167 & 0.9932 \end{pmatrix}, \quad S = \begin{pmatrix} 0.9242 & -0.9900 \\ 0.0000 & 1.127 \end{pmatrix},$$

Again, $J = TST^{-1}$.

With the help of matrix T, we are able to define new auxiliary variables $\tilde{k}_t$ and $\tilde{c}_t$:

$$\begin{pmatrix} \tilde{k}_t \\ \tilde{c}_t \end{pmatrix} = T^{-1} \begin{pmatrix} k_t - k \\ c_t - c \end{pmatrix}. \tag{2.23}$$

such that our system of difference equations (2.21) can be rewritten as:

$$\begin{pmatrix} \tilde{k}_{t+1} \\ \tilde{c}_{t+1} \end{pmatrix} = S \begin{pmatrix} \tilde{k}_t \\ \tilde{c}_t \end{pmatrix} = \begin{pmatrix} \rho_1 & -0.9900 \\ 0.0000 & \rho_2 \end{pmatrix} \begin{pmatrix} \tilde{k}_t \\ \tilde{c}_t \end{pmatrix}. \tag{2.24}$$

Consider the second equation of (2.24):

$$\tilde{c}_{t+1} = \rho_2 \tilde{c}_t.$$

We can rearrange the equation to obtain

$$\tilde{c}_t = \frac{1}{\rho_2} \tilde{c}_{t+1}.$$

Iterating this equation forward and substituting it into itself, we derive

$$\tilde{c}_t = \lim_{i \to \infty} \frac{1}{\rho_2^i} \tilde{c}_{t+i}.$$

Assuming that $\tilde{c}_{t+i}$ remains bounded for $i \to \infty$, and with $\rho_2 > 1$, it follows that

$$\tilde{c}_t = 0.$$

[26]For those readers interested in numerical linear algebra, a Givens rotation is represented by a matrix transformation. In our problem, we search for a matrix

$$G = \begin{pmatrix} d & -e \\ e & d \end{pmatrix}$$

that helps to transform $\tilde{S}$ into $S = G\tilde{S}$.

Since T^{-1} is given by

$$T^{-1} = \begin{pmatrix} 0.9932 & 0.1167 \\ -0.1167 & 0.9932 \end{pmatrix},$$

we find that

$$\tilde{c}_t = -0.1167(k_t - k) + 0.9932(c_t - c) = 0,$$

or

$$c_t - c = 0.1175(k_t - k). \tag{2.25}$$

Notice that the unstable eigenvalue $\rho_2 = 1.1271$ helps us to pin down the jump variable c_t as a function of the sluggish variable k_t!

Next, we need to determine $k_{t+1} - k$ as a function of $k_t - k$. Therefore, we use the first-difference equation from (2.24):

$$\begin{aligned}
\tilde{k}_{t+1} &= 0.9242\tilde{k}_t - 0.9900\tilde{c}_t \\
&= 0.9242\,[0.9932(k_t - k) + 0.1167(c_t - c)] \\
0.9932(k_{t+1} - k) + 0.1167(c_{t+1} - c) &= 0.9242\,[0.9932(k_t - k) + 0.1167(c_t - c)] \\
0.9932(k_{t+1} - k) + 0.1167 \cdot 0.1175(k_{t+1} - k) &= 0.9242\,[0.9932(k_t - k) + 0.1167 \cdot 0.1175(k_t - k)],
\end{aligned}$$

or[27]

$$k_{t+1} - k = 0.9242(k_t - k). \tag{2.26}$$

Clearly, this equation is stable, and k_t converges to k for every value $k_0 > 0$. To derive this stability, we needed an eigenvalue ρ_1 with an absolute value smaller than one.

Given the initial condition for k_0, we can compute c_0 with the help of (2.25) and k_1 with the help of (2.26). Iterating forward for $t = 1, \ldots,$ we can derive the complete time path for $\{k_t, c_t\}_{t=0}^{\infty}$ (or, as in our numerical example, $\{k_t, c_t\}_{t=13}^{100}$).

Figure 2.8 illustrates the dynamics that result from (1) the solution with the direct computation (directly solving the non-linear system of equations) by the solid line and (2) from using the linearization method by the broken line. In each case, we use the initial condition $k_{13} = 5.877$, when Z_t is falling back to its steady-state value, as the initial value for the capital stock. The results are close, but not identical. In particular, the speed of adjustment is higher in the case of direct computation.

[27] Notice that the coefficient of the first-order difference equation is equal to the stable root of the Jacobian, $\rho_1 = 0.9242$.

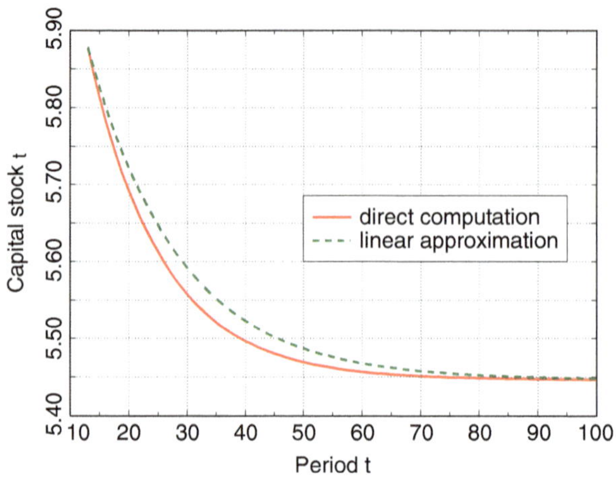

Fig. 2.8 Direct computation versus linear approximation of the saddle path in the numerical example

2.2.3 Elastic Labor

In business cycle theory or in the study of tax policies, we are interested in the behavior of employment. For example, we would like to analyze the cyclical fluctuation of employment or the effect of an income tax on the labor supply. For this reason, we introduce elastic labor supply into the Ramsey model in the following.

The number of households N_t is still growing at rate n. The household is endowed with one unit of time that it spends either on work L_t or leisure $1 - L_t$. The instantaneous utility function depends on both leisure $1 - L$ and consumption c according to:

$$u(c, 1-L) = \frac{\left(c^{\iota}(1-L)^{1-\iota}\right)^{1-\sigma} - 1}{1-\sigma}, \tag{2.27}$$

where ι and $1 - \iota$ denote the weights of consumption and leisure in utility.

The intertemporal utility of the individual household in period $t = 0$ is represented by

$$U = \sum_{t=0}^{\infty} \beta^t u(c_t, 1 - L_t). \tag{2.28}$$

Aggregate labor $N_t L_t$ is given by the product of the number of households N_t and their labor supply L_t. We, again, assume that the production function is Cobb-Douglas:

$$Y_t = Z_t K_t^{\alpha} (N_t L_t)^{1-\alpha}, \tag{2.29}$$

meaning that per capita production is presented by

$$y_t \equiv \frac{Y_t}{N_t} = Z_t k_t^{\alpha} L_t^{1-\alpha}, \tag{2.30}$$

where we define $k_t \equiv K_t/N_t$.

With this definition, the resource constraint is still given by (2.8), and the first-order conditions of the central planner's optimization problem follow from the derivation of the Lagrangian

$$\mathscr{L} = \sum_{t=0}^{\infty} \beta^t \left[\frac{\left(c_t^{\iota}(1-L_t)^{1-\iota}\right)^{1-\sigma} - 1}{1-\sigma} + \lambda_t \left(Z_t k_t^{\alpha} L_t^{1-\alpha} + (1-\delta)k_t - c_t - (1+n)k_{t+1} \right) \right] \tag{2.31}$$

with respect to c_t, L_t, and k_{t+1}. In particular, the first-order conditions are given by:

$$\lambda_t = \iota c_t^{\iota(1-\sigma)-1}(1-L_t)^{(1-\iota)(1-\sigma)}, \tag{2.32a}$$

$$\lambda_t(1-\alpha)Z_t k_t^{\alpha} L_t^{-\alpha} = (1-\iota)c_t^{\iota(1-\sigma)}(1-L_t)^{(1-\iota)(1-\sigma)-1}, \tag{2.32b}$$

$$(1+n)\lambda_t = \lambda_{t+1}\beta \left[1 + \alpha Z_{t+1} k_{t+1}^{\alpha-1} L_{t+1}^{1-\alpha} - \delta \right]. \tag{2.32c}$$

Equations (2.32a) and (2.32b) can be combined to yield:

$$(1-\alpha)Z_t k_t^{\alpha} L_t^{-\alpha} = \frac{1-\iota}{\iota} \frac{c_t}{1-L_t}. \tag{2.33}$$

The Euler equation is derived from inserting (2.32a) into (2.32c):

$$\left(\frac{c_t}{c_{t+1}}\right)^{\iota(1-\sigma)-1} \left(\frac{1-L_t}{1-L_{t+1}}\right)^{(1-\iota)(1-\sigma)} = \frac{\beta}{1+n} \left[1 + \alpha Z_{t+1} k_{t+1}^{\alpha-1} L_{t+1}^{1-\alpha} - \delta \right]. \tag{2.34}$$

In steady state, $Z_t = Z = 1$, $c_t = c_{t+1} = c$, $k_t = k_{t+1} = k$, and $L_t = L_{t+1} = L$, and thus, the Euler equation simplifies to

$$\frac{1+n}{\beta} - 1 + \delta = \alpha k^{\alpha-1} L^{1-\alpha}. \tag{2.35}$$

We will study a numerical example and, for this reason, calibrate the model as in the previous section. Accordingly, we set $\beta = 0.96$, $\sigma = 2.0$, $\alpha = 0.36$, $\delta = 0.08$, and $n = 0$. The only new parameter in the model with elastic labor supply is the utility parameter ι, which determines the weights of consumption c_t and leisure $1 - L_t$ in utility. We will set the parameter such that the steady-state labor supply is equal to 30% of the available time, which is broadly consistent with empirical observations.

Therefore, we set $L = 0.3$. From the steady-state Euler equation (2.35), we can determine the steady-state capital intensity k:

$$k = \left(\frac{\alpha}{\frac{1+n}{\beta} - 1 + \delta} \right)^{\frac{1}{1-\alpha}} L.$$

For our calibration, $k = 1.634$, and therefore, $y = Y/N = k^{\alpha} L^{1-\alpha} = 0.552$. From the resource constraint, we can derive $c = y - (\delta + n)k = 0.422$. Finally, we use (2.33) to calibrate $\iota = 0.338$.

We repeat our analysis of the dynamics as in the previous section and consider a temporary increase in productivity Z_t from 1 to 1.1 during periods $t = 10, 11, 12$. In comparison to the present case, the dynamic system (2.14) needs to be adjusted to allow for elastic labor supply in production. In addition, we have to add a new equation for the optimal labor supply, and thus, we have to solve a non-linear system of equations in 200 variables k_t and L_t, $t = 1, \ldots, 100$ with $k_0 = k_{101} = k$ and $L_0 = L_{101} = L$:

$$\frac{\beta}{1+n} \left[1 + \alpha Z_t k_t^{\alpha-1} L_t^{1-\alpha} - \delta \right] = \left(\frac{Z_t k_t^{\alpha} L_t^{1-\alpha} + (1-\delta)k_t - (1+n)k_{t+1}}{Z_{t-1} k_{t-1}^{\alpha} L_{t-1}^{1-\alpha} + (1-\delta)k_{t-1} - (1+n)k_t} \right)^{1-\iota(1-\sigma)} \cdot \left(\frac{1 - L_{t-1}}{1 - L_t} \right)^{(1-\iota)(1-\sigma)}, \tag{2.36a}$$

$$(1-\alpha) Z_t k_t^{\alpha} L_t^{-\alpha} = \frac{1-\iota}{\iota} \frac{c_t}{1 - L_t}. \tag{2.36b}$$

The dynamics of the model are illustrated in Fig. 2.9.[28] First consider the case of an unexpected increase in Z_t, which is illustrated by the broken lines. As the technology level Z_t rises by 10% in period $t = 10$, output increases, and therefore, both consumption and investment increase as in the case with exogenous labor. Higher consumption increases the marginal utility of leisure, meaning that this income effect also tends to increase leisure. However, the marginal product of labor $Z_t(1-\alpha)k_t^{\alpha} L_t^{-\alpha}$ also increases, and this substitution effect dominates the income effect such that labor increases and leisure declines.

In the case of an expected increase in the technology level Z_t in period $t = 10$, the household already increases both consumption and leisure prior to period $t =$

[28] For a better illustration of the dynamics during the early periods 1–20, I only used 40 periods for the number of transition periods. Although the adjustment is not complete after 40 periods, the approximation is close during periods when the technology shock increases to $Z_t = 1.1$. The MATLAB/Gauss program `Ch2_ramsey2.m`/`Ch2_ramsey2.g` computes the solution presented in Fig. 2.9.

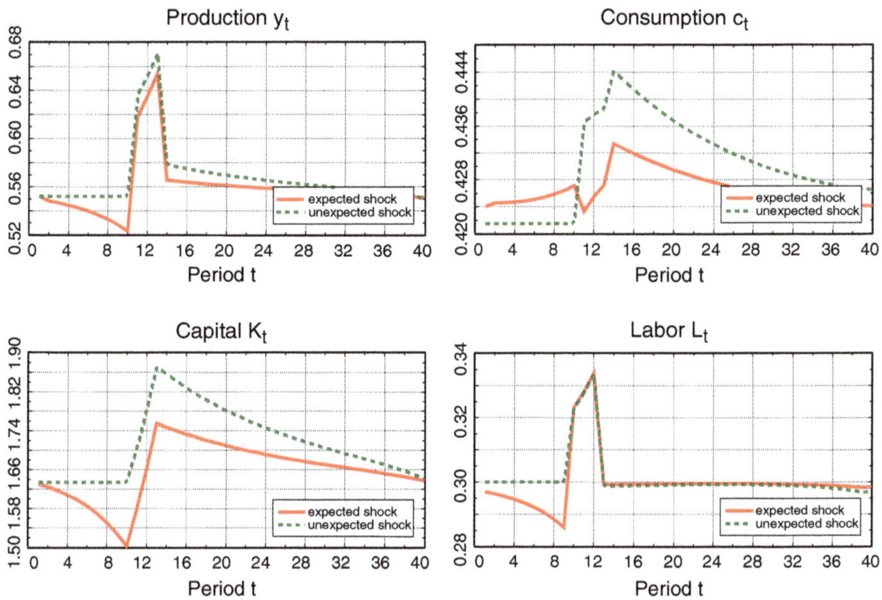

Fig. 2.9 Equilibrium dynamics in the Ramsey model with elastic labor

10 to intertemporally smooth utility. The time paths of output y_t, consumption c_t, capital k_t, and labor L_t are illustrated by the solid line in Fig. 2.9. Since both labor and capital decline prior to period $t = 10$, output also falls in these periods before the technology level Z_t rises above its steady-state value.

The behavior of the variables output, consumption, investment, and labor and the dynamics of these variables that result from a temporary increase (or decrease) in the technology level are at the heart of modern business cycle models. In these models, the technology level Z_t is stochastic, and a rise or decline in its level above or below its steady-state value results in intertemporal substitution of labor and consumption at the household level. As a consequence, these variables fluctuate and are subject to characteristic co-movements that have been the subject of the RBC theory, as initiated by Kydland and Prescott (1982) and Long and Plosser (1983).[29] Surprisingly, an exogenous AR(1) process for technology helps to mimic the empirical time-series behavior of these variables quite well. We will return to this point in Sect. 2.4.

[29]In 2004, Finn E. Kydland and Edward C. Prescott received the Nobel prize for their research on RBCs.

2.3 Decentralized Economy

In this section, we turn our attention from a centralized economy with a benevolent central planner to the case of a decentralized market economy. Households own the capital stock and supply both capital and labor to the market. The firms demand both inputs and produce a homogenous output good. The price mechanism establishes equilibrium in goods and factor markets such that demand equals supply.

2.3.1 Households

Again, the number of households is equal to N_t, and the growth rate of population is denoted by n. All households are identical, meaning that $k_t = K_t/N_t$ is both the average and the individual capital stock per household. Since households are identical, we can study the behavior of all households by means of the individual household and simply multiply individual labor supply L_t and capital k_t by the population size N_t to derive aggregate quantities.

The individual household maximizes lifetime utility (2.28), where instantaneous utility is specified as in (2.27). The household receives two types of income: income from labor and income from capital. A household's labor income $w_t L_t$ is given by the product of the real wage rate w_t and labor supply L_t, while its capital income $r_t k_t$ is equal to the product of the real interest rate r_t and capital supply k_t. The household can spend its income on either consumption c_t or savings s_t:

$$w_t L_t + r_t k_t = c_t + s_t.$$

Savings s_t increase the wealth of the household. Since the only type of asset in the economy is capital, household wealth accumulates according to

$$(1+n)k_{t+1} = (1-\delta)k_t + s_t,$$

where we assume that capital depreciates at rate δ. Notice that the LHS includes a multiplicative term $(1+n)$, which reflects the fact that each individual has $(1+n)$ descendants in the next period, and the assets are distributed equally among the descendants.

Combining the last two equations, we derive the household's budget constraint

$$w_t L_t + r_t k_t + (1-\delta)k_t = c_t + (1+n)k_{t+1}. \tag{2.37}$$

The first-order conditions of the household's optimization problem follow from the derivation of the Lagrangian

$$\mathscr{L} = \sum_{t=0}^{\infty} \beta^t \left[\frac{\left(c_t^{\iota}(1-L_t)^{1-\iota}\right)^{1-\sigma} - 1}{1-\sigma} + \lambda_t \left(w_t L_t + r_t k_t + (1-\delta)k_t - c_t - (1+n)k_{t+1}\right) \right] \tag{2.38}$$

with respect to c_t, L_t, and k_{t+1}. In particular, the first-order conditions are given by:

$$\lambda_t = \iota c_t^{\iota(1-\sigma)-1}(1-L_t)^{(1-\iota)(1-\sigma)}, \tag{2.39a}$$

$$\lambda_t w_t = (1-\iota)c_t^{\iota(1-\sigma)}(1-L_t)^{(1-\iota)(1-\sigma)-1}, \tag{2.39b}$$

$$(1+n)\lambda_t = \lambda_{t+1}\beta\left[1+r_{t+1}-\delta\right]. \tag{2.39c}$$

Equation (2.39b) governs the household's decision with respect to the optimal labor supply. A marginal increase in labor supply by 1 h increases consumption by w_t units. Since the marginal utility from one unit of consumption amounts to λ_t, the LHS of the equation, $\lambda_t w_t$, describes the utility gained from working one additional hour. The RHS presents the marginal disutility from reducing leisure $1-L_t$ by 1 h. At the optimum, the marginal increase in utility stemming from higher consumption must be equal to the marginal decrease in utility stemming from reduced leisure. If this were not the case, the household could increase its total utility by re-balancing its allocations to consumption and leisure. Equation (2.39c), again, describes the optimal intertemporal consumption or savings decision, which is identical to that in the centralized economy described in the previous section.

Equations (2.39a) and (2.39b) can be combined to yield

$$w_t = \frac{1-\iota}{\iota}\frac{c_t}{1-L_t}. \tag{2.40}$$

The Euler equation is derived from inserting (2.39a) into (2.39c):

$$\left(\frac{c_t}{c_{t+1}}\right)^{\iota(1-\sigma)-1}\left(\frac{1-L_t}{1-L_{t+1}}\right)^{(1-\iota)(1-\sigma)} = \frac{\beta}{1+n}\left[1+r_{t+1}-\delta\right]. \tag{2.41}$$

2.3.2 Production

We assume that the goods and factor markets are subject to perfect competition. A representative firm produces with the help of labor and capital according to the production function (2.29) and maximizes profits

$$\Pi_t = Y_t - w_t N_t L_t - r_t K_t = Z_t K_t^{\alpha}(N_t L_t)^{1-\alpha} - w_t N_t L_t - r_t K_t. \tag{2.42}$$

Profit maximization with respect to labor $N_t L_t$ and capital K_t results in the following first-order conditions:

$$w_t = (1-\alpha)Z_t K_t^{\alpha}(N_t L_t)^{-\alpha}, \tag{2.43a}$$

$$r_t = \alpha Z_t K_t^{\alpha-1}(N_t L_t)^{1-\alpha}, \tag{2.43b}$$

which establishes the well-known result that in competitive (factor and goods) markets, the production factors are rewarded by their marginal products. Since the production function is linear-homogenous, Euler's theorem applies, and total production is equal to total costs:

$$Y_t = w_t N_t L_t + r_t K_t = Z_t K_t^{\alpha} (N_t L_t)^{1-\alpha}$$

and profits are zero, $\Pi_t = 0$.

Inserting (2.43) into the budget constraint of the household (2.37), we derive the resource constraint of the economy:

$$y_t = c_t + (1+n)k_{t+1} - (1-\delta)k_t. \tag{2.44}$$

Accordingly, total production is equal to consumption plus investment.

2.3.3 First Fundamental Theorem of Welfare Economics

Inserting (2.43) into (2.39), we observe that we derive exactly the same equilibrium conditions (2.32) as in the case of the central planner. Hence, the allocation in the market economy with perfect competition is equal to the central planner's solution. Since the allocation of the benevolent central planner who maximizes household utility subject to the resource constraint is Pareto-efficient, the allocation in the competitive equilibrium (with no externalities) is also Pareto-efficient. This result is known as the *first fundamental theorem of welfare economics*.[30]

In addition, the equilibrium dynamics can be computed with the help of (2.36), and the dynamics for the equilibrium variables y_t, k_t, L_t, and c_t are identical in the decentralized and centralized economies. In particular, Fig. 2.9 also describes the dynamics in the market economy that is subject to the technology process (2.13).

2.4 The Stochastic Ramsey Model

In this section, I present the stochastic Ramsey model, which forms the basic building block of modern business cycle models.[31] In these models, shocks hit the economy in every period t, and the households and firms optimally adjust their behavior over time. More specifically, the households form *rational expectations* meaning that the agents inside the model know the model and take the model's

[30] The *second theorem of welfare economics* states that any efficient allocation can be sustained by a competitive equilibrium and, thus, constitutes the converse of the first theorem.

[31] Two basic types of business cycle models are presented by RBC models, in which only real variables enter the model, and New Keynesian models, in which nominal variables enter the model and prices and/or wages are sticky. Both types of business cycle models are described in greater detail in Heer and Maußner (2009), McCandless (2008), and Cooley (1995).

prediction as valid. In other words, the agents behave model-consistent.[32] The use of the mathematical expectations operator $\mathbb{E}_t\{\}$ in the following reflects this assumption. Agents form expectations with respect to the probability distribution of random variables in the form of the total factor productivity shocks conditional on the information available at period t.

The stochastic Ramsey model is used for business cycle analysis concentrating on the study of short-term fluctuations. Since most empirical time series for variables such as consumption, output, or investment are available on a quarterly basis,[33] we henceforth also consider a period length of one quarter.

2.4.1 The Model

In the standard RBC model, the technology level Z_t is the exogenous stochastic variable and usually follows a first-order autoregressive (AR1) process[34]:

$$\ln Z_t = \rho^Z \ln Z_{t-1} + \epsilon_t, \quad \epsilon_t \sim N(0, \sigma^Z), \tag{2.45}$$

where the innovation ϵ_t is normally distributed with standard deviation σ^Z.

Empirically, these innovations ϵ_t are identified with the help of a time series on the technology level Z_t. Since technology is not directly observable, Z_t can be evaluated as residual with the help of observations of output Y_t, capital K_t, and labor L_t from the following equation:

$$Z_t = \frac{Y_t}{K_t^\alpha L_t^{1-\alpha}}.$$

The growth rate of total factor productivity is also called the *Solow residual*. It is the part of output growth that is not accounted for by measures of input (labor, capital)

[32]The notation of rational expectation was originally introduced by Muth (1961).

[33]Some time series are also available as monthly data, e.g., industrial production and employment. Some other economic variables such as distributional measures of income and consumption concentration in the form of their Gini coefficients, however, are only available on an annual basis, rendering the analysis of the short-term distributional effects of economic policy more difficult.

[34]In some studies, the technology level follows a unit root process with trend

$$\ln Z_t = \ln Z_{t-1} + a + \epsilon_t, \quad \epsilon_t \sim N(0, \sigma^Z),$$

where a denotes the drift or growth rate of total factor productivity. The modeling of the technology process (and, more generally, time series of macroeconomic variables) is not an innocuous assumption and affects business-cycle results. For example, Cogley and Nasan (1995) demonstrate that if pre-filtered series are first-order integrated, then HP-filtering of the series may result in business cycles that do not exist in the original pre-filtered data.

growth.[35] We use the autoregressive parameter $\rho^Z = 0.95$ and a standard deviation of the innovation equal to $\sigma^Z = 0.007$, as estimated by Cooley and Prescott (1995) with the help of quarterly data for the US economy.[36]

The rest of the model is identical to that of the decentralized economy in the previous section. Due to the stochastic nature of technology and, hence, wages and interest rates, the household maximizes expected utility in period 0

$$U = \mathbb{E}_0 \sum_{t=0}^{\infty} \beta^t u(c_t, 1 - L_t), \tag{2.46}$$

subject to the budget constraint (2.37). Instantaneous utility $u(c, 1-L)$ is a function of consumption c and leisure $1-L$, as presented by (2.27). The Lagrangian function is represented by

$$\mathscr{L} = \mathbb{E}_0 \sum_{t=0}^{\infty} \beta^t \left[\frac{\left(c_t^{\iota}(1-L_t)^{1-\iota}\right)^{1-\sigma} - 1}{1-\sigma} + \lambda_t \left(w_t L_t + r_t k_t + (1-\delta)k_t - c_t - (1+n)k_{t+1}\right) \right]. \tag{2.47}$$

The static first-order condition with respect to leisure is still described by (2.33), while the dynamic Euler equation is adjusted to account for the stochastic nature of the next-period real interest rate r_{t+1}[37]:

$$\frac{1+n}{\beta} = \mathbb{E}_t \left\{ \left(\frac{c_{t+1}}{c_t}\right)^{\iota(1-\sigma)-1} \left(\frac{1-L_{t+1}}{1-L_t}\right)^{(1-\iota)(1-\sigma)} [1 + r_{t+1} - \delta] \right\}. \tag{2.48}$$

[35]Basu, Fernald, and Kimball (2006) construct a measure of technology change in the presence of variable capacity utilization and imperfect competition.

[36]The calibration of RBC models with respect to the characteristics of other industrialized countries employs similar values, e.g., Heer and Maußner (2009) estimate $\rho^Z = 0.90$ and $\sigma^Z = 0.0072$ for the German economy.

[37] Notice that we interchanged the derivative and the expectational operator to derive the first-order conditions using:

$$\frac{d}{dx}\mathbb{E} f(x, Z) = \mathbb{E}\frac{d}{dx} f(x, Z).$$

This condition holds if $f(x, Z)$ is integrable for all x and f is differentiable with respect to x. Furthermore, the expected value of Z is finite, $\mathbb{E}(Z) < \infty$. The above equation is a special application of the *Leibniz integral rule* according to which

$$\frac{d}{dx}\left(\int_{a(x)}^{b(x)} f(x, Z)\, dZ\right) = f(x, b(x)) \cdot \frac{d}{dx} b(x) - f(x, a(x)) \cdot \frac{d}{dx} a(x) + \int_{a(x)}^{b(x)} \frac{\partial}{\partial x} f(x, Z)\, dZ.$$

The firm maximizes profits (2.42) such that factor prices (2.43) are equal to their marginal products. In addition, the goods market clears, meaning that the resource constraint of the economy is represented by (2.44). The equilibrium conditions of the RBC model are summarized by the following eight equations in the eight variables k_t, Z_t, c_t, L_t, y_t, w_t, r_t, and i_t:

$$\frac{1+n}{\beta} = \mathbb{E}_t \left\{ \left(\frac{c_{t+1}}{c_t}\right)^{\iota(1-\sigma)-1} \left(\frac{1-L_{t+1}}{1-L_t}\right)^{(1-\iota)(1-\sigma)} [1+r_{t+1}-\delta] \right\}, \tag{2.49a}$$

$$y_t = c_t + (1+n)k_{t+1} - (1-\delta)k_t, \tag{2.49b}$$

$$\ln Z_t = \rho^Z \ln Z_{t-1} + \epsilon_t, \tag{2.49c}$$

$$w_t = \frac{1-\iota}{\iota}\frac{c_t}{1-L_t}, \tag{2.49d}$$

$$w_t = (1-\alpha)Z_t k_t^\alpha L_t^{-\alpha}, \tag{2.49e}$$

$$r_t = \alpha Z_t k_t^{\alpha-1} L_t^{1-\alpha}, \tag{2.49f}$$

$$y_t = Z_t k_t^\alpha L_t^{1-\alpha}, \tag{2.49g}$$

$$i_t = (1+n)k_{t+1} - (1-\delta)k_t. \tag{2.49h}$$

In the deterministic steady state, the variables are constant, with $\epsilon_t \equiv 0$ and $Z_t \equiv 1$. The steady-state values are therefore identical to those in the decentralized model in the previous section.

To compute the equilibrium dynamics of the stochastic model, we use numerical methods. All parameters are set as in Sect. 2.2.3 except for $\beta = 0.99$ and $\delta = 2.0\%$, which are adjusted for a period length of one quarter (rather than 1 year). In particular, we use a linear approximation of the system of equations around the steady state.[38] Optimal consumption and labor are expressed in the form of the following policy functions that describe the percentage deviation from the steady state[39]:

$$\begin{pmatrix} \hat{c}_t \\ \hat{L}_t \end{pmatrix} = \begin{pmatrix} 0.4946 \\ -0.1706 \end{pmatrix} \hat{k}_t + \begin{pmatrix} 0.4908 \\ 0.6457 \end{pmatrix} \hat{Z}_t. \tag{2.50}$$

[38]Therefore, our approximation is fairly close to the steady state but becomes increasingly inaccurate with increasing distance from the steady state. Linear approximation is a useful technique for the behavior of economies during tranquil times. During periods of severe crisis such as the Great Recession of 2007–2008, one should instead apply global approximation methods, as described in Chapters 5 and 6 in Heer and Maußner (2009).

[39]The computation of the policy functions is described in greater detail in Appendix 2.3.

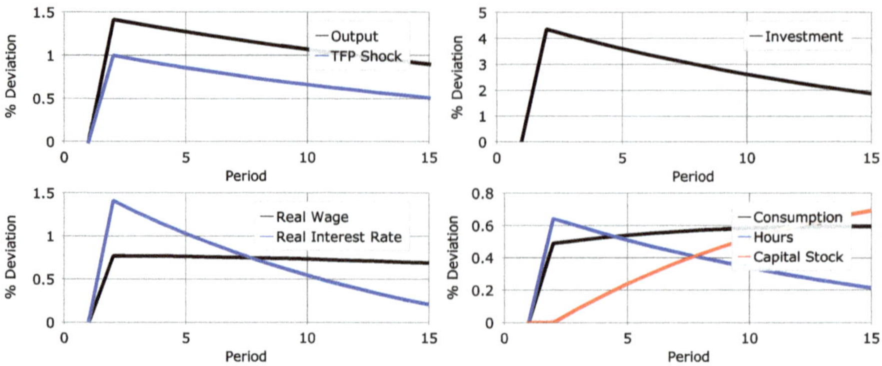

Fig. 2.10 Impulse responses to a technology shock in the RBC benchmark model

Accordingly, consumption increases by 0.49% if the technology level is equal to $Z_t = 1.01$, ceteris paribus (and, hence, 1.0% above its steady-state level, $\hat{Z}_t = 0.01$). Similarly, consumption increases if capital is above its steady-state value due to the wealth effect. Labor supply, however, declines with a higher capital stock and increases with technology. A one percent increase in Z_t results in an increase in labor L of 0.65%.

As our solution for the dynamics of the capital stock k_t, we find

$$\hat{k}_{t+1} = 0.9695\hat{k}_t + 0.0869\hat{Z}_t. \tag{2.51}$$

The dynamics of the model are best explained by means of impulse response functions, which we describe in the next section.

2.4.2 Impulse Responses

The behavior of the policy functions can be illustrated with the help of the impulse response functions displayed in Fig. 2.10.[40] These functions present the percentage deviation of policy functions from the steady state if the economy is in steady state in period $t = 1$ and is hit by a shock $\epsilon_2 = 0.01$ in period $t = 2$. In the next period, $t = 3$, the shock falls back to zero, meaning that $\epsilon_t = 0$ for $t = 3, \ldots$. Since technology Z_t is modeled as an autoregressive process with autoregressive parameter $\rho^Z = 0.95$, technology increases by 0.01 in period $t = 2$ and slowly decreases in the periods thereafter, e.g., $Z_3 = 0.0095$ and $Z_4 = 0.0090$. The dynamics for the technology level Z_t are illustrated by the blue line in the upper-left panel of Fig. 2.10.

[40]The MATLAB and Gauss programs, `Ch2_rbc.m` and `Ch2_rbc.g`, compute the policy functions, the impulse responses, and the time series statistics.

With a higher technology level Z_t, productivity and, hence, wages w_t and the interest rate r_t also increase, and thus, the income of the household increases. Therefore, the household is able to increase both consumption c_t and investment i_t, which rise by 0.48% and 4.3%, respectively. In its effort to smooth utility over time, the household saves a large share of its additional income such that it is also able to increase consumption over the following periods. Therefore, consumption remains 0.5% above the steady-state level for the next 4–5 years (corresponding to 16–20 quarters).

Since the wage rate increases, the household also increases its labor supply. Again, the substitution effect is larger than the income effect, as in the deterministic model in the previous section. Therefore, output, illustrated by the black line in the upper-left panel of Fig. 2.10, increases more strongly than technology and rises by 1.4% on impact.[41]

While the interest rate r_t increases on impact and falls immediately afterward, the wage w_t remains at a higher level over a long period. This is a consequence of the behavior of the technology level Z_t and the capital stock k_t. While technology Z_t falls gradually, the capital stock k_t accumulates slowly over time. The former effect reduces the wage, while the latter increases labor productivity. Consequently, the dynamics of the wage w_t are much more sluggish than those of the interest rate r_t, where the growth in the capital stock k_t and the decline in the technology level Z_t both reduce the marginal product of capital and, hence, r_t.

2.4.3 Time Series Behavior

To compare the behavior of the RBC model with empirical evidence, we employ statistical methods from time series analysis. In particular, we generate a series of T normally distributed random variables that we use as inputs into our model as time series for the stochastic shock $\{\epsilon_t\}_{t=1}^T$. Often, the number of observations T is chosen in accordance with the number of empirical observation periods so that the two time series have equal length. Since our empirical data will have a length of 238 quarters, we also choose $T = 238$. In period $t = 0$, we set all state variables in the model equal to their steady-state values, $Z_0 = 1$ and $k_0 = k$. With the help of the time series $\{\epsilon_t\}_{t=1}^T$, we are able to compute a time series for $\{Z_t\}_{t=0}^T$. We use these values for the exogenous variable Z_t and the starting value $\hat{k}_0 = 0$ to compute time series for $\hat{k}_t$ with the help of (2.51). Given the series for the two state variables $\hat{k}_t$ and $\hat{Z}_t$, we are able to compute the policy functions for the other variables $\hat{c}_t$, $\hat{L}_t$, $\hat{i}_t$,

[41] Empirical studies such as Galí (1999) and Basu, Fernald, and Kimball (2006) find that a positive technology shock led to a contraction of labor inputs. The standard RBC model is inconsistent with this observation. In Sect. 4.5.2, we present a New Keynesian model with sticky prices and adjustment costs of capital that is able to account for this fact.

$\hat{w}_t$, and $\hat{r}_t$. We repeat this simulation 500 times for different random samples for the shocks and use these realizations to compute the averages of the second moments.[42]

Since we are interested in the short-run behavior of the series that is characteristic of the business cycle, we attempt to eliminate cycles with low frequency. For this reason, we apply the same filter to both the empirical and the model series. In RBC models, the most commonly used filter is the HP filter of Hodrick and Prescott (1997).[43] Let $\{y_t\}_{t=1}^T$ denote the log of a time series that may be considered a realization of a non-stationary stochastic process. The growth component $\{g_t\}_{t=1}^T$ of this series as defined by the HP filter is the solution to the following minimization problem[44]:

$$\min_{\{g_t\}_{t=1}^T} \sum_{t=1}^{T} (y_t - g_t)^2 + \lambda \sum_{t=2}^{T-1} [(g_{t+1} - g_t) - (g_t - g_{t-1})]^2. \tag{2.52}$$

If $\lambda = 0$, the solution $y_t = g_t$ is simply the original series. For $\lambda \to \infty$, the series g_t is chosen so that the growth rate is constant. Thus, with a choice of λ, the filter returns a series $\{y_t\}_{t=1}^T$ that approximates the series but is either closer to the original series or a linear trend. For quarterly data, as in the present case, the usual filter weight chosen is $\lambda = 1600$.[45]

Figure 2.11 presents the log of quarterly US GDP[46] during the period 1947:Q1–2015:Q2 (solid red line) and the fitted HP trend (broken green line). GDP displays an upward trend, and we can clearly recognize periods of recession during 1982 and the Great Recession during the period 2007–2008. To better distinguish the trend and the original series from one another, Fig. 2.12 displays the two series for a shorter time period, 1947–1959.

The cyclical components for both HP-filtered GDP (solid red line) and hours (broken green line) that are computed as the deviation of the original series from the HP trend are presented in Fig. 2.13. Obviously, the cyclical components of GDP and output move very closely with one another. In addition, the variances of the two series are also similar. Notice that both series are presented as percentage deviations because we first logged the original series before we applied the HP filter.

---

[42]Sometimes, the researcher also cuts the first 50 periods or similarly from the simulation so that the initialization of the state variables in the first period with their steady state values does not have any effect on the results.

[43]The paper had already circulated as a discussion paper two decades earlier and was then introduced as a working paper by Hodrick and Prescott in 1980 that was published in 1997.

[44]For a more detailed description of this filter and its computation, see for example, Chapter 12.4 in Heer and Maußner (2009).

[45]Ravn and Uhlig (2001) and Baxter and King (1999) propose values of λ equal to 6.5 and 10 for annual data, respectively.

[46]The data are described in more detail in Appendix 2.4.

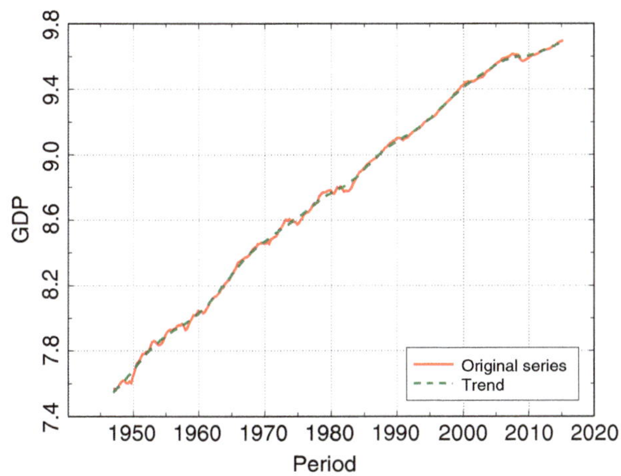

Fig. 2.11 Quarterly US GDP, 1947–2014: original series and trend

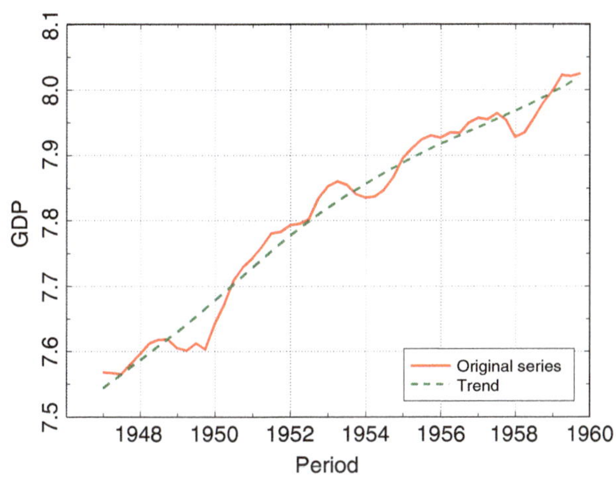

Fig. 2.12 Quarterly US GDP, 1947–1959: original series and trend

The upper half of Table 2.1 presents the second moments for US time series data.[47] We restrict our attention to the period 1953:Q1–2015:Q2. Our main reason for the shorter time period is the behavior of fiscal policy during the period 1947–1952 that we will consider in Chap. 4. During this period, government expenditures

[47]The statistics are computed with the help of the MATLAB or Gauss programs `Ch2_data.m` and `Ch2_data.g`.

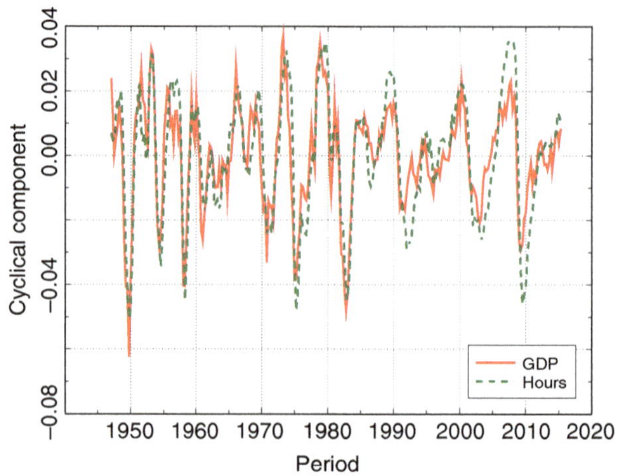

Fig. 2.13 HP-filtered cyclical component of quarterly US GDP and hours, 1947–2014

Table 2.1 US statistics 1953–2014 and RBC model

Variable	s_x (in %)	s_x/s_Y	r_{xY}	r_{xL}
US statistics 1953–2014				
Output Y	1.51	1.00	1.00	0.87
Hours L	1.86	1.24	0.87	1.00
Consumption C	1.21	0.80	0.88	0.76
Investment I	4.54	3.01	0.68	0.86
Wage w	1.02	0.67	−0.27	−0.26
Risk-free rate r^f	0.29	0.19	0.43	0.60
Equity return r^e	6.71	4.44	−0.27	−0.27
RBC model				
Output Y	1.32	1.00	1.00	0.99
Hours L	0.68	0.51	0.99	1.00
Consumption C	0.38	0.29	0.95	0.90
Investment I	4.39	3.33	1.00	1.00
Real wage w	0.65	0.49	0.99	0.97
Real interest rate r	1.35	1.02	0.98	1.00

Notes: $s_x :=$ Standard deviation of time series x in percent, where $x \in \{Y, L, I, C, w, r^f, r^e\}$. Empirical time series were HP filtered with weight 1600. $s_x/s_Y :=$ standard deviation of the variable x relative to the standard deviation of output Y. $r_{xY} :=$ Cross-correlation of the variable with output. $r_{xL} :=$ Cross-correlation of the variable with labor

were determined by the Korean war rather than by cyclical fiscal policy, and we therefore exclude these periods from our observations.[48]

[48]For the risk-free rate and the real equity return rate, we restrict our attention to the period 1959:Q2–2015:Q2. To construct the inflation rate for the computation of the real return, we use

Empirically, the volatility of output Y (as measured by the standard deviation of HP-filtered log GDP) is equal to 1.51%. Hours, as measured by working hours L, are approximately as volatile as output, while private consumption C is less volatile. Notice that investment I is three times as volatile as output Y. In addition, output Y, consumption C, investment I, and labor L are highly correlated, but not perfectly. The empirical correlations of these three variables are in the range 0.68–0.88.

The dynamics of the factor prices are less pro-cyclical than those of the aggregate demand components. In particular, wages are negatively correlated with output and labor. To determine the behavior of the interest rate r, we examine two different rates. First, we report the statistics for the US treasury bill rate, a variable that is often used to measure the risk-free rate in finance. During the period 1947–2014, the average annualized real T-Bill rate was 1.56%. As our second measure, we compute the real equity return of the S&P-500 index in the US stock market, which exhibited an annual average of 7.26%.[49] Accordingly, the equity premium during the period 1947–2014 that measures the difference between the two returns was 5.7%. In both cases, we computed real rates of return by subtracting the inflation rate.[50] Notice that the real equity return is more than five times as volatile as the real risk-free rate. In addition, the risk-free rate is pro-cyclical, while the equity return has a low, negative correlation with both GDP and labor.

How does this time series behavior compare with that implied by the standard RBC model? The lower half of Table 2.1 presents the second moments from the 500 simulations of our RBC model. The RBC model replicates the fact that consumption is less volatile than output, while labor is not volatile enough. In addition, the correlation of output, consumption, and hours is too high compared with the data.

With respect to the factor prices, the basic RBC model is unable to replicate the dynamics of wages and interest rates. Both factor prices are much too pro-cyclical, which is not surprising. A positive productivity shock increases output and employment and, hence, the marginal product of both labor and capital. Consequently, all variables move closely together. The RBC literature has proposed many model extensions that help to considerably improve the fit of the model with respect to empirical observations, including sticky wages, labor market frictions (such as search unemployment), and the introduction of other shocks (such as demand shocks). For instance, we will consider government demand shocks in Chap. 4.

Another basic criticism of RBC theory is based upon the experience of the recent financial crisis. During the period 2008–2009, output and hours fell dramatically, while labor productivity rose. Hence, the standard RBC model that is based on

---

the price index for Private Consumption Expenditures (Excluding Food and Energy), which is only available during the period 1959:Q1–2015:Q2.

[49]The parameter $\beta = 0.99$ in the RBC model is often calibrated to imply an annual real interest rate of 4%, which is a midpoint between the real returns of T-Bills and US equity.

[50]In the case of the interest rate or equity return, which are already measured in percentage points, we do not take the log but rather apply the HP filter to the original series.

shocks to total factor productivity should have difficulty explaining this episode. McGrattan and Prescott (2014) show that there is no inconsistency if intangible capital is included in the analysis.

Appendix 2.1: Intertemporal Elasticity of Substitution and Savings

Derivation of the Intertemporal Elasticity of Substitution

In this appendix, we derive an expression for the intertemporal elasticity of substitution (IES), $1/\sigma$, in a simplified two-period model and study how savings depend on σ. In addition, we analyze the effects of non-capital income in the second period when a change in the interest rate also entails an endowment effect. For this reason, assume that a household lives for 2 periods and maximizes utility

$$U = u(c_1) + \beta u(c_2),$$

where β denotes the discount factor, and c_t represents consumption in period $t = 1, 2$. The household receives the wage incomes y_1 and y_2 in periods 1 and 2, respectively, and thus, the intertemporal budget constraint is given by:

$$y_1 + \frac{y_2}{1+r} = c_1 + \frac{c_2}{1+r},$$

where r denotes the real rate of interest.

The first-order condition for household utility maximization follows from the derivation of the Lagrangian

$$\mathscr{L} = u(c_1) + \beta u(c_2) + \lambda \left[y_1 + \frac{y_2}{1+r} - c_1 - \frac{c_2}{1+r} \right]$$

with respect to c_1 and c_2:

$$u'(c_1) = \lambda, \tag{2.53a}$$

$$\beta u'(c_2) = \frac{\lambda}{1+r}. \tag{2.53b}$$

The first-order conditions of the household can be summarized by the following Euler equation:

$$R \equiv 1 + r = \frac{u'(c_1)}{\beta u'(c_2)},$$

where the right-hand side of the equation is equal to the marginal rate of intertemporal substitution in consumption.

Taking logarithms of both sides and using the approximation $\ln(1+r) \approx r$,[51] we derive:

$$r = -\ln\left(\frac{u'(c_2)}{u'(c_1)}\right) - \ln\beta. \tag{2.54}$$

For small values, logarithms are very close to percentage changes, and thus, we can interpret r as the real interest rate.

The IES is defined as the percentage change in consumption growth for a one-percentage-point increase in the real interest rate:

$$\frac{\partial \ln \frac{c_2}{c_1}}{\partial r}.$$

By substituting (2.54) into the definition of the elasticity above, we can see that the definition of the IES is equivalent to the elasticity of consumption growth with respect to marginal utility growth:

$$\frac{\partial \ln \frac{c_2}{c_1}}{\partial r} = -\frac{\partial \ln \frac{c_2}{c_1}}{\partial \ln\left(\frac{u'(c_2)}{u'(c_1)}\right)}. \tag{2.55}$$

Let utility of consumption be given by

$$u(c) = \begin{cases} \frac{c^{1-\sigma}-1}{1-\sigma} & \sigma \neq 1, \\ \ln c & \sigma = 1. \end{cases}$$

This utility function belongs to the family of the so-called 'Constant-Relative-Risk-Aversion' (CRRA) utility functions and has marginal utility $u'(c) = c^{-\sigma}$. σ is equal to the elasticity of the marginal utility of consumption with respect to consumption and is also called the coefficient of relative risk aversion. Thus,

$$\ln\left(\frac{u'(c_2)}{u'(c_1)}\right) = -\sigma \ln\left(\frac{c_2}{c_1}\right),$$

[51] This approximation follows from a first-order Taylor series expansion

$$f(x) \approx f'(x_0)(x - x_0)$$

with $f(x) = \ln(1+x)$ and $x_0 = 0$ implying:

$$\ln(1+x) \approx \frac{1}{1+x_0}(x - x_0) = x.$$

or

$$\ln\left(\frac{c_2}{c_1}\right) = -\frac{1}{\sigma}\ln\left(\frac{u'(c_2)}{u'(c_1)}\right).$$

Hence, applying this result to expression (2.55), we obtain

$$-\frac{\partial \ln \frac{c_2}{c_1}}{\partial \ln\left(\frac{u'(c_2)}{u'(c_1)}\right)} = -\left(-\frac{1}{\sigma}\right) = \frac{1}{\sigma}.$$

Accordingly, the IES is equal to $1/\sigma$ and, hence, equal to the reciprocal of the elasticity of the marginal utility of consumption with respect to consumption.

CES Utility and Savings

Let us consider another example of a utility function with a constant IES:

$$U = U(c_1, c_2) = \left[(c_1)^\rho + \beta(c_2)^\rho\right]^{\frac{1}{\rho}}. \tag{2.56}$$

In contrast to (2.1), this function is not characterized by additive separability of the period's instantaneous utility functions. It is easy to show that (2.56) is characterized by a constant IES equal to

$$1/\sigma = \frac{1}{1-\rho}.$$

As a numerical example, we assume that the household has non-capital incomes y_1 and y_2 in periods 1 and 2, respectively, and thus, the budget restriction is given by:

$$y_1 + \frac{y_2}{1+r} = c_1 + \frac{c_2}{1+r}.$$

In the initial steady state, we consider the values $y_1 = 100$, $y_2 = 0$, and $r = 5\%$ with discount factor $\beta = 1/1.1$. The values of first-period consumption c_1 and savings s are summarized in columns three and four and in the first and third entry rows of Table 2.2 for two different values of $\sigma \in \{0.5, 1.5\}$.

Table 2.2 Savings and interest rates

$1/\sigma$	r (%)	c_1	s
0.5	5	51.80	48.20
0.5	10	52.38	47.62
1.5	5	52.96	47.04
1.5	10	52.38	47.62

Table 2.3 Savings and interest rates with $y_2 = 90$

$1/\sigma$	r (%)	c_1	s
0.5	5	96.20	3.80
0.5	10	95.24	4.76
1.5	5	98.36	1.64
1.5	10	95.24	4.76

How does an increase in r from 5% to 10% affect savings $s = y_1 - c_1$? Clearly, savings s decrease (increase) if the IES is below one, $1/\sigma < 1$ (above one, $1/\sigma > 1$).

If we also consider non-capital income in the second-period, e.g., $y_2 = 90$, there is an additional endowment effect. If r increases, the discounted value of the non-capital income, $y_1 + \frac{y_2}{1+r}$, decreases, and consequently, this reduces consumption in both periods. If c_1 declines, savings s increase due to this effect. Due to this endowment effect, savings even increase in response to an interest rate rise for $\sigma = 0.5$ in the present example. The effect of increasing the interest rate from 5% to 10% with $y_1 = 100$ and $y_2 = 90$ is summarized in Table 2.3.[52]

As a consequence of this finding, we often have a "normal" reaction of savings—i.e. a rise in savings s if the interest rate r increases—in large-scale OLG models in which households receive income in many periods, such as the Auerbach-Kotlikoff model that we will consider in Chap. 6. In these models, instability is less of a problem. The intuition is simple: If agents accumulate more savings and the capital stock increases, the interest rate will decrease and, eventually, will prevent households from accumulating additional savings.

Appendix 2.2: Solving Non-linear Equations Numerically

In many applications in this book, we need to find the solution x to a non-linear equation:

$$f(x) = 0 \tag{2.57}$$

In this appendix, we will focus on presenting only the main idea of the solution method that we predominantly apply hereinafter. For a more comprehensive presentation of this problem for economists, see Chapter 5 in Judd (1998) or Chapter 3 in Miranda and Fackler (2002).

One of the most widely applied and successful methods to solve non-linear equation (2.57) is the *Newton-Rhapson method* also know as Newton's method.

[52] In Problem 2.6, you are asked to analytically demonstrate this result.

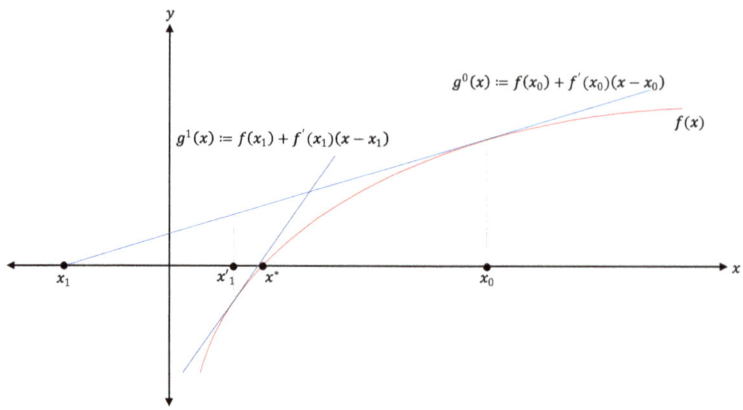

Fig. 2.14 Newton-Rhapson method

The idea is illustrated in Fig. 2.14. The function $f(x)$ has a so-called *root* at a point $x = x^*$ that we attempt to locate. However, since the function is non-linear and is given in implicit form, we may not be able to directly solve for x but instead have to provide a guess x_0 and evaluate the function at this point, $f(x_0)$, and attempt to get as close to the solution, x^*, as possible.

The idea of the Netwon-Rhapson method is to use an iterative scheme, where, starting with an initial guess x_0, we can compute x_{s+1} with the help of x_s using successive linearization around x_s. Starting at our initial guess x_0, the linear approximation of $f : [a, b] \to \mathbb{R}$ at x_0 is given by:

$$g^0(x) := f(x_0) + f'(x_0)(x - x_0),$$

where x is supposed to be the root that we are attempting to locate. Since $g^0(x) = 0$, we obtain:

$$0 = f(x_0) + f'(x_0)(x_1 - x_0) \quad \Rightarrow \quad x_1 = x_0 - \frac{f(x_0)}{f'(x_0)}.$$

This iteration step takes us to point x_1 in Fig. 2.14. Since we have not yet found the solution, we need to continue to iterate forward:

$$x_{s+1} = x_s - \frac{f(x_s)}{f'(x_s)}.$$

When should we stop the algorithm?

1. We may stop if we are close to the solution, x^*. Therefore, we stop if $|f(x_s)| < \epsilon$, e.g., if $|f(x)| < 10^{-5}$,

2. or we may stop the algorithm if successive values of x_s no longer change, e.g., if x_s and x_{s+1} are close to one another:

$$\frac{|x_s - x_{s+1}|}{1 + |x_s|} \leq \epsilon, \quad \epsilon \in \mathbb{R}_{++}. \tag{2.58}$$

Therefore, we examine the percentage change in x_s. In the denominator, the number 1 is added to $|x_s|$ to ensure that we do not encounter the problem of numerical inaccuracies if x_s is small in absolute value. To understand this problem, note that the computer has a given machine accuracy, e.g., it can only store and compute numbers with an accuracy of 1e-16. Let us assume that x_s =1.0e-15. In this case, we cannot determine the solution with an accuracy of 1% because the computer is not able to distinguish between 1.01e-16 and 1.0e-16. Therefore, (2.58) would not be applicable if we did not add the number one to the denominator.

The Newton-Rhapson algorithm is summarized in Algorithm 2.1.

Algorithm 2.1 (Newton-Raphson)

Purpose: *Solve* $f(x) = 0$, *where* $x \in \mathbb{R}$.

Steps:

Step 1: Initialize: choose x_0.
Step 2: Compute $f(x_0)$ *and* $f'(x_0)$ *and iterate on the sequence:*

$$x_{s+1} = x_s - \frac{f(x_s)}{f'(x_s)}.$$

Step 3: Check for convergence: If $\|f(x_s)\|_\infty < \epsilon$ *and/or* $|x_{s+1} - x_s|/(1 + |x_s|) \leq \epsilon$ *for a given tolerance* $\epsilon \in \mathbb{R}_{++}$, *stop; otherwise return to step 2.*

There are two major problems with the Newton-Rhapson method:

1. x_{s+1} may not be defined. For example, in Fig. 2.14, x_1 lies outside the definition area of $f(.)$, e.g., utility from negative consumption cannot be evaluated (if the computer has to evaluate $\ln(x)$ for $x \leq 0$, the computation breaks down). We need to choose a different value, x_1', as the starting point in the second iteration.
2. It may be impossible to derive $f'(.)$.

To circumvent these problems, we modify the Newton-Rhapson algorithm as follows:

1. We backtrack x_{s+1}' along the direction of $f'(x_s)$ to a point x_{s+1} at which f can be evaluated. For example, we could simply take the midpoint between x_s and x_{s+1}. If this point $x_{s+1}' = (x_s + x_{s+1})/2$ is still not admissible, we take the midpoint

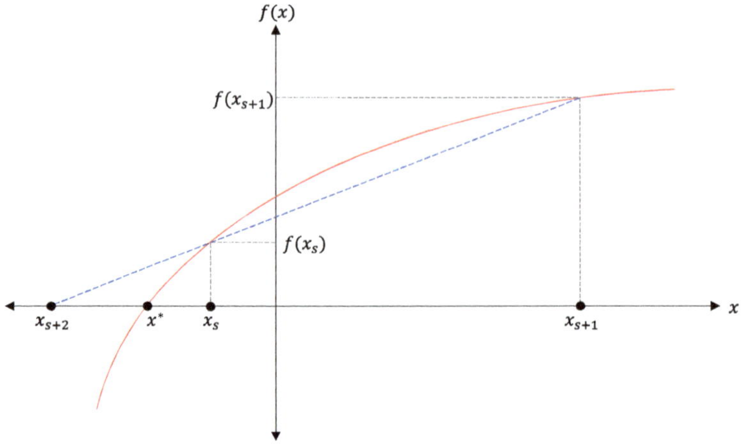

Fig. 2.15 Secant method

between x_s and x'_{s+1}. We continue in this fashion until we are able to evaluate $f(.)$.

2. Instead of $f'(.)$, we use the slope of the secant rather than the derivative to compute the derivative. Therefore, the Newton-Rhapson step is changed to:

$$x_{s+2} = x_{s+1} - \frac{x_{s+1} - x_s}{f(x_{s+1}) - f(x_s)} f(x_{s+1}). \tag{2.59}$$

To compute the secant, we need two former iteration points, x_s and x_{s+1}, and, consequently, two initial points for this method. The secant method is illustrated in Fig. 2.15.

The altered algorithm is called the *Modified* or *Quasi-Newton Method*. In the following, we will state the algorithm for the case in which we do not have to solve one single non-linear equation, $f(x) = 0$, but a system of system of n non-linear equations in the unknowns $x = [x_1, x_2, \ldots, x_n]$:

$$\left.\begin{aligned} 0 &= f^1(x_1, x_2, \ldots, x_n), \\ 0 &= f^2(x_1, x_2, \ldots, x_n), \\ \vdots &= \vdots \\ 0 &= f^n(x_1, x_2, \ldots, x_n), \end{aligned}\right\} \Longleftrightarrow 0 = f(x). \tag{2.60}$$

The equivalent to $f'(x)$ in the multi-variable case is the Jacobian matrix $J(x)$ of partial derivatives of $f = [f^1, f^2, \ldots, f^n]'$ with respect to $x_i, i = 1, 2, \ldots n$. We use the notation

$$f_j^i := \frac{\partial f^i(x)}{\partial x_j}.$$

The Jacobian is defined by:

$$J(x) := \begin{bmatrix} f_1^1 & f_2^1 & \ldots & f_n^1 \\ f_1^2 & f_2^2 & \ldots & f_n^2 \\ \vdots & \vdots & \ddots & \vdots \\ f_1^n & f_2^n & \ldots & f_n^n \end{bmatrix}. \tag{2.61}$$

Algorithm 2.2 (Modified Newton-Raphson)

Purpose: *Solve* $f(x) = 0$, *where* $f = [f^1, f^2, \ldots, f^n]'$ *and* $x \in \mathbb{R}^n$.

Steps:

Step 1: *Initialize: Choose* x_0.
Step 2: *Compute* $J(x_0)$ *the Jacobian of* f *at* x_0, *and solve* $J(x_0)dx = -f(x_0)$. *If* $x_1 = x_0 + dx$ *is not an admissible point, choose* $\lambda \in (0, 1)$ *such that* $x_2 = x_0 + \lambda dx$ *is admissible, and set* $x_1 = x_2$.
Step 3: *Check for convergence: If* $\|f(x_1)\|_\infty < \epsilon$ *and/or* $|x_i^1 - x_i^0|/(1+|x_i^0|) \leq \epsilon \, \forall i$ *for a given tolerance* $\epsilon \in \mathbb{R}_{++}$, *stop; otherwise set* $x_0 = x_1$, *and return to step 2.*

This algorithm is implemented in the numerical Gauss routine *FixVMN1(.)* that we use in our programs.[53] In MATLAB, the command *fsolve* computes the solution to non-linear equations.

The most pressing remaining problem for the researcher who wants to apply the Newton-Rhapson algorithm and the routine *FixVMN1(.)* is to come up with a good initial guess that is close to the true solution. If the guess is not close enough, the sequence x_s, $s = 1, 2, \ldots$, might not converge. Possible methods to find a good initial guess are as follows:

1. Grid search over an interval. Of course, in this case, we have to ensure that the algorithm may not have to evaluate the function at a point that results in a breakdown (e.g., $1/0$ or $(-2.5)^{1/2}$).

[53] The toolbox is available as the Gauss source file *toolbox.src* from my homepage.

2. Educated guess.
3. Genetic search.
4. Backstepping.

A diligent discussion of these problems is provided in Chapter 11.5 of Heer and Maußner (2009). We will also discuss this problem in the upcoming applications in this book if appropriate.

Appendix 2.3: Solving the Benchmark RBC Model

In this appendix, we compute the solution of the benchmark real business cycle (RBC) model described in Sect. 2.4. To solve it we use a linear approximation of the system of difference equations at the steady state. The algorithm is based upon the pioneering work of Blanchard and Kahn (1980) and follows the approach of King and Watson (2002).[54]

Our model is described by the first-order condition of the household, (2.36), and the resource constraint (2.44). After substitution of the factor prices from (2.43) and the production function, $y_t = Z_t k_t^\alpha L_t^{1-\alpha}$, the equilibrium can be presented by the following equations in k_t, λ_t, c_t, and L_t:

$$\lambda_t = \iota c_t^{\iota(1-\sigma)-1}(1-L_t)^{(1-\iota)(1-\sigma)}, \tag{2.62a}$$

$$\lambda_t(1-\alpha)Z_t k_t^\alpha L_t^{-\alpha} = (1-\iota)c_t^{\iota(1-\sigma)}(1-L_t)^{(1-\iota)(1-\sigma)-1}, \tag{2.62b}$$

$$(1+n)\lambda_t = \mathbb{E}_t \lambda_{t+1}\beta\left[1+\alpha Z_{t+1}k_{t+1}^{\alpha-1}L_{t+1}^{1-\alpha}-\delta\right], \tag{2.62c}$$

$$Z_t k_t^\alpha L_t^{1-\alpha} = c_t + (1+n)k_{t+1} - (1-\delta)k_t. \tag{2.62d}$$

We have to distinguish four types of variables in the stochastic system of difference equations (2.62). First, we have the predetermined variables x_t. In our case, the only predetermined variable in period t is the capital stock k_t because investment $i_{t-1} = k_t - (1-\delta)k_{t-1}$ was already chosen at the end of period $t-1$. The second set of variables are the so-called costate variables λ_t. The number of these variables must be equal to the number of dynamic equations (those equations that contain variables from both periods t and $t+1$) minus the number of predetermined variables. Therefore, we need one costate variable. In addition, the costate variables must be dynamic, i.e., these variables must appear in the system of equations with time index $t+1$. In our case, the Lagrange multiplier λ_t is the costate variable. The third type of variables are control variables u_t, which are the remaining variables

[54] This appendix is intended to offer a short introduction to the ideas of solution methods using a simple example. A much more detailed technical description with a generalization to multi-dimensional problems is provided in Chapter 2.4 of Heer and Maußner (2009) or in Chapter 6.8 of McCandless (2008).

in the model that are either chosen by the household and/or the firm, e.g., labor supply, or that are determined by equilibrium conditions, e.g., the prices that equate demand and supply. In our case, the control variables are consumption c_t and labor L_t. Finally, the fourth type of variable is the exogenous state variable, which is the technology shock Z_t.

To solve the system of equations (2.62), we linearize it around the deterministic steady state in a first step and solve the linearized system in a second step. Therefore, we first successively linearize each of the equations in (2.62) starting with (2.62a). We take the logarithm of (2.62a):

$$\ln \lambda_t = \ln \iota - [1 - \iota(1-\sigma)] \ln c_t + (1-\iota)(1-\sigma) \ln(1 - L_t).$$

and compute the total differential of this equation in steady state with $Z_t = Z$, $k_t = k$, $c_t = c$, and $L_t = L$:

$$\frac{d\lambda_t}{\lambda} = -[1 - \iota(1-\sigma)] \frac{dc_t}{c} - (1-\iota)(1-\sigma)\frac{L}{1-L}\frac{dL_t}{L}.$$

Using the notation $\hat{x}_t \equiv dx_t/x$ for the percentage deviation of the variable $x \in \{Z, k, c, L, \lambda\}$, we obtain:

$$\hat{\lambda}_t = -[1 - \iota(1-\sigma)]\,\hat{c}_t - (1-\iota)(1-\sigma)\frac{L}{1-L}\hat{L}_t.$$

Similarly, we log-linearize (2.62b):

$$\alpha\hat{k}_t + \hat{\lambda}_t + \hat{Z}_t = \iota(1-\sigma)\hat{c}_t + \underbrace{\left[1 - (1-\iota)(1-\sigma)\frac{L}{1-L} + \alpha\right]}_{\zeta}\hat{L}_t.$$

Rearranging terms, we find

$$\begin{pmatrix} -[1-\iota(1-\sigma)] & -(1-\iota)(1-\sigma)\frac{L}{1-L} \\ \iota(1-\sigma) & \zeta \end{pmatrix}\begin{pmatrix} \hat{c}_t \\ \hat{L}_t \end{pmatrix} = \begin{pmatrix} 0 & 1 \\ \alpha & 1 \end{pmatrix}\begin{pmatrix} \hat{k}_t \\ \hat{\lambda}_t \end{pmatrix} + \begin{pmatrix} 0 \\ 1 \end{pmatrix}\hat{Z}_t, \tag{2.63}$$

or, evaluated in steady state,

$$\underbrace{\begin{pmatrix} -1.3375 & 0.2839 \\ -0.3375 & 1.0725 \end{pmatrix}}_{C_u}\begin{pmatrix} \hat{c}_t \\ \hat{L}_t \end{pmatrix} = \underbrace{\begin{pmatrix} 0.0000 & 1.0000 \\ 0.3600 & 1.0000 \end{pmatrix}}_{C_{x\lambda}}\begin{pmatrix} \hat{k}_t \\ \hat{\lambda}_t \end{pmatrix} + \underbrace{\begin{pmatrix} 0.0000 \\ 1.0000 \end{pmatrix}}_{C_Z}\hat{Z}_t. \tag{2.64}$$

More generally, we can write (2.64) in the following form:

$$C_u u_t = C_{x\lambda} \begin{bmatrix} x_t \\ \Lambda_t \end{bmatrix} + C_z z_t, \tag{2.65}$$

where we define the variables u_t, x_t, Λ_t, and $z_t$as follows:

$$u_t \equiv \begin{pmatrix} \hat{c}_t \\ \hat{L}_t \end{pmatrix}, \quad x_t \equiv \hat{k}_t, \quad \Lambda_t \equiv \hat{\lambda}_t, \quad z_t \equiv \hat{Z}_t.$$

Next we consider (2.62c), substitute $r_t = \alpha Z_t (L_t/k_t)^{1-\alpha}$ and differentiate totally in steady state:

$$d\lambda_t = \mathbb{E}_t \left[d\lambda_{t+1} + \frac{\lambda \beta r}{1+n} \left(\frac{dZ_{t+1}}{Z} - (1-\alpha)\frac{dk_{t+1}}{k} + (1-\alpha)\frac{dL_{t+1}}{L} \right) \right]$$

and, after dividing by λ:

$$\frac{(1-\alpha)\beta r}{1+n} \mathbb{E}_t \hat{k}_{t+1} - \mathbb{E}_t \hat{\lambda}_{t+1} + \hat{\lambda}_t = \frac{(1-\alpha)\beta r}{1+n} \mathbb{E}_t \hat{L}_{t+1} + \frac{\beta r}{1+n} \mathbb{E}_t \hat{Z}_{t+1}. \tag{2.66}$$

Finally, the resource constraint of the economy is presented by (2.62d). Taking the total differential in steady state yields

$$\hat{Z}_t + \left[\alpha + (1-\delta)\frac{k}{y} \right] \hat{k}_t + (1-\alpha)\hat{L}_t = \frac{c}{y}\hat{c}_t + (1+n)\frac{k}{y}\hat{k}_{t+1}.$$

Rearranging the last two log-linearized equations, we obtain

$$\begin{aligned} &\begin{pmatrix} \frac{(1-\alpha)\beta r}{1+n} & -1 \\ (1+n)\frac{k}{y} & 0 \end{pmatrix} \mathbb{E}_t \begin{pmatrix} \hat{k}_{t+1} \\ \hat{\lambda}_{t+1} \end{pmatrix} + \begin{pmatrix} 0 & 1 \\ -\left[\alpha + (1-\delta)\frac{k}{y}\right] & 0 \end{pmatrix} \begin{pmatrix} \hat{k}_t \\ \hat{\lambda}_t \end{pmatrix} \\ &= \begin{pmatrix} 0 & \frac{(1-\alpha)\beta r}{1+n} \\ 0 & 0 \end{pmatrix} \mathbb{E}_t \begin{pmatrix} \hat{c}_{t+1} \\ \hat{L}_{t+1} \end{pmatrix} + \begin{pmatrix} 0 & 0 \\ -\frac{c}{y} & (1-\alpha) \end{pmatrix} \begin{pmatrix} \hat{c}_t \\ \hat{L}_t \end{pmatrix} \\ &+ \begin{pmatrix} \frac{\beta r}{1+n} \\ 0 \end{pmatrix} \mathbb{E}_t \hat{Z}_{t+1} + \begin{pmatrix} 0 \\ 1 \end{pmatrix} \hat{Z}_t. \end{aligned} \tag{2.67}$$

Computing the matrices of this system of equations by plugging in the parameter and equilibrium values of the model, we derive

$$\underbrace{\begin{pmatrix} 0.0191 & -1.0000 \\ 11.96 & 0.0000 \end{pmatrix}}_{D_{x\lambda}} \mathbb{E}_t \begin{pmatrix} \hat{k}_{t+1} \\ \hat{c}_{t+1} \end{pmatrix} + \underbrace{\begin{pmatrix} 0 & 1.0000 \\ -12.0805 & 0 \end{pmatrix}}_{F_{x\lambda}} \begin{pmatrix} \hat{k}_t \\ \hat{c}_t \end{pmatrix}$$

$$= \underbrace{\begin{pmatrix} 0 & 0.0191 \\ 0 & 0 \end{pmatrix}}_{D_u} \begin{pmatrix} \hat{c}_{t+1} \\ \hat{L}_{t+1} \end{pmatrix} + \underbrace{\begin{pmatrix} 0 & 0 \\ -0.7608 & 0.6400 \end{pmatrix}}_{F_u} \begin{pmatrix} \hat{c}_t \\ \hat{L}_t \end{pmatrix} + \underbrace{\begin{pmatrix} 0.0298 \\ 0 \end{pmatrix}}_{D_z} \mathbb{E}_t \hat{Z}_{t+1}$$

$$+ \underbrace{\begin{pmatrix} 0 \\ 1.000 \end{pmatrix}}_{F_z} \hat{Z}_t.$$

More generally, this equation can be written as:

$$D_{x\lambda} \mathbb{E}_t \begin{bmatrix} x_{t+1} \\ \Lambda_{t+1} \end{bmatrix} + F_{x\lambda} \begin{bmatrix} x_t \\ \Lambda_t \end{bmatrix} = D_u \mathbb{E}_t u_{t+1} + F_u u_t + D_z \mathbb{E}_t z_{t+1} + F_z z_t. \tag{2.68}$$

Collecting all dynamic equations characterizing our linearized stochastic system of difference equations, we derive (2.69):

$$C_u u_t = C_{x\lambda} \begin{bmatrix} x_t \\ \Lambda_t \end{bmatrix} + C_z z_t, \tag{2.69a}$$

$$D_{x\lambda} \mathbb{E}_t \begin{bmatrix} x_{t+1} \\ \Lambda_{t+1} \end{bmatrix} + F_{x\lambda} \begin{bmatrix} x_t \\ \Lambda_t \end{bmatrix} = D_u \mathbb{E}_t u_{t+1} + F_u u_t \tag{2.69b}$$

$$+ D_z \mathbb{E}_t z_{t+1} + F_z z_t.$$

In addition, we assume that Z_t is governed by the AR(1)-process (2.45):

$$\hat{Z}_t = \rho^Z \hat{Z}_{t-1} + \epsilon_t.$$

To reduce the system, we assume that the first equation can be solved for the vector u_t:

$$u_t = C_u^{-1} C_{x\lambda} \begin{bmatrix} x_t \\ \Lambda_t \end{bmatrix} + C_u^{-1} C_z z_t. \tag{2.70}$$

In our case,

$$\begin{pmatrix} \hat{c}_t \\ \hat{L}_t \end{pmatrix} = \begin{pmatrix} 0.0764 & -0.5891 \\ 0.3597 & 0.7470 \end{pmatrix} \begin{pmatrix} \hat{k}_t \\ \hat{\lambda}_t \end{pmatrix} + \begin{pmatrix} 0.2120 \\ 0.9992 \end{pmatrix} \hat{Z}_t. \tag{2.71}$$

Shifting the time index one period into the future and taking expectations conditional on information as of period t yields:

$$\mathbb{E}_t u_{t+1} = C_u^{-1} C_{x\lambda} \mathbb{E}_t \begin{bmatrix} x_{t+1} \\ \Lambda_{t+1} \end{bmatrix} + C_u^{-1} C_z \mathbb{E}_t z_{t+1}. \tag{2.72}$$

The solutions (2.70) and (2.72) allow us to eliminate u_t and $\mathbb{E}_t u_{t+1}$ from (2.69b):

$$\begin{aligned} \left(D_{x\lambda} - D_u C_u^{-1} C_{x\lambda}\right) \mathbb{E}_t \begin{bmatrix} x_{t+1} \\ \Lambda_{t+1} \end{bmatrix} = &- \left(F_{x\lambda} - F_u C_u^{-1} C_{x\lambda}\right) \begin{bmatrix} x_t \\ \Lambda_t \end{bmatrix} \\ &+ \left(D_z + D_u C_u^{-1} C_z\right) \mathbb{E}_t z_{t+1} \\ &+ \left(F_z + F_u C_u^{-1} C_z\right) z_t. \end{aligned}$$

Assume that the matrix $D_{x\lambda} - D_u C_u^{-1} C_{x\lambda}$ is invertible. Furthermore, (2.45) implies $\mathbb{E}_t z_{t+1} = \rho^Z z_t$. Consequently, we obtain the reduced dynamic system:

$$\begin{aligned} \mathbb{E}_t \begin{bmatrix} x_{t+1} \\ \Lambda_{t+1} \end{bmatrix} &= W \begin{bmatrix} x_t \\ \Lambda_t \end{bmatrix} + R z_t, \\ W &= -\left(D_{x\lambda} - D_u C_u^{-1} C_{x\lambda}\right)^{-1} \left(F_{x\lambda} - F_u C_u^{-1} C_{x\lambda}\right), \\ R &= \left(D_{x\lambda} - D_u C_u^{-1} C_{x\lambda}\right)^{-1} \\ &\quad \times \left[\left(D_z + D_u C_u^{-1} C_z\right) \rho^Z + \left(F_z + F_u C_u^{-1} C_z\right)\right]. \end{aligned} \tag{2.73}$$

In our example,

$$\mathbb{E}_t \begin{bmatrix} \hat{k}_{t+1} \\ \hat{\lambda}_{t+1} \end{bmatrix} = \underbrace{\begin{pmatrix} 1.0245 & 0.0774 \\ 0.0123 & 0.9869 \end{pmatrix}}_{W} \begin{bmatrix} \hat{k}_t \\ \hat{\lambda}_t \end{bmatrix} + \underbrace{\begin{pmatrix} 0.1236 \\ -0.0443 \end{pmatrix}}_{R} \hat{Z}_t, \tag{2.74}$$

Next, we proceed in the same way as in the derivation of our stability result in Sect. 2.2.2. In particular, we use the Schur factorization of W:

$$S = T^{-1} W T,$$

with

$$S = \begin{pmatrix} 0.9695 & 0.0651 \\ 0.0000 & 1.0419 \end{pmatrix}, \quad T = \begin{pmatrix} 0.8154 & 0.5789 \\ -0.5789 & 0.8154 \end{pmatrix}.$$

We can reformulate the equation in the transformed variables

$$\begin{bmatrix}\tilde{k}_t \\ \tilde{\lambda}_t\end{bmatrix} \equiv T^{-1}\begin{bmatrix}\hat{k}_t \\ \hat{\lambda}_t\end{bmatrix} = \begin{bmatrix}0.8154 & -0.5789 \\ 0.5789 & 0.8154\end{bmatrix}\begin{bmatrix}\hat{k}_t \\ \hat{\lambda}_t\end{bmatrix}. \tag{2.75}$$

Multiplying (2.74) by T^{-1}, we obtain

$$\mathbb{E}_t\begin{bmatrix}\tilde{k}_{t+1} \\ \tilde{\lambda}_{t+1}\end{bmatrix} = \begin{pmatrix}0.9695 & 0.0651 \\ 0.0000 & 1.0419\end{pmatrix}\begin{bmatrix}\tilde{k}_t \\ \tilde{\lambda}_t\end{bmatrix} + \begin{pmatrix}0.1264 \\ 0.03545\end{pmatrix}\hat{Z}_t.$$

We solve this difference equation line by line, starting with the last line. Accordingly,

$$\mathbb{E}_t\tilde{\lambda}_{t+1} = 1.0419\tilde{\lambda}_t + 0.03545\hat{Z}_t.$$

Rearranging this equation, we obtain

$$\begin{aligned}
\tilde{\lambda}_t &= \frac{1}{1.0419}\mathbb{E}_t\tilde{\lambda}_{t+1} \underbrace{-0.0340}_{\rho_1}\hat{Z}_t \\
&= \frac{1}{1.0419}\mathbb{E}_t\left(\frac{1}{1.0419}\mathbb{E}_{t+1}\tilde{\lambda}_{t+2} + \rho_1\hat{Z}_{t+1}\right) + \rho_1\hat{Z}_t \\
&= \ldots \\
&= \lim_{i\to\infty}\mathbb{E}_t\frac{1}{1.0419^i}\tilde{\lambda}_{t+i} + \rho_1\left(\hat{Z}_t + \frac{\rho^Z}{1.0419}\hat{Z}_t + \ldots\right) \\
&= \frac{\rho_1}{1-\frac{\rho^Z}{1.0419}}\hat{Z}_t \\
&= -0.3854\hat{Z}_t,
\end{aligned}$$

where we have used the assumptions that (1) $\mathbb{E}_t\hat{Z}_{t+i} = (\rho^Z)^i\hat{Z}_t$ and (2) $\lim_{i\to\infty}\tilde{\lambda}_{t+i} < \infty$. Noticing that $\tilde{\lambda}_t = 0.5789\hat{k}_t + 0.8154\hat{\lambda}_t$, we derive

$$\hat{\lambda}_t = -\frac{0.5789}{0.8154}\hat{k}_t - \frac{0.3854}{0.8154}\hat{Z}_t = -0.7099\hat{k}_t - 0.4727\hat{Z}_t. \tag{2.76}$$

Therefore, we have derived the policy function for the marginal utility of consumption, $\hat{\lambda}_t$, given the values of the state variables $\hat{k}_t$ and $\hat{Z}_t$.

Next, we have to solve the first line of the difference equation

$$\tilde{k}_{t+1} = 0.9695\tilde{k}_t + 0.0651\tilde{\lambda}_t + 0.1264\hat{Z}_t.$$

Substituting for $\tilde{k}_t$ and $\tilde{\lambda}_t$ from (2.75) and for $\hat{\lambda}_t$ from (2.76), we find that[55]

$$\hat{k}_{t+1} = 0.9695\hat{k}_t + 0.0870\hat{Z}_t. \tag{2.77}$$

With the help of the policy function for $\hat{\lambda}_t$, we can also compute the policy function for $\hat{c}_t$ and $\hat{L}_t$ from (2.71):

$$\begin{pmatrix} \hat{c}_t \\ \hat{L}_t \end{pmatrix} = \begin{pmatrix} 0.4946 \\ -0.1706 \end{pmatrix} \hat{k}_t + \begin{pmatrix} 0.4908 \\ 0.6457 \end{pmatrix} \hat{Z}_t.$$

Similarly, we can compute the policy function for output, wages, and the interest rate with the help of $\hat{L}_t$ from

$$\begin{aligned} \hat{y}_t &= \hat{Z}_t + \alpha\hat{k}_t + (1-\alpha)\hat{L}_t, \\ \hat{r}_t &= \hat{Z}_t - \alpha\hat{k}_t + (1-\alpha)\hat{L}_t, \\ \hat{w}_t &= \hat{Z}_t + \alpha\hat{k}_t - (1-\alpha)\hat{L}_t. \end{aligned}$$

Appendix 2.4: Data Sources

Most data are taken from the FRED database of the Federal Reserve Bank of St. Louis at `https://research.stlouisfed.org/fred2/`, where you can download many US macroeconomic time series (Accessed on 28 October 2015). The time series that we use in this chapter are also attached as a separate ASCII file "*Fred_Data.txt*" to my Matlab/Gauss programs.

- **Output** Gross Domestic Product. Bureau of Economic Analysis (BEA), retrieved from FRED, Federal Reserve Bank of St. Louis. Series ID: GDPA.
- **Consumption** Real Personal Consumption Expenditures, Billions of Chained 2009 Dollars, Quarterly, Seasonally Adjusted Annual Rate. Series ID: PCECC96.
- **Investment** Private Nonresidential Fixed Investment. BEA, retrieved from FRED, Federal Reserve Bank of St. Louis. Series ID: PNFIA.
- **Labor Supply** Nonfarm Business Sector: Hours of All Persons. US. Bureau of Labor Statistics, retrieved from FRED, Federal Reserve Bank of St. Louis. Series ID: HOANBS.

[55]In (2.77), we report the result from the computation with the program *Ch2_rbc.g*, where we used an accuracy of eight digits. If you compute the result by hand using the numerical values from Eqs. (2.75) and (2.76), the result might diverge by an order of 10^{-3} due to rounding errors.

- **Wages** Nonfarm Business Sector—Compensation Per Hour. BLS, retrieved from FRED, Federal Reserve Bank of St. Louis. Series ID: COMPNFB.
- **Nominal Interest Rate for the Risk-free Rate** 3-Month Treasury Bill: Secondary Market Rate, Percent, Quarterly, Not Seasonally Adjusted. Series ID: TB3MS.
- **Nominal Equity Return** Standard & Poor's 500 Total Return, Yield, Percent, Quarterly, Not Seasonally Adjusted, own calculation.
- **Inflation Rate** Personal Consumption Expenditures: Chain-type Price Index Less Food and Energy, Index 2009 = 100, Quarterly, Seasonally Adjusted, retrieved from FRED, Federal Reserve Bank of St. Louis. Series ID: JCXFE.

Problems

2.1. Compute the dynamics for the centralized deterministic Ramsey model with labor supply as presented in Fig. 2.9 for the case of a CES production function (2.6). Use the value $\sigma_p = 3/4$ for the production substitution elasticity as in Heer and Schubert (2012). For all other parameters, use the values provided in Sect. 2.2.

2.2. Compute the Jacobian (2.16) at the steady state $k_t = k$, $x_t = k$.

2.3. Consider the market economy described in Sect. 2.3. Assume that the government has to finance exogenous expenditures G_t that do not affect utility or production (in Chap. 4, we will modify this assumption). Consider two cases:

1. Government expenditures are financed by a lump-sum tax T_t such that each household pays T_t/N_t.
2. Government expenditures are financed by a tax on wage income such that the individual household pays $\tau_t w_t L_t$ in taxes.

Is the allocation in the economy in both cases Pareto-optimal?

2.4. Consider the market economy described in Sect. 2.3. Assume that the initial capital stock per capita amounts to 50% of its steady state value.

1. Compute the transition dynamics to the steady state. Assume that the transition is completed after 50 periods. How does the savings rate behave during the transition? Contrast your finding in the Ramsey model with the Solow model where the savings rate is constant.
2. Introduce capital adjustment costs into the Ramsey model. For this reason, assume that the household faces the following costs to increase its capital stock:

$$k_{t+1} = (1-\delta)k_t + \Phi\left(\frac{i_t}{k_t}\right),$$

where i_t denotes investment and $\Phi(.)$ is a concave function that takes the following form:

$$\Phi\left(\frac{i_t}{k_t}\right) = \frac{a_1}{1-\zeta}\left(\frac{i_t}{k_t}\right)^{1-\zeta} + a_2.$$

Calibrate the parameter of the adjustment cost function $\Phi(.)$ as follows: $\zeta = 1/0.23$ is taken from Jermann (1998). Assume that adjustment costs play no role in steady state, meaning that $i = \delta k$, and that the multiplier of the adjustment cost constraint in the Lagrangian,

$$\mathscr{L} = \sum_{t=0}^{\infty} \beta^t \{u(c_t, L_t) + \lambda_t [w_t L_t + r_t k_t - c_t - i_t]$$
$$+ q_t \left[k_{t+1} - (1-\delta)k_t - \Phi\left(\frac{i_t}{k_t}\right)\right]\}$$

is equal to one in steady state, $q = 1$, implying

$$a_1 = \delta^{\zeta},$$

and

$$a_2 = \frac{-\zeta}{1-\zeta}\delta.$$

How do capital adjustment costs affect the savings rate during the transition?

3. Compute the impulse responses and the time series statistics for the RBC model with capital adjustment costs.

2.5. Two-Sector Model (follows Heer, Maußner, and Süssmuth 2018) Consider a two-sector economy in which a consumption and an investment good are produced in separate production sectors. The consumption goods sector (subscript C) employs the technology

$$C_t = Z_{Ct} L_{Ct}^{1-\alpha} K_{Ct}^{\alpha}, \quad \alpha \in (0, 1),$$

where L_{Ct} and K_{Ct} denote labor and capital employed in this sector. Z_{Ct} denotes the total factor productivity (TFP). We assume that the log of Z_{Ct} follows a random walk with drift parameter a_C and (possibly) autocorrelated innovations ϵ_{Ct}:

$$\ln Z_{Ct} = \ln Z_{Ct-1} + a_C + \epsilon_{Ct},$$
$$\epsilon_{Ct} = \rho^C \epsilon_{Ct-1} + \eta_{Ct}, \quad \eta_{Ct} \text{ iid } \mathscr{N}(0, \sigma^C).$$

The investment goods sector (subscript I) uses the production technology

$$I_t = Z_{It} L_{It}^{1-\alpha} K_{It}^{\alpha},$$

so that I_t is the amount of investment goods in period t which sell at the relative price p_t. The process for total factor productivity in the investment sector is also difference stationary, yet with a different drift rate a_I:

$$\begin{aligned} \ln Z_{It} &= \ln Z_{It-1} + a_I + \epsilon_{It}, \\ \epsilon_{It} &= \rho^I \epsilon_{It-1} + \eta_{It}, \quad \eta_{It} \text{ iid } \mathcal{N}(0, \sigma^I). \end{aligned}$$

The economy's output in units of the consumption good is equal to

$$Y_t = C_t + p_t I_t.$$

Total labor and capital in the economy equal

$$\begin{aligned} L_t &= L_{Ct} + L_{It}, \\ K_t &= K_{Ct} + K_{It}. \end{aligned}$$

The firms in the two sector maximize profits

$$\begin{aligned} \Pi_t^C &= C_t - w_{Ct+s} L_{Ct} - p_t I_{Ct}, \\ \Pi_t^I &= p_t I_t - w_{It} L_{It} - p_t I_{It}. \end{aligned}$$

A representative household supplies labor L_{Ct} to the consumption goods sector and L_{It} to the investment goods sector. The respective wage rates are w_{Ct} and w_{It} and are equal to each other in equilibrium. The household maximizes intertemporal utility

$$\max \mathbb{E}_t \sum_{s=0}^{\infty} \beta^s u\left(C_{t+s}, L_{t+s}\right),$$

where its instantaneous utility function u depends on consumption C_t and labor L_t:

$$u(C_t, N_t) = \ln(C_t) - \frac{\nu_0}{1+\nu_1} L_t^{1+1/\nu_1}, \quad \nu_0, \nu_1 > 0.$$

The household also owns the capital stock which is subject to adjustment costs:

$$K_{Xt+1} = \Phi_X(I_{Xt}/K_{Xt})K_{Xt} + (1-\delta)K_{Xt}, \quad \delta \in (0,1], \quad X \in \{C, I\},$$

where δ denotes the rate of capital depreciation and the adjustment cost functions Φ_X, $X \in \{C, I\}$, have the same functional forms as the one specified in Problem 2.4 above.

1. Derive the equilibrium conditions of the model.
2. Reformulate the equilibrium equations in stationary variables. Show that, in the long-run, the relative price of the two goods p_t is driven by the different rates of technological progress.
3. Compute the model and its impulse responses to a productivity shock in the two sectors. Use the calibration of Heer, Maußner, and Süssmuth (2018): $\beta = 0.994$, $\nu_1 = 0.3$, $\alpha = 0.36$, $\delta = 0.021$, $\zeta_C = 6.0$, $\zeta_I = 2.0$, $a_C = 0.00054$, $a_I = 0.0077$, $\rho^C = 0$, $\rho^I = 0.28$, $\sigma^C = 0.0070$, $\sigma^I = 0.0084$, and the steady state value of labor $L = 0.33$ for the calibration of ν_0. How do the responses differ between the productivity shocks to the consumption and investment goods sectors?

2.6. For the two-period model in Appendix 2.1, show that the optimal consumption in period 1, c_1, is given by:

$$c_1 = \frac{y_1 + \frac{y_2}{1+r}}{1 + \beta^{1/\sigma}\left[1+r\right]^{1/\sigma - 1}}. \tag{2.78}$$

For $y_2 = 0$, it follows immediately that $\frac{\partial c_1}{\partial r} > 0$ if and only if $1/\sigma > 1$. Recompute the values in the Tables 2.2 and 2.3 in Appendix 2.1 with the help of Gauss (use the *proc()* command to compute the value of the function (2.78)) or MATLAB.

References

Attanasio, O., & Low, H. (2004). Estimating Euler equations. *Review of Economic Dynamics, 7*, 406–435.

Auerbach, A. J., & Kotlikoff, L. (1987). *Dynamic fiscal policy*. New York: Cambridge University Press.

Azariadis, C. (1993). *Intertemporal macroeconomics*. Oxford: Blackwell.

Basu, S., Fernald, J. G., & Kimball, M. S. (2006). Are technology improvements contractionary?. *American Economic Review, 96*, 1418–1448.

Baxter, M., & King, R. G. (1999). Measuring business cycles: Approximate band-pass filters for economic time series. *Review of Economics and Statistics, 81*(4), 575–593.

Beaudry, P., & Portier, F. (2004). An exploration into Pigou's theory of cycles. *Journal of Monetary Economics, 51*(6), 1183–1216.

Beaudry, P., & Portier, F. (2006). Stock prices, news, and economic fluctuations. *American Economic Review, 596*(4), 1293–1307.

Blanchard, O. J., & Kahn, C. M. (1980). The solution of linear difference models under rational expectations. *Econometrica, 48*, 1305–1311.

Cass, D. (1965). Optimum growth in an aggregative model of capital accumulation. *Review of Economic Studies, 32*, 233–240.

Cogley, T., & Nasan, J. M. (1995). Output dynamics in real-business-cycle models. *American Economic Review, 85*, 492–511.

Cooley, T. F. (1995). *Frontiers of business cycle research*. Princeton: Princeton University Press.
Cooley, T. F., & Prescott, E. C. (1995). Economic growth and business cycles. In T. F. Cooley (Ed.), *Frontiers of business cycle research* (pp. 1–64). Princeton: Princeton University Press.
deJong, D. N., & Dave, C. (2011). *Structual macroeconometrics* (2nd ed.). Princeton: Princeton University Press.
Galí, J. (1999). Technology, employment, and the business cycle: Do technology shocks explain aggregate fluctuations?. *American Economic Review, 89*, 249–271.
Gandolfo, G. (2009). *Economic dynamics*. Berlin/Heidelberg: Springer.
Heer, B., & Maußner, A. (2009). *Dynamic general equilibrium modeling: Computational methods and applications* (2nd ed.). Heidelberg: Springer.
Heer, B., Maußner, A., & Süssmuth, B. (2018). Cyclical asset returns in the consumption and investment goods sector. *Review of Economic Dynamics, 28*, 51–70.
Heer, B., & Schubert, S. F. (2012). Unemployment and debt dynamics in a highly indebted small open economy. *Journal of International Money and Finance, 31*, 1392–1413.
Hodrick, R. J., & Prescott, E. C. (1997). Postwar US business cycles: An empirical investigation. *Journal of Money, Credit, and Banking, 29*, 1–16.
Jermann, U. J. (1998). Asset pricing in production economies. *Journal of Monetary Economics, 41*, 257–275.
Judd, K. L. (1998). *Numerical methods in economics*. Cambridge: MIT Press.
King, R. G., & Watson, M. W. (2002). System reduction and solution algorithms for singular difference systems under rational expectations. *Computational Economics, 20*, 57–86.
Koopmans, T. C. (1965). On the concept of optimal economic growth. In *The econometric approach to development planning* (pp. 225–287). Amsterdam: North-Holland.
Kydland, F. E., & Prescott, E. C. (1982). Time to build and agregate fluctuations. *Econometrica, 50*, 1345–1370.
Long, J. B., & Plosser, C. I. (1983). Real business cycles. *Journal of Political Economy, 91*, 39–69.
McCandless, G. (2008). *The ABCs of RBCs*. Cambridge: Harvard University Press.
McGrattan, E. R., & Prescott, E. C. (2014). A reassessment of real business cycle theory. *American Economic Review, 104*, 177–182.
Mehra, R., & Prescott, E. C. (1985). The equity premium: a puzzle. *Journal of Monetary Economics, 15*, 145–161.
Miranda, M. J., & Fackler, P. L. (2002). *Applied computational economics and finance*. Cambridge: MIT Press.
Muth, J. F. (1961). Rational expectations and the theory of price movements. *Econometrica, 29*, 315–335.
Prescott, E. C. (1986). Theory ahead of business-cycle measurement. *Carnegie-Rochester Conference on Public Policy, 24*, 11–44.
Ramsey, F. (1928). A mathematical theory of saving. *Economic Journal, 38*, 543–559.
Ravn, M. O., & Uhlig, H. (2001). *On adjusting the HP-filter for the frequency of observations* (CEPR Discussion Paper, 2858).

The Overlapping Generations Model 3

3.1 Introduction

This chapter investigate the standard overlapping generations (OLG) model with two periods. It serves as one of the main tools to study problems in modern public finance such as pensions, unemployment insurance, and debt. The standard OLG model will be shown to be possibly Pareto-inefficient. In addition, we discuss the problem of stability and note that it is less relevant for the large-scale OLG models that we consider in later chapters than for simple two-period models. With the help of an example, we show that the transition in the OLG model might take place over very long time horizons exceeding several decades. In addition, important extensions of the standard two-period OLG model such as bequests and growth are introduced.

3.2 The Model

In this section, we exmine the two-period model developed by Diamond (1965) and Samuelson (1958). We begin by describing the demographics in the model, before we turn to the household utility maximization problem. Household labor supply is assumed to be inelastic (an assumption that we will dispense with in later chapters), and consumption and savings depend on the factor prices, the real wage rate w_t, and the real interest rate r_t. Production is characterized by constant returns to scale, and we assume an infinite number of representative firms of measure one to allow us to study the problem with the help of a representative firm. We will first study the decentralized competitive equilibrium before we compare it to the command optimum.

B. Heer, *Public Economics*, Springer Texts in Business and Economics,
https://doi.org/10.1007/978-3-030-00989-2_3

3.2.1 Demographics

Each period has two living generations, a young and an old generation. In each generation, the cohort members are identical, which allows us to study their behavior by means of a representative member that we refer to as the young or the old, respectively.

The population of the young generation is denoted by N_t and grows at rate n:

$$N_t = (1+n)N_{t-1}. \tag{3.1}$$

As a consequence, the total population $\tilde{N}_t = N_t + N_{t-1}$ also grows at rate n:

$$\frac{\tilde{N}_{t+1}}{\tilde{N}_t} = \frac{N_{t+1} + N_t}{N_t + N_{t-1}} = \frac{(1+n)N_t + N_t}{(1+n)N_{t-1} + N_{t-1}} = \frac{N_t}{N_{t-1}} = 1+n. \tag{3.2}$$

3.2.2 Household Utility Maximization

The young supplies one unit of labor inelastically in period t, $l_t^1 = 1$, while the old is retired and does not work, $l_t^2 = 0$. Here, and in the following, the superscript $i \in \{1, 2\}$ denotes the age of the generation, while the subscript t denotes the period. The generation born in period t consumes c_t^1 in period t and c_{t+1}^2 in period $t+1$, while the generation born in period $t-1$ consumes c_t^2 in period t. Total consumption C_t, therefore, amounts to

$$C_t = N_t c_t^1 + N_{t-1} c_t^2.$$

The utility of the young generation is assumed to be additively separable in the utilities of the two periods:

$$U_t = u(c_t^1) + \beta u(c_{t+1}^2). \tag{3.3}$$

Instantaneous utility is assumed to be concave:

$$u'(.) > 0, \quad u''(.) < 0, \quad \lim_{c \to 0} u'(c) = \infty.$$

In addition, we assume that the marginal utility of consumption approaches infinity if consumption goes to zero.

The young generation discounts future utility with the discount factor $\beta > 0$. Contrary to the Ramsey model, the assumption that $\beta < 1$ is not necessary to solve the maximization problem here. For finite instantaneous utility $u(.)$, lifetime utility

U_t is bounded given a discount factor $\beta > 1$, meaning that we can compare and rank different lifetime consumption paths $\{c_t^1, c_{t+1}^2\}$.[1]

Individuals are born without any assets (we will introduce bequests later in this chapter) and work one unit of time during their first period of life, earning the real wage w_t. Therefore, wages are the only income that they either consume or save, meaning that savings are given by (since only the young save, the superscript is dropped):

$$s_t = w_t - c_t^1. \tag{3.4}$$

In the second period, agents retire and consume their savings

$$c_{t+1}^2 = (1 + r_{t+1})s_t, \tag{3.5}$$

where r_{t+1} denotes the real interest rate over the period $t + 1$. Equation (3.4) can be substituted into (3.5) to yield the intertemporal budget constraint:

$$c_t^1 + \frac{c_{t+1}^2}{1 + r_{t+1}} = w_t. \tag{3.6}$$

Let us take a more detailed look at the timing. At the beginning of period t, agents begin to provide labor services to the firms and work for the full length of the period. They receive their wage w_t at the end of the period when the production of the good is complete, and they also consume c_t^1 at this time. Savings s_t are invested in the capital market at the end of period t (which is also that at the beginning of period $t + 1$) and earn interest r_{t+1} over period $t + 1$. At the end of period $t + 1$, savings plus interest is remunerated to the old agents, and consumption of amount c_{t+1}^2 takes place.

Accordingly, the young generation household's optimization problem consists in maximizing (3.3) subject to (3.6). The Lagrangian function is represented by

$$\mathscr{L} = u(c_t^1) + \beta u(c_{t+1}^2) + \lambda \left[w_t - c_t^1 - \frac{c_{t+1}^2}{1 + r_{t+1}} \right].$$

Differentiating this expression with respect to c_t^1 and c_{t+1}^2 implies the first-order condition:

$$u'(c_t^1) = \beta(1 + r_{t+1})u'(c_{t+1}^2). \tag{3.7}$$

[1]In later chapters, we will study different applications with $\beta > 1$. Hurd (1989) shows that empirical discount factors are above one if one accounts for mortality risk and bequests.

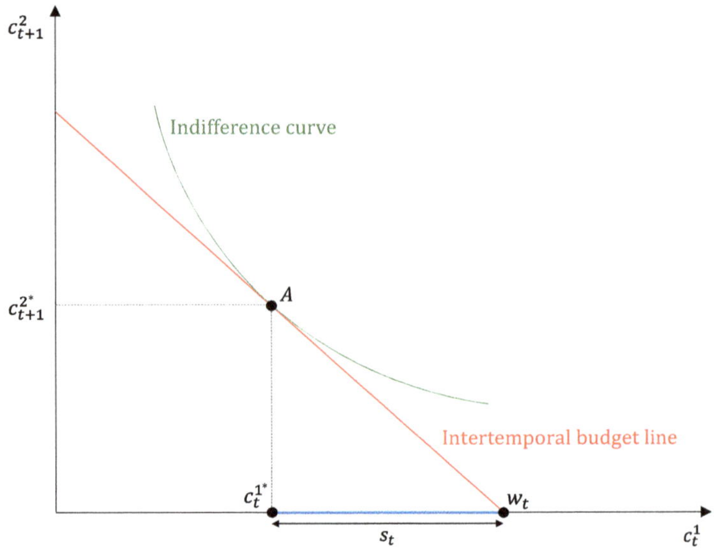

Fig. 3.1 Savings decision in the OLG model

Substituting (3.4) and (3.5) for c_t^1 and c_{t+1}^2 in (3.7)

$$u'(w_t - s_t) = \beta(1 + r_{t+1})u'((1 + r_{t+1})s_t), \tag{3.8}$$

implies savings as an implicit function of the wage rate in period t, w_t, and the interest rate in period $t + 1$, r_{t+1}:

$$s_t = s(w_t, r_{t+1}). \tag{3.9}$$

The intertemporal optimization problem of the young is illustrated in Fig. 3.1. The optimal intertemporal consumption allocation $\{c_t^{1*}, c_{t+1}^{2*}\}$ is represented by point A, where the indifference curve $U_t(c_t^1, c_{t+1}^2)$ is tangent to the intertemporal budget constraint (3.6). Savings are equal to wages w_t minus consumption c_t^1 and are represented by the length of the blue line in Fig. 3.1.

3.2.2.1 Properties of the Savings Function $s(w_t, r_{t+1})$

In the following, we briefly show that the optimization problem has a unique solution for savings $s_t = s(w_t, r_{t+1})$ and that while savings always increase with the wage rate w_t, the response of savings to an increase in the interest rate depends on the magnitude of the elasticity of intertemporal substitution. For an elasticity

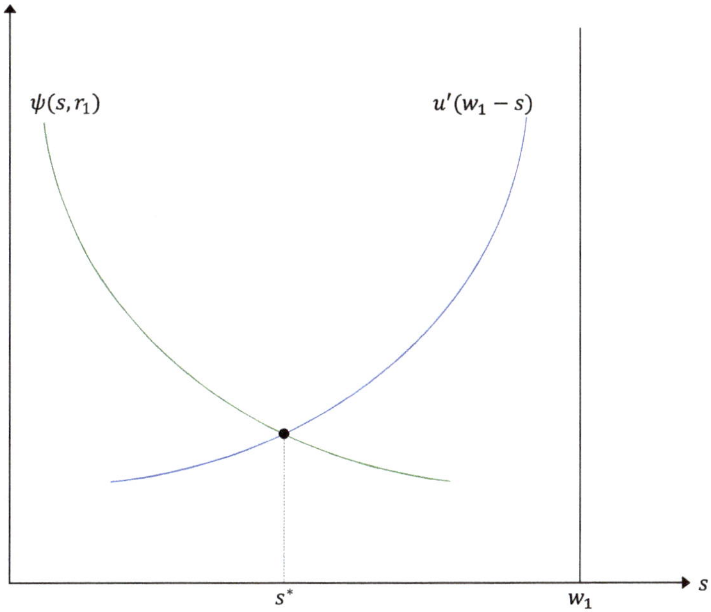

Fig. 3.2 Equilibrium savings in the OLG model

above (below) one, savings will increase (decrease) with higher interest rates. The elasticity of intertemporal substitution, therefore, will be crucial for the analysis in later chapters.[2]

To derive these properties of the savings functions (3.9), let us restate (3.7) as follows:

$$u'(w - s) = \beta u' \left[(1 + r)s\right](1 + r) \equiv \psi(s, r).$$

The left-hand side (LHS) of this equation is a strictly increasing function of savings s on the interval $[0, w]$ with domain $[u'(w), \infty]$, while the right-hand side (RHS) is a strictly decreasing function of s, denoted by $\psi(s, r)$, on the interval $[0, w]$ with domain $\left[u'\left((1 + r)w\right)(1 + r), \infty\right]$. As a consequence, there is one unique solution $s(w, r)$. For illustration, the two sides of the equations are graphed in Fig. 3.2. The upward-sloping blue curve displays the increasing LHS of the equation, while the downward-sloping green curve $\psi(s, r)$ displays its RHS for given wages $w_t = w_1$ and interest rates $r_{t+1} = r_1$. For $s \to w_1$, the LHS $u'(w_1 - s)$ approaches infinity.

[2]The concept of the elasticity of intertemporal substitution is reviewed in Appendix 2.1.

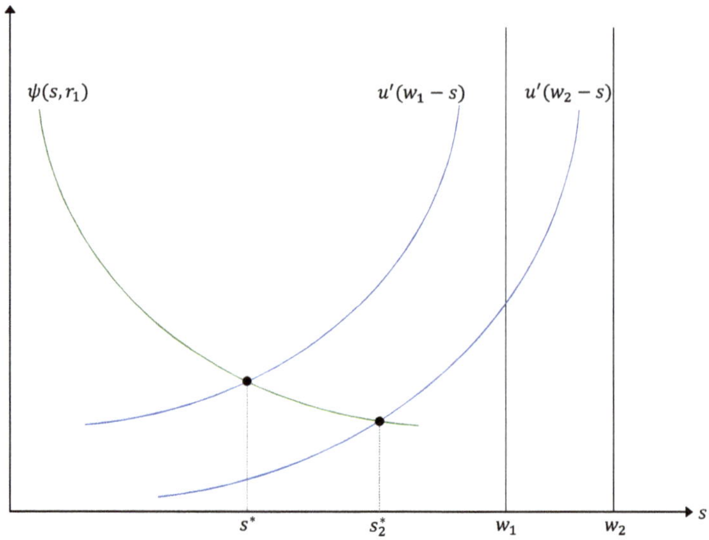

Fig. 3.3 Equilibrium savings for $w_2 > w_1$

Increasing w from w_1 to w_2 results in a downward shift of the LHS $u'(w - s)$, as illustrated in Fig. 3.3, because marginal utility $u'(.)$ is a monotone decreasing function of consumption. As a consequence, savings always increase.[3]

To derive the comparative statics of savings $s(.)$ with respect to the interest rate r_{t+1}, we study the effect of a tax change on the location of the two curves in Fig. 3.4. Increasing r from r_1 to r_2 does not affect the location of the LHS of the equation, while it shifts the curve $\psi(s, r)$ outwards, meaning that savings increase if and only if $\partial\psi/\partial r > 0$:

$$\begin{aligned}\frac{\partial\psi(s,r)}{\partial r} &= \beta u''\left[(1+r)s\right](1+r)s + \beta u'\left[(1+r)s\right] \\ &= \beta u'(c^2)\left[\frac{u''(c^2)}{u'(c^2)}c_2 + 1\right] \\ &= \beta u'(c^2)\left[1-\sigma\right],\end{aligned}$$

where

$$\sigma \equiv -\frac{u''(c)}{u'(c)}c$$

[3]You are encouraged to analytically derive this result by taking the total differential of (3.9).

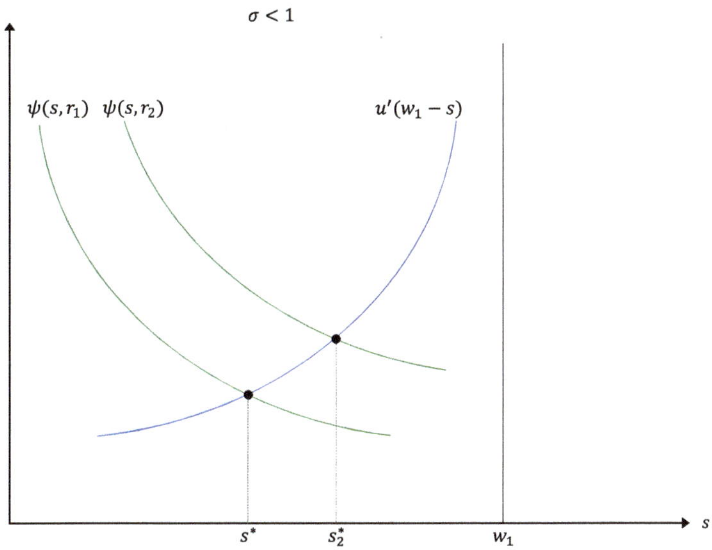

Fig. 3.4 Equilibrium savings for $r_2 > r_1$ if $\sigma < 1$

is the absolute value of the elasticity of marginal utility with respect to consumption (coefficient of relative risk aversion in the economy with uncertainty). Therefore, $\frac{\partial s}{\partial r} > 0$ iff $\sigma < 1$. This case is depicted in Fig. 3.4.

In Appendix 2.1, we show that the elasticity of intertemporal substitution is equal to the reciprocal of the elasticity of marginal utility with respect to consumption and, therefore, equal to $1/\sigma$. Accordingly, we can restate our result above such that an increase in the interest rate increases (decreases) savings if the elasticity of intertemporal substitution is larger (smaller) than one.[4]

3.2.3 Firms

The economy comprises a large number of identical firms that produce output Y_t using capital K_t and labor L_t as inputs:

$$Y_t = F(K_t, L_t). \tag{3.10}$$

[4]In Appendix 2.1, we also show that the result depends on the assumption that non-capital income in period 2 is zero.

For notational simplicity, we assume that $F(.,.)$ is net production such that depreciation is already accounted for. Production is characterized by constant returns to scale, implying

$$\frac{Y_t}{L_t} = F\left(\frac{K_t}{L_t}, 1\right) \equiv f(k_t),$$

where $k \equiv K/L$ denotes the capital-labor ratio or, equivalently, the capital intensity.

The representative firm pays factor prices w_t and r_t to the workers and capital owners, and thus, profits are given by

$$\Pi_t = Y_t - w_t L_t - r_t K_t. \tag{3.11}$$

Assuming perfect competition in both goods and factor markets, profit maximization (3.11) subject to (3.10) implies the first-order conditions[5]:

$$w_t = \frac{\partial F(K_t, L_t)}{\partial L_t} = f(k_t) - k_t f'(k_t), \tag{3.12a}$$

$$r_t = \frac{\partial F(K_t, L_t)}{\partial K_t} = f'(k_t). \tag{3.12b}$$

According to (3.12a) and (3.12b), the wage rate and the interest rate are equal to the marginal products of labor and capital. Notice that both factor prices w_t and r_t are functions of only one variable, capital intensity k_t. This is a very convenient property of the constant-returns-to-scale production function because it will allow us to derive the equilibrium dynamics of the model as the solution of a dynamic equation in k_t.

3.2.4 Equilibrium

In general equilibrium, the labor supply of the young is equal to the labor demanded by firms:

$$N_t l_t^1 = N_t = L_t.$$

The aggregate capital stock K_t increases with investment:

$$K_{t+1} - K_t = I_t. \tag{3.13}$$

[5]To understand the RHS of (3.12a), notice that $\frac{\partial F(K_t, L_t)}{\partial L_t} = \frac{\partial L_t f(K_t/L_t)}{\partial L_t}$.

In addition, investment is equal to aggregate savings:

$$I_t = N_t s_t - K_t. \tag{3.14}$$

At the end of period t, the old consume (or dissave) all their accumulated savings, which is equal to the aggregate capital stock K_t (recall that the young are born without assets). The young invest all their savings $N_t s_t$ in the capital stock of the firms K_{t+1}. Therefore, in capital market equilibrium

$$K_{t+1} = N_t s_t \tag{3.15}$$

or, after dividing by N_t,

$$(1+n)k_{t+1} = s_t = s(w_t, r_{t+1}). \tag{3.16}$$

Since $w_t = w_t(k_t)$ and $r_{t+1} = r_{t+1}(k_{t+1})$ according to (3.12), Eq. (3.16) constitutes a dynamic equation in k_t that describes the equilibrium dynamics of the economy for a given k_0:

$$k_{t+1} = g(k_t). \tag{3.17}$$

3.2.4.1 Steady State

In steady state, the capital stock is constant, $k_t = k_{t+1} = k$, implying:

$$(1+n)k = s(w(k), r(k)). \tag{3.18}$$

For given functional forms of the utility function and production function, this equation can be solved for k.

3.2.5 Existence and Stability

There is no guarantee that there exists a unique stable steady state for the dynamic equation $k_{t+1} = g(k_t)$ in (3.17). Figure 3.5 displays three different forms for $g(k_t)$. For the case depicted by the lower red curve, there exists no steady state with strictly positive production (only the trivial steady state with $k = 0$). The green line illustrates the case in which there are two steady states, A and C, one locally stable (A) and one unstable (C). If there is a small deviation of k_t from the point C, the capital stock converges to point A or to infinity over time. Only for the case described by the blue line does there exist a unique and stable steady state, at point B.

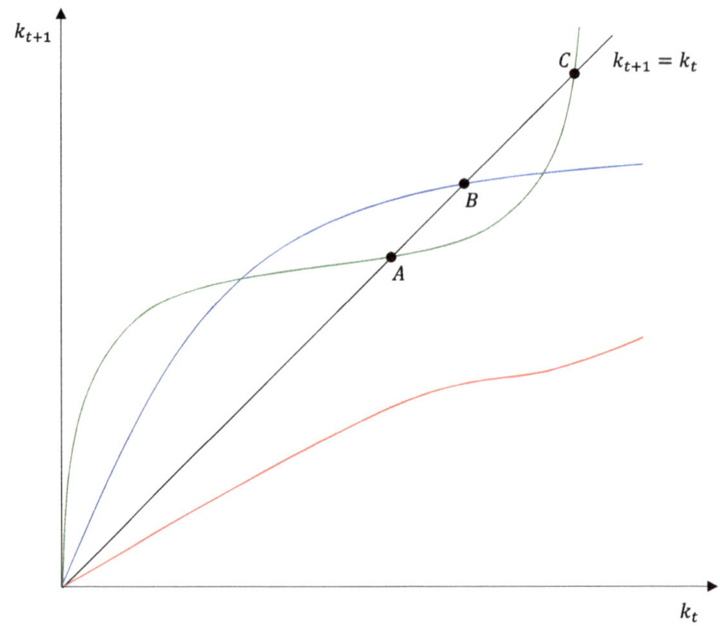

Fig. 3.5 Uniqueness and existence of the steady state

How can instability, e.g., at point C in Fig. 3.5 arise?[6] The economic intuition is straightforward: For a high capital stock k_t, the interest rate is low. For instability to occur, a perpetually increasing capital stock that leads to perpetually decreasing interest rates, savings must continue to rise. Of course, this requires that the first partial derivative of savings with respect to the interest rate must be smaller than zero, $s_r < 0$.[7]

Let us formally derive this result with the help of $k_{t+1} = g(k_t)$:

$$\begin{aligned} k_{t+1} &= \frac{s\left[w(k_t), r(k_{t+1})\right]}{1+n} \\ &= \frac{s\left[f(k_t) - k_t f'(k_t), f'(k_{t+1})\right]}{1+n}. \end{aligned}$$

[6]The argument follows Blanchard and Fischer (1989), Chapter 2.

[7]In Appendix 2.1, we demonstrate that this is the case for CES utility functions with an elasticity of intertemporal substitution below one. We also argue that stability is a less important problem in the large-scale OLG models with many periods that we will consider in later chapters.

Total Differentiation implies:

$$\frac{dk_{t+1}}{dk_t} = \frac{-s_w(k_t)k_t f''(k_t)}{1+n-s_r(k_{t+1})f''(k_{t+1})}.$$

Assuming that there exists a unique equilibrium, $k > 0$, this derivative evaluated around the steady state is equal to:

$$\frac{dk_{t+1}}{dk_t} = \frac{-s_w(k)kf''(k)}{1+n-s_r(k)f''(k)}.$$

Local stability requires

$$\left|\frac{-s_w(k)kf''(k)}{1+n-s_r(k)f''(k)}\right| < 1.$$

The numerator is positive since $f'' < 0$ and $s_w > 0$ (assuming that consumption c^2 in period 2 is a normal good). The above expression for the RHS increases with smaller values of s_r. Therefore, the condition is more likely to fail if high interest rates reduce savings, $s_r < 0$.

3.2.6 A Numerical Example

In the following, we present a very simple, but instructive, numerical example to illustrate the dynamics in the OLG model and the solution methods. The example will also be helpful in understanding the more elaborate models in Chaps. 6 and 7.

Since we analyze two periods of life, working and retirement, it makes sense to choose a period length equal to 30 years. Individuals begin working at real-life age 25 and retire at age 55. They die at age 84. Of course, this is a very crude approximation of real life and suffers from the model assumption that the two periods have equal length. We will extend the standard OLG model with two periods to one with many periods in later chapters such that the working period is longer than the retirement period. Population grows at rate $n = 0.1$.

Next, we have to choose functional forms for the utility and production functions. We choose a utility function that is characterized by a constant elasticity of intertemporal substitution, $1/\sigma = 1$[8]:

$$U_t = \ln(c_t^1) + \beta \ln(c_{t+1}^2).$$

[8]Compare Appendix 2.1 for the analysis of the utility function which is characterized by a constant intertemporal elasticity of substitution.

With this choice, an increase in the interest rate r_t has no effect on savings because the substitution and income effect cancel one another out.[9] Production is described by a Cobb-Douglas function:

$$Y_t = K_t^{\alpha} L_t^{1-\alpha},$$

implying (using (3.12))

$$\frac{Y_t}{L_t} = \frac{Y_t}{N_t} = y_t = f(k_t) = \left(\frac{K_t}{N_t}\right)^{\alpha} = k_t^{\alpha}, \tag{3.19a}$$

$$w_t = (1-\alpha)k_t^{\alpha}, \tag{3.19b}$$

$$r_t = \alpha k_t^{\alpha-1}. \tag{3.19c}$$

We also need to choose values for the parameters β and α. We choose $\beta = 0.4 \approx 0.97^{30}$. To motivate this choice of parameter value, consider that the young attempts to smooth intertemporal consumption as in the Ramsey model. Accordingly, c^1 and c^2 will be approximately equal,[10] meaning that we can assume that the marginal utilities $u'(c^1)$ and $u'(c^2)$ are also equal to one another. From (3.7), it follows that $1/\beta = 1 + r$. Hence, if we assume an annual interest rate r equal to 3% and a model period length of 30 years, $\beta = 0.40$ serves as an initial approximation. As in Chap. 2, we set $\alpha = 0.36$ equal to one minus the labor income share, which, empirically, amounts to approximately 64% in modern industrialized countries.

For this choice of the functional forms, we can compute optimal savings (3.9) with the help of (3.7) as follows:

$$s_t = s_t(w_t, r_{t+1}) = \frac{\beta}{1+\beta} w_t. \tag{3.20}$$

In this special case with $1/\sigma = 1$, as noted above, the savings function does not depend on the interest rate. With $w_t = (1-\alpha)k_t^{\alpha}$, the goods market equilibrium condition (3.16) becomes

$$k_{t+1} = \frac{s_t}{1+n} = \frac{\beta}{1+\beta}\frac{1-\alpha}{1+n} k_t^{\alpha} = g(k_t), \tag{3.21}$$

which describes equilibrium dynamics with the help of a first-order difference equation in k_t. In steady state, $k_t = k_{t+1}$, and thus, (3.21) can be solved for the steady-state capital intensity k:

$$k = \left(\frac{\beta}{1+\beta}\frac{1-\alpha}{1+n}\right)^{\frac{1}{1-\alpha}} = 0.0606,$$

[9]For this result, we also assume that the so-called endowment effect is equal to zero. This will be the case if the household receives zero non-capital income in the second period of life. See also Appendix 2.1.

[10]In the steady state of the Ramsey model in Chap. 2, consumption is constant over the lifetime.

and therefore,

$$
\begin{aligned}
w &= (1-\alpha)k^{\alpha} = 0.233,\\
r &= \alpha k^{\alpha-1} = 2.166,\\
s &= \frac{\beta}{1+\beta} w = (1+n)k = 0.0666,\\
y &= k^{\alpha} = 0.364,\\
c^1 &= w - s = 0.167,\\
c^2 &= (1+r)s = 0.211.
\end{aligned}
$$

Notice that consumption is increasing over the lifetime, meaning that the Euler condition

$$
\frac{c^2}{c^1} = (1+r)\beta > 1
$$

implies that the discount rate $\theta = 1/\beta - 1$ is smaller than the interest rate r. In addition, the steady-state capital stock k is smaller than in the corresponding Ramsey model (with equal functional forms for utility and production and the same parameters α, β, and n). Notice, however, that this need not be true in general. Wickens (2011) compares the OLG model with the representative agent model[11] and shows that it is not clear in which model the capital stock will be larger, which depends on the parameters θ, α, and the depreciation rate δ. In addition, it is also not possible to determine whether consumption c^1 in the OLG model is smaller or larger than the steady-state consumption in the representative agent model.

For given initial per capita capital stock in period 0, e.g., $k_0 = \frac{k}{3}$, we can compute the dynamics from (3.21). For the solution of the numerical problem, we use the Gauss program `Ch3_olg_dyn1.g`. The convergence of the capital intensity is displayed in Fig. 3.6. The capital intensity k_t approaches its long-run steady state within 3–4 periods, which corresponds to 90–120 years. Compared to the Ramsey model from Chap. 2, the OLG model is characterized by a much longer length of the transition period. We will find this result to hold in many other applications in this book. In addition, we find that the dynamic system displays stability, i.e., it converges to the long-run equilibrium.[12] For the solution to be (locally) stable, the following condition must hold:

$$
\left| \frac{dk_{t+1}}{dk_t} \right| < 1.0. \tag{3.22}
$$

[11] See Section 6.3.4 in Wickens (2011).

[12] To be more precise, (3.16) has two solutions for the steady state, $k_{t+1} = k_t = k$: (1) $k = 0$, which is unstable, and (2) $k = \left(\frac{\beta}{1+\beta}\frac{1-\alpha}{1+n}\right)^{\frac{1}{1-\alpha}}$, which is stable.

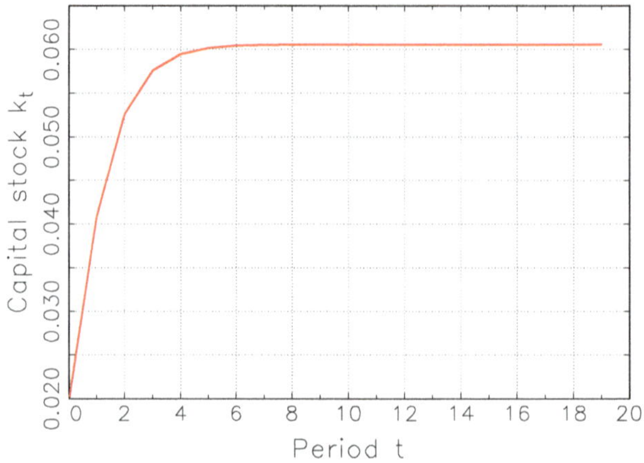

Fig. 3.6 Dynamics of the capital stock k_t in the numerical example of an OLG model

In the above example,

$$\frac{dk_{t+1}}{dk_t} = \alpha \frac{\beta}{1+\beta}\frac{1-\alpha}{1+n} k_t^{\alpha-1}.$$

Evaluated for $k = 0$, the derivative goes to infinity. Evaluated for $k = \left(\frac{\beta}{1+\beta}\frac{1-\alpha}{1+n}\right)^{\frac{1}{1-\alpha}}$, the absolute value of the derivative is smaller than 1 for $\alpha < 1$:

$$\begin{aligned}\frac{dk_{t+1}}{dk_t} &= \alpha \frac{\beta}{1+\beta}\frac{1-\alpha}{1+n} k_t^{\alpha-1} \\ &= \alpha \frac{\beta}{1+\beta}\frac{1-\alpha}{1+n}\left(\frac{\beta}{1+\beta}\frac{1-\alpha}{1+n}\right)^{\frac{\alpha-1}{1-\alpha}} \\ &= \alpha.\end{aligned}$$

The dynamics for a given initial capital stock k_0 are illustrated in Fig. 3.7. Starting at k_0 in period $t = 0$, we find k_1 on the blue curve as $k_1 = \frac{\beta}{1+\beta}\frac{1-\alpha}{1+n}k_0^\alpha$ using (3.21). The capital stock k_1 is graphed on the ordinate k_{t+1}. With the help of the diagonal $k_{t+1} = k_t$ (the black line in Fig. 3.7), the capital stock k_{t+1} is mirrored to the abscissa k_t. Iterating t one period ahead, k_1 becomes the present-period capital stock k_t, and k_2 can be allocated with the help of the function $g(k_t)$. Continuing in this fashion over periods $t = 2, \ldots$, the complete transition path can be traced. In Fig. 3.7, this is illustrated by the arrows.

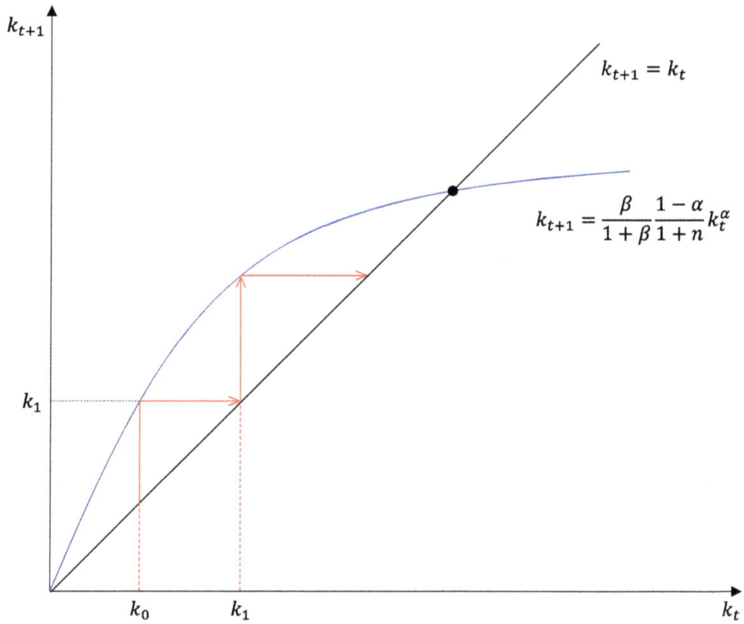

Fig. 3.7 Stability in the numerical example of an OLG model

The main lines of the GAUSS program `Ch3_olg_dyn1.g` are as follows:

```
alpha=0.36; // production elasticity of capital
beta=0.40; // discount factor
n=0.1; // population growth rate
k=( beta/(1+beta) * (1-alpha)/(1+n) )^(1/(1-alpha));
k0=k/3; // initial capital stock

tt=seqa(1,1,20); // periods
kt=zeros(20,1); // time series for capital stock

kt[1]=k0;
i=1;
do until i==20;
i=i+1;
kt[i]=kdyn(kt[i-1]);
endo;

proc kdyn(x);
retp( beta/(1+beta) * (1-alpha)/(1+n) * x^(alpha) );
endp;
```

In the first 3 lines, the parameter values are set. In line 4, the stable steady-state value is computed. The initial capital stock is set to one-third of the steady-state value in line 5. Assuming a transition length of 20 periods, the time path for k_t is initialized at zero in lines 5–6. The iteration over k_{t+1} given k_t is performed in lines 7–12 using the procedure `kdyn(x)` in lines 13–15 for the computation of $g(k_t)$.[13]

3.3 The Command Optimum

3.3.1 Modified Golden Rule

In this section, we study the question of how a benevolent dictator would allocate consumption over time and between generations. Therefore, we need to choose how the central planner weights the utility of the different generations. We will postulate that he discounts the utility of future generations at the common rate R. Accordingly, the social welfare function is

$$\mathscr{U} = \beta u(c_0^2) + \sum_{t=0}^{T-1} \frac{1}{(1+R)^t} \left[u(c_t^1) + \beta u(c_{t+1}^2) \right]. \tag{3.23}$$

In particular, we assume that the central planner cares only about the utility of the $T+1$ current and future generations. For $R=0$, the central planner cares equally about all generations.

We assume that the production side is identical to that in the previous section. The young generation supplies labor inelastically, $l_t^s = 1$, meaning that total labor L_t is equal to the size of the young population N_t. Therefore, the central planner faces the following resource constraint:

$$K_t + F(K_t, N_t) = K_{t+1} + N_t c_t^1 + N_{t-1} c_t^2, \tag{3.24}$$

or, in per capita variables,

$$k_t + f(k_t) = (1+n)k_{t+1} + c_t^1 + \frac{1}{1+n} c_t^2. \tag{3.25}$$

This maximization problem can be solved by differentiating the Lagrange function

$$\begin{aligned}\mathscr{L} = \beta u(c_0^2) + \sum_{t=0}^{T-1} \frac{1}{(1+R)^t} \Bigg\{ & \left[u(c_t^1) + \beta u(c_{t+1}^2) \right] \\ & + \lambda_t \left[k_t + f(k_t) - (1+n)k_{t+1} - c_t^1 - \frac{1}{1+n} c_t^2 \right] \Bigg\}\end{aligned}$$

[13] If you are using a different computer language, e.g., MATLAB, you are encouraged to translate this little simple program into the language that you are using.

with respect to the variables c_t^1, c_{t+1}^2, and k_{t+1}, implying the first-order conditions

$$\lambda_t = u'(c_t^1), \tag{3.26a}$$

$$\beta u'(c_{t+1}^2) = \frac{1}{1+R}\frac{\lambda_{t+1}}{1+n}, \tag{3.26b}$$

$$\lambda_t = \frac{1}{1+R}\frac{\lambda_{t+1}}{1+n}\left[1 + f'(k_{t+1})\right]. \tag{3.26c}$$

Combining (3.26a), (3.26b), and (3.26c), we can derive the counterpart to the first-order condition of the household (3.7) in the centralized economy:

$$u'(c_t^1) = \beta\left[1 + f'(k_{t+1})\right] u'(c_{t+1}^2).$$

In steady state, $c_t^1 = c_{t+1}^1 = c^1$, $c_t^2 = c_{t+1}^2 = c^2$, and $k_t = k$:

$$\beta u'(c^2) = \frac{1}{1+R}\frac{u'(c^1)}{1+n}, \tag{3.27a}$$

$$1 + f'(k) = (1+R)(1+n). \tag{3.27b}$$

As a result, the steady state satisfies the *modified golden rule* (using the approximation that $Rn \approx 0$ for small R and n):

$$f'(k) = R + n.$$

Accordingly, if the central planner attaches equal weight to each generation ($R = 0$), the marginal product of capital (equal to the real interest rate in decentralized economy) in the steady state is equal to the population growth rate n.

3.3.2 Pareto-Efficiency

3.3.2.1 Efficiency in the Command Optimum

In the following, we study the question of whether the steady state in the command optimum is efficient. To do so, let us reconsider the steady-state condition (3.27b):

$$1 + f'(k) = (1+n)(1+R).$$

Let $c = c^1 + \frac{c^2}{1+n}$ denote total consumption in each period of the steady state divided by the number of young households, $c_t = C_t/N_t$. Then, (3.25) implies

$$f(k) - nk = c.$$

A decrease in k increases steady-state consumption if and only if

$$\frac{dc}{dk} = f'(k) - n < 0.$$

If $f'(k) < n$ (i.e., for $R < 0$), everyone can be made better off if the capital stock is decreased. Obviously, this condition holds in steady state. Moreover, this result holds not only for the new steady state with lower capital $\tilde{k}$ and higher consumption $\tilde{c}$ but also during the transition. During the transition, consumption c_t may even be higher than in the new long-run equilibrium, $\tilde{c}$. At a minimum, however, it is guaranteed that c_t during the transition is higher than the old steady-state c.

To understand this result, assume that, prior to period t, we are in steady state with capital stock k and consumption c as given by (3.27b). In period t, a small change $dk < 0$ is initiated such that total consumption in periods t and $t+1$ changes as follows (assuming that the change in savings, $dk < 0$, is completely effectuated in period t)[14]:

$$dc_t = -(1+n)dk > 0, \tag{3.28a}$$

$$dc_{t+1} = (f' - n)dk > 0. \tag{3.28b}$$

Therefore, the total consumption and, hence, utility of all generations could be increased if $f'(k) < n$. Consequently, the OLG model—in contrast to the Ramsey model—may exhibit *dynamic inefficiency*.[15]

In the case in which

$$\frac{dc}{dk} = f'(k) - n > 0,$$

steady-state consumption could be increased if the capital stock were to grow. However, in this case, the generations during the transition have to sacrifice consumption to accumulate additional savings. In particular, if the change, $dk > 0$, is completely enacted in period t, total consumption in period t and in period $t+1$ adjust according to:

$$dc_t = -(1+n)dk < 0,$$

$$dc_{t+1} = (f' - n)dk > 0.$$

[14] Notice that (3.25) implies that $c_t = k_t - (1+n)k_{t+1} + f(k_t)$. For $dk_{t+1} = dk < 0$, $dc_t = -(1+n)dk_{t+1} > 0$.

[15] Compare this with the Pareto efficiency of the Ramsey model in Sect. 2.3.3.

3.3.2.2 Efficiency in the Decentralized Economy

We can extend the same argument for the command economy from above to the case of a competitive economy. The steady state in the market economy is described by (3.16), which we restate for the readers' convenience:

$$(1+n)k_{t+1} = s_t = s(w_t, r_{t+1}).$$

For the special case of Cobb-Douglas production and logarithmic utility, we have shown (3.21):

$$k_{t+1} = \frac{s_t}{1+n} = \frac{\beta}{1+\beta}\frac{1-\alpha}{1+n}k_t^{\alpha}$$

implying that

$$k = \left(\frac{\beta}{1+\beta}\frac{1-\alpha}{1+n}\right)^{\frac{1}{1-\alpha}}.$$

For the steady state, $k = k_t = k_{t+1}$, it is not a priori evident whether

$$r = f'(k) \gtrless n$$

From $f'(k) = \alpha k^{\alpha-1}$, we derive that the condition is equivalent to

$$\frac{\alpha}{1-\alpha}\frac{1+\beta}{\beta} \gtrless \frac{n}{1+n}.$$

For this equation to hold with the '$<$' sign, population growth needs to be positive, $n > 0$. Let us consider empirical values, e.g., $\alpha = 0.36$ and $\beta = 0.97$ with annual periods. To have dynamic inefficiency, we would need an annual population growth rate in excess of 114%. Thus, for empirically relevant numbers, dynamic inefficiency seems less relevant. In fact, in all the applications in this book, dynamic inefficiency does not arise and is a merely theoretical phenomenon.

Why is the consideration of Pareto efficiency in the market economy important? Imagine you would like to conduct a welfare analysis of an economically relevant public policy in an OLG model. For example, you might consider a policy in the form of higher taxes on capital income that will lead to a reduction in savings. Such a policy—despite distorting the intertemporal allocation of consumption—may increase welfare in a dynamically inefficient economy but not in a dynamically efficient economy.

3.3.3 Altruism

In this section, we show that the decentralized economy is equivalent to the command economy if both generations are altruistic and the bequest motive is operative. First, we review some empirical evidence from the US economy on bequests, before we implement bequests in the standard OLG model.

In this section, all bequests are planned and non-accidental. In Sect. 6.4, we will add accidental bequests. In this case, survival from age j to age $j+1$ will be stochastic, and in the event of death, the household may leave unintended bequests. In general equilibrium, therefore, we have to model what happens to accidental bequests.[16]

3.3.3.1 Bequests

Bequests are important in real life. They contribute a large share of the total wealth that households accumulate over their life-cycle. Using US data, Kotlikoff and Summers (1981) estimate the contribution of intergenerational transfers to aggregate capital accumulation. They find that intergenerational transfers account for the largest share of aggregate US capital formation; only a small share of capital formation results from the life-cycle saving motive. Gale and Scholz (1994) find that 3.7% of households in the US Survey of Consumer Finances (SCF) report receiving an inheritance between 1983 and 1985. Conditional on receipt, the mean inheritance amounts to \$42,729, while aggregate inheritance amount to 2.65% of GNP. Using PSID data, Hendricks (2007) estimates a lower value of aggregate inheritance amounting to 1.85% of aggregate output. He finds that households inherit 2.4 times their gross mean annual income or \$55,000 in 1994. However, because it is difficult to distinguish between intended and unintended bequests, Gale and Schulz also analyze the transfers between generations (between parent and child) during a parent's lifetime and estimate that intended transfers account for at least 20% of net worth. In this regard, Skinner and Zeldes (2002) provide evidence that most bequests are accidental, rather than intended. Only 8% of all households state inheritance as an operative motive for accumulating savings.

Therefore, both an operative bequest motive and accidental bequests may help to solve a number of savings puzzles that are obtained in heterogeneous-agent OLG models.[17] In these models:

1. the wealth concentration is smaller than observed empirically. This puzzle has been studied by Huggett (1996), among others.

---

[16]There will be three options: (i) We may introduce a parent-child link. (ii) The government may confiscate accidental bequests. Alternatively, (iii), we may assume a perfect annuities market such that financial intermediaries invest the funds on behalf of the household, which receives a higher return in the event of survival. Otherwise, the financial intermediary will receive the assets.

[17]Typically, these models consider households with different income and/or individual productivity levels. We will analyze these types of models in later chapters.

2. Rich households have high saving rates (see, for example, Dynan, Skinner, and Zeldes 2004).

A parent that leaves intended bequests to his child may have various motives for doing so. One is gift exchange, e.g., the parent promises the child higher bequests if the child spends more time with him or takes care of him. Another motive is pure altruism, on which we focus hereinafter. Let the parent's utility be given by

$$V_t = u(c_t^1) + \beta u(c_{t+1}^2) + \frac{1}{1+R} V_{t+1}. \tag{3.29}$$

Parents discount the lifetime utility of their children at rate $R > 0$. Equation (3.29) can be solved recursively forward to yield:

$$V_t = \sum_{i=0}^{\infty} \frac{1}{(1+R)^i} \left[u(c_{t+i}^1) + \beta u(c_{t+i+1}^2) \right].$$

Evidently, this expression is equivalent to that of the central planner (with possibly different discount rates R, however, and an infinite planning horizon T) provided in (3.23). The budget constraints of the cohort born in period t (while young) and period t (while old) are represented by:

$$c_t^1 + s_t = w_t + beq_t, \tag{3.30a}$$

$$c_{t+1}^2 + (1+n)beq_{t+1} = (1+r_{t+1})s_t. \tag{3.30b}$$

where $beq_t \geq 0$ denotes the bequest received by each member of generation t. Notice that bequests cannot be negative: You cannot leave your debt to your children. These two budgets can be combined into the following intertemporal budget constraint:

$$c_t^1 + \frac{c_{t+1}^2 + (1+n)beq_{t+1}}{1+r_{t+1}} = w_t + beq_t. \tag{3.31}$$

Maximizing (3.29) subject to (3.31) results in the following first-order conditions[18]:

$$\frac{\partial}{\partial c_t^1} : \lambda_t = u'(c_t^1), \tag{3.32a}$$

$$\frac{\partial}{\partial c_{t+1}^2} : \frac{\lambda_t}{1+r_{t+1}} = \beta u'(c_{t+1}^2), \tag{3.32b}$$

[18] For details, see Appendix 3.1.

$$\frac{\partial}{\partial beq_{t+1}} : \lambda_t \frac{1+n}{1+r_{t+1}} \geq \frac{\lambda_{t+1}}{1+R}. \tag{3.32c}$$

where λ_t denotes the Lagrange multiplier of the generation-t budget constraint.

Therefore,

$$u'(c_t^1) = \beta u'(c_{t+1}^2) \left[1 + r_{t+1}\right], \tag{3.33a}$$

$$u'(c_t^1) \frac{1+n}{1+r_{t+1}} \begin{cases} = \frac{u'(c_{t+1}^1)}{1+R} & \text{if } beq_{t+1} > 0, \\ \geq \frac{u'(c_{t+1}^1)}{1+R} & \text{if } beq_{t+1} = 0. \end{cases} \tag{3.33b}$$

Let us consider the steady state so that we can drop the time index from the variables. Evidently, if $beq > 0$ and for $r = f'(k)$, these conditions coincide with those of the command optimum, and consequently, the economy with altruism is Pareto-efficient if $f'(k) > n$. Moreover, in steady state, $c_t^1 = c_{t+1}^1$, and for the case in which $beq > 0$, the modified golden rule follows:

$$(1+r) = (1+n)(1+R).$$

For the case in which $beq = 0$, however, the modified golden rule does not hold:

$$(1+r) \leq (1+n)(1+R).$$

Is the assumption of parents' altruism realistic? This question will become important again in Chap. 7, where we study Ricardian equivalence, i.e., whether it matters if additional government expenditures are financed by means of debt or lump-sum taxes. In the case of altruistic households, parents consider the effect of higher debt on future taxes and, therefore, the utility of their children. As a consequence, parents act as if they themselves would have to pay the taxes in the future. Without altruism, they do not increase their savings to compensate their children for the loss in net income. Therefore, altruism is a pre-condition for Ricardian equivalence. However, Altonij, Hayashi, and Kotlikoff (1997) show that the implications of altruism for intergenerational risk-sharing behavior are rejected empirically.

3.3.4 Dynamics in the Command Optimum

In the following, we examine the dynamics in the economy with a central planner. The example is instructive in the sense that it shows (1) that the adjustment dynamics in the OLG model often take longer than in the Ramsey model, which will be an important result for the applications in later chapters on pensions and the demographic transition, and (2) how this problem can be solved with simple computational methods, i.e., the solution of a non-linear function in one variable.

First, we need to specify functional forms for the production and utility functions, before we select the parameter values. We choose the same specification as in the example in Sect. 3.2.6: Production is characterized by a Cobb-Douglas function, $y = f(k) = k^\alpha$ with $\alpha = 0.36$, and instantaneous utility is assumed to be logarithmic, $u(c) = \ln c$. Utility in old age is discounted with the factor $\beta = 0.40$, while the population grows at rate $n = 0.1$. The central planner equally weights the utilities of the different generations, $R = 0$. Furthermore, the period length is equal to 30 years.

We study the dynamics over a time horizon of 20 periods. Denote the steady-state value of the capital stock by $k_t = k$. The initial and the terminal capital stock, k_0 and k_{21}, are given by $k_0 = k_{21} = 0.7k$. We know from the Ramsey model (and the Turnpike theorem) that capital should rise rapidly from its initial value to k and rest there almost until the end of the time horizon.

To compute the dynamics, we first have to solve for the steady state:

$$k = \left(\frac{\alpha}{n}\right)^{\frac{1}{1-\alpha}} = 7.400,$$

implying $k_0 = k_{21} = 0.7k = 5.180$.

To derive the dynamics, we have to insert the functional specifications of production $f(.)$ and instantaneous utility $u(.)$ into (3.26a), (3.26b), and (3.26c), yielding:

$$\frac{\beta}{c_t^2} = \frac{1}{1+n}\frac{1}{c_t^1}, \tag{3.34a}$$

$$\frac{1}{c_t^1} = \frac{1}{1+n}\left[1 + \alpha k_{t+1}^{\alpha-1}\right]\frac{1}{c_{t+1}^1}, \tag{3.34b}$$

$$k_t + k_t^\alpha = (1+n)k_{t+1} + c_t^1 + \frac{c_t^2}{1+n}. \tag{3.34c}$$

After some algebra, (3.34) can be solved for a second-order difference equation in k_t:

$$k_{t+2} = \frac{k_{t+1} + k_{t+1}^\alpha}{1+n} - \frac{1 + \alpha k_{t+1}^{\alpha-1}}{(1+n)^2}\left[k_t + k_t^\alpha - (1+n)k_{t+1}\right]. \tag{3.35}$$

The solution to this first-order condition (an initial and a final value problem) is provided in Fig. 3.8.

3.3.4.1 Numerical Computation

The solution of the problem consists in finding the sequence of $T = 20$ capital stocks $\{k_1, \ldots, k_T\}$, Formally, this is a non-linear equations problem in T variables k_t, $t = 1, \ldots, T$, i.e., we have to solve the 20 equations (3.35) for $t = 0, \ldots, 19$ given the initial and final values of the capital stock, k_0 and k_{21}.

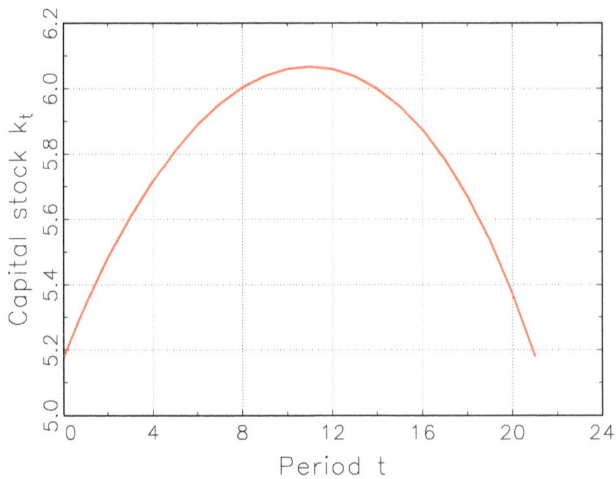

Fig. 3.8 Dynamics of the capital stock k_t in the command optimum

Solving large-scale non-linear equation problems, however, can be quite cumbersome, as we argued in Appendix 2.2. However, we can conveniently transform the problem of 20 non-linear equations with 20 unknowns into finding the solution to a non-linear equation in *one* variable, namely k_1, by making use of the recursive nature of the problem. In fact, given a guess $\tilde{k}_1$ for k_1, we can compute a guess for $\tilde{k}_2$ for k_2 with the help of (3.35) for period $t = 0$. Given the values $\tilde{k}_1$ and $\tilde{k}_2$, we can compute $\tilde{k}_3$ from (3.35) with $t = 1$. We can recursively continue in this fashion for $t = 2, \ldots, 18$. Having finally computed $\tilde{k}_{19}$ and $\tilde{k}_{20}$ by this recursive procedure, we can also compute $\tilde{k}_{21}$ from (3.35) for period $t = 19$. If this value is equal to the given final capital stock, k_{21}, we are done. Otherwise, we need to search again and find a new guess for k_1. This method is called *forward shooting*.

Therefore, we only solve the one non-linear equation $g(\tilde{k}_1) = \tilde{k}_{21} - k_{21}$. Numerically, this recursive function is implemented in the Gauss program `Ch3_turnpike.g` and MATLAB program `Ch3_turnpike.m`. The non-linear function is implemented in the procedure *kdyn()*:

```
proc kdyn(x);
local i,kt;
    kt=zeros(bigt+2,1); // time series for capital stock
kt[1]=k0;
kt[2]=x;
i=2;
do until i==bigt+2;
i=i+1;
kt[i]=(kt[i-1]+kt[i-1]^(alpha) ) / (1+n)
- (1+alpha*kt[i-1]^(alpha-1)) / (1+n)^2 *
( kt[i-2]+kt[i-2]^(alpha) -(1+n)*kt[i-1] );
endo;
retp(kt[bigt+2]-kfinal);
endp;
```

In the procedure, *k0* and *kfinal* denote k_0 and $k_{T+1} = k_{21}$, while *bigt* denotes the number of transition periods T. In lines 9–11, (3.35) is implemented, and the iteration is carried out in lines 6–12. The procedure returns the value $\tilde{k}_{21} - k_{21}$.

With regard to the present computation, some comments are in order:

- Finding the solution to a non-linear equation is almost as much an *art* as a science.
- In the present problem, Alfred Maußner's routine *FixVMN1* that we use will be able to find the solution for a given initial value, while the GAUSS-provided non-linear equation solver *eqSolve* will not (in Gauss 11.0). You may want to test this by substituting the command *eqsolve(.)* in the present program.
- Finding the solution depends on good initial guesses.
- In our problem, if we set $T = 100$, we cannot find a solution for the given initial guess for k_1.
- Given these difficulties in solving non-linear problems, you often have to experiment with:
  1. different solution algorithms and
  2. initial guesses, or
  3. you may look for a reformulation of the non-linear equation. Let us consider the example of a typical first-order condition for the household in the Ramsey model:

$$u'(c(k_t)) = \beta u'(c(k_{t+1}))\left[1 + f'(k_{t+1})\right],$$
$$k_{t+1} + c_t(k_t) = f(k_t) + k_t.$$

     It is possible that the non-linear equation solver will not find a solution $\{k_t^*, k_{t+1}^*\}$ for this system. In this case, one possible remedy to this problem is to reformulate the above equations:

$$1 = \beta \frac{u'(c(k_{t+1}))}{u'(c(k_t))}\left[1 + f'(k_{t+1})\right],$$
$$k_{t+1} + c_t(k_t) = f(k_t) + k_t.$$

     Due to the reformulation, the Jacobian of the system and therefore the Newton step described in the Algorithm 2.1 change.
  4. In the present problem, we began with a guess for k_1 and iterated forward (and computed k_2 from (3.35)). Alternatively, one could attempt to iterate backward in time and provide a guess $\tilde{k}_T$ for k_T and compute $\tilde{k}_{T-1}$ from (3.35). This method is called *backward shooting*.
  5. In the present case of a non-linear system of equations in one variable, it is always a good idea to graph the non-linear equation as a function over a grid and choose one value that is close to the solution $g(\tilde{k}_1) \approx 0$. In Fig. 3.9, $kdyn(k_1)$ is presented for different values of k_1, and an initial guess $\tilde{k}_1 = 5.35$

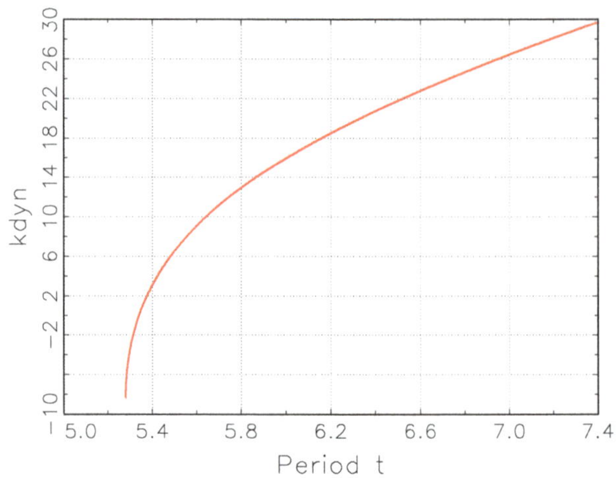

Fig. 3.9 Values of the residual function *kdyn()* over a grid of k_1

appears to be close to the solution. If you graph the function with the help of numerical software, you need to be careful with values of k_1 that produce a missing value, e.g., if the GAUSS program has to execute a command such as k_t^α with $k_t < 0$.[19]

3.4 The Two-Period OLG Model with Technological Progress

In OLG models, we often need to consider economic growth for the following reasons, among others:

1. Growth helps to alleviate the pressure on the pension system during the demographic transition.
2. Indebted governments hope to grow out of their debt.
3. Growth may create jobs.

In the following, we will extend the two-period OLG model for growth. The young population supplies one unit of labor inelastically and grows at rate n:

$$N_t = (1+n)N_{t-1}.$$

[19]Study how we implemented these conditions in the Gauss program *Ch3_turnpike.g*.

Again, total labor supply is equal to N_t. Instantaneous utility is assumed to be logarithmic so that lifetime utility is presented by[20]:

$$U_t = \ln(c_t^1) + \beta \ln(c_{t+1}^2) \tag{3.36}$$

Production is described by a Cobb-Douglas function:

$$Y_t = K_t^{\alpha} \left(A_t N_t\right)^{1-\alpha}, \tag{3.37}$$

where we have introduced labor-augmenting technological progress that grows at the exogenous rate γ:

$$A_t = (1+\gamma) A_{t-1}. \tag{3.38}$$

Accordingly, the household now receives wage rate w_t per *efficiency unit* A_t:

$$s_t = w_t A_t - c_t^1, \tag{3.39a}$$

$$c_{t+1}^2 = (1 + r_{t+1}) s_t. \tag{3.39b}$$

The first-order condition of the household optimization problem is given by:

$$s_t = \frac{\beta}{1+\beta} A_t w_t.$$

Therefore, savings s_t are proportional to the wage per efficiency unit w_t.

Firms rent savings s_t in the form of capital from the young agents at the end of period t, use it in production during period $t+1$ and repay the complete amount of savings at the end of period $t+1$. Therefore, they also have to bear the cost of depreciated capital when they repay the savings. Firms maximize profits:

$$\Pi_t = K_t^{\alpha} \left(A_t N_t\right)^{1-\alpha} - w_t A_t N_t - r_t K_t - \delta K_t,$$

implying the first-order conditions

$$w_t = (1-\alpha) \left(\frac{K_t}{A_t N_t}\right)^{\alpha} = (1-\alpha) k_t^{\alpha}, \tag{3.40a}$$

$$r_t = \alpha k_t^{\alpha - 1} - \delta, \tag{3.40b}$$

[20]You need to be careful to specify the utility functions in OLG models with economic growth. For an intertemporal elasticity of consumption that is not equal to one, it makes a difference whether utility is specified as a function of individual consumption or individual consumption per efficiency unit. For details see Appendix 3.2.

where $k = K/AN$ is defined as capital per efficiency unit of labor. For convenience, we will use the terms capital and capital stock per efficiency unit of labor interchangeably to refer to k.

In capital market equilibrium, total capital demand is equal to total savings:

$$N_t s_t = K_{t+1}.$$

Therefore,

$$N_t s_t = N_t \frac{\beta}{1+\beta} A_t w_t = K_{t+1},$$

or, after dividing by efficient labor $A_t N_t$ on both sides

$$k_{t+1}(1+\gamma)(1+n) = \frac{\beta}{1+\beta}(1-\alpha)k_t^{\alpha}. \tag{3.41}$$

3.4.1 Steady State

In steady state, the capital stock per efficiency unit of labor is constant, implying:

$$k = \left(\frac{\beta}{1+\beta} \frac{(1-\alpha)}{(1+\gamma)(1+n)} \right)^{\frac{1}{1-\alpha}}.$$

In steady state, the per capita capital stock grows at rate γ. The real interest rate r is given by:

$$r = \alpha k^{\alpha-1} - \delta = \frac{1+\beta}{\beta} \frac{\alpha}{1-\alpha}(1+\gamma)(1+n) - \delta.$$

3.4.2 Pareto Efficiency and the Golden Rule

To study Pareto efficiency in the model with growth, we consider the benevolent central planner's problem. The central planner's resource constraint is given by:

$$(1-\delta)K_t + F(K_t, A_t N_t) = K_{t+1} + N_t c_t^1 + N_{t-1} c_t^2.$$

Defining c_t as above and dividing by $A_t N_t$, we obtain:

$$(1-\delta)k_t + f(k_t) = (1+n)(1+\gamma)k_{t+1} + c_t,$$

with

$$c_t \equiv \frac{N_t c_t^1 + N_{t-1} c_t^2}{A_t N_t},$$

In steady state, $k_t = k$, and thus,

$$f(k) = (n + \gamma + n\gamma + \delta)k + c.$$

If n and γ are small, we can drop the term $n\gamma$. Notice, however, that if a period corresponds to 30 years, $n\gamma$ might not be quantitatively negligible; therefore, we retain the term in the following. Differentiating the above equation with respect to k, we obtain

$$\frac{dc}{dk} = f'(k) - \delta - (n + \gamma + n\gamma).$$

The economy, therefore, is dynamically inefficient ($dc/dk < 0$) if

$$r = f'(k) - \delta < n + \gamma + n\gamma.$$

Accordingly, the economy is not Pareto-efficient if the interest rate is smaller than the sum of the population growth rate n and the economic growth rate γ (and the joint product $n\gamma$). In comparison with the case without growth, $\gamma = 0$, this condition is more likely to occur. The real interest rate and the growth rate in the US postwar period have been close to one another, depending on how one measures the real interest rate, i.e., what asset one considers (government bonds, stocks). Mehra and Prescott (2003) find that the average (annual) real return of US government bonds (US T-Bills) amounted to 1.19% during the period 1889–2000, while US stocks had an average (annual) return of 8.06% during this period.[21] In comparison, the average (annual) growth rate of real income in the US during the period 1950–2010 was equal to 2.2% (using Penn Word Tables, Version 6-1, by Summers, Heston and Aten).

Appendix 3.1: Kuhn-Tucker First-Order Conditions in the Model with Altruistic Bequests

The model with altruistic bequests involves the constraint

$$beq_{t+1} \geq 0.$$

[21]Jagannathan, McGrattan, and Scherbina (2000) note that the so-called equity premium, the difference between the two returns, averaged approximately 7% points during the period 1926–1970 and only approximately 0.7 of percentage points thereafter.

Households cannot leave negative bequests.

To be more precise, the optimization problem needs to be solved using the *Kuhn-Tucker* method. The Lagrange function is given by

$$\mathscr{L} = \sum_{t=0}^{\infty} \frac{1}{(1+R)^t} \left\{ \left[u(c_t^1) + \beta u(c_{t+1}^2) \right] + \right.$$
$$\left. \lambda_t \left[w_t + beq_t - c_t^1 - \frac{c_{t+1}^2 + (1+n)beq_{t+1}}{1+r_{t+1}} \right] + \mu_t beq_{t+1} \right\}$$

with the first-order conditions (with respect to beq_{t+1}):

$$\frac{1}{1+R}\lambda_{t+1} = \lambda_t \frac{1+n}{1+r_{t+1}} - \mu_t, \tag{3.42a}$$

$$\mu_t \cdot beq_{t+1} = 0, \tag{3.42b}$$

$$\mu_t \geq 0. \tag{3.42c}$$

Two cases can be distinguished:

1. $beq_{t+1} > 0$: In this case, $\mu_t = 0$, and

$$\frac{1}{1+R}\lambda_{t+1} = \lambda_t \frac{1+n}{1+r_{t+1}}.$$

2. $beq_{t+1} = 0$: $\mu_t \geq 0$, and

$$\frac{1}{1+R}\lambda_{t+1} \leq \lambda_t \frac{1+n}{1+r_{t+1}}.$$

The conditions are restated in (3.33).

Appendix 3.2: Utility Function and Economic Growth

In the OLG model with economic growth, we assumed the functional form of lifetime utility to be represented by (3.36). In this appendix, we study the sensitivity of savings with respect to the choice of the arguments c_t^1 and c_{t+1}^2 in the utility function.

There are basically two definitions of the variables c_t^1 and c_t^2, we might choose from. (1) We can define c_t^1 and c_{t+1}^2 as the per capita consumption of the generation born in period t, $c_t^1 = C_t^1/N_t$ and $c_{t+1}^2 = C_{t+1}^2/N_t$, where C_t^s denotes total consumption of the s-year old household in period t. This seems to be a natural formulation of preferences. (2) We can employ the arguments $\tilde{c}_t^1 \equiv C_t^1/(A_t N_t)$ and $\tilde{c}_{t+1}^2 \equiv C_{t+1}^2/(A_{t+1}N_t)$. We can interpret this alternative behavior in the sense that consumption habits adjust to technological change. The newest computer in 2000 provides the current household with the same utility as the newest computer in 2020 does. As an implication, households do not grow happier over time, which is in accordance with the so-called *Easterlin paradox*. According to this paradox, a higher level of a country's per capita gross domestic product does not correlate with greater self-reported levels of happiness among its citizens.[22]

Next, we turn to the question how these two formulations of preferences affect the savings behavior of the household. For this reason, let us generalize life-time utility (3.36) to the case with a constant intertemporal elasticity of substitution $1/\sigma$ and (1) let us use individual consumption (or, equally, consumption per household member) as its arguments:

$$U_t = \frac{\left(C_t^1/N_t\right)^{1-\sigma} - 1}{1-\sigma} + \beta \frac{\left(C_{t+1}^2/N_t\right)^{1-\sigma} - 1}{1-\sigma}. \tag{3.43}$$

Total savings of the young household with N_t members is represented by

$$S_t = A_t N_t w_t - C_t^1,$$

and total consumption of the household in period 2 is equal to savings plus interest earnings

$$C_{t+1}^2 = (1 + r_{t+1})S_t,$$

such that the intertemporal budget constraint can be formulated in terms of individual consumption $c_t^1 \equiv C_t^1/N_t$ and $c_{t+1}^2 \equiv C_{t+1}^2/N_t$:

$$c_t^1 + \frac{c_{t+1}^2}{1 + r_{t+1}} = A_t w_t. \tag{3.44}$$

Accordingly, we can formulate the Lagrange function of the household as follows:

$$\mathscr{L} = \frac{\left(c_t^1\right)^{1-\sigma} - 1}{1-\sigma} + \beta \frac{\left(c_{t+1}^2\right)^{1-\sigma} - 1}{1-\sigma} + \lambda_t \left[A_t w_t - c_t^1 - \frac{c_{t+1}^2}{1 + r_{t+1}} \right].$$

[22]However, the validity of the Easterlin paradox is not undisputed; see, for example, Clark, Frijters, and Shields (2008).

The first-order conditions of the household's maximization problem follow from the derivation of the above Lagrangean with respect to c_t^1 and c_{t+1}^2:

$$\lambda_t = \left(c_t^1\right)^{-\sigma}, \tag{3.45a}$$

$$\frac{\lambda_t}{1+r_{t+1}} = \beta\left(c_{t+1}^2\right)^{-\sigma}, \tag{3.45b}$$

and, therefore,

$$\left(\frac{c_{t+1}^2}{c_t^1}\right)^{\sigma} = \beta(1+r_{t+1}). \tag{3.46}$$

Inserting this first-order condition in the intertemporal budget constraint (3.44), we derive optimal individual consumption

$$c_t^1 = \frac{A_t w_t}{1+\beta^{\frac{1}{\sigma}}(1+r_{t+1})^{\frac{1}{\sigma}-1}} \tag{3.47}$$

and, hence, savings

$$\begin{aligned} s_t &\equiv \frac{S_t}{A_t N_t} \\ &= w_t - \frac{c_t^1}{A_t} \\ &= w_t - \frac{w_t}{1+\beta^{\frac{1}{\sigma}}(1+r_{t+1})^{\frac{1}{\sigma}-1}} \\ &= \left[1 - \frac{1}{1+\beta^{\frac{1}{\sigma}}(1+r_{t+1})^{\frac{1}{\sigma}-1}}\right] w_t. \end{aligned}$$

Notice that we would have derived the same amount of optimal savings if we had used the arguments $c_t^1 \equiv C_t^1/(A_t N_t)$ and $c_{t+1}^2 \equiv C_{t+1}^2/(A_t N_t)$. In Chaps. 6 and 7, we will use this notation in the life-time utility of the household in a growing economy.

(2) Let us consider the second specification with lifetime utility

$$\tilde{U}_t = \frac{\left(C_t^1/(A_t N_t)\right)^{1-\sigma} - 1}{1-\sigma} + \beta\frac{\left(C_{t+1}^2/(A_{t+1} N_t)\right)^{1-\sigma} - 1}{1-\sigma}. \tag{3.48}$$

In this case, consumption habits adjust to the level of the current technology A_t. Let us now define $\tilde{c}_t^1 \equiv C_t^1/(A_t N_t)$ and $\tilde{c}_{t+1}^2 \equiv C_{t+1}^2/(A_{t+1}N_t)$. With this definition, we can formulate the budgets at age 1 and 2 as follows (using $A_{t+1}/A_t = 1+\gamma$):

$$s_t = w_t - \tilde{c}_t^1$$

$$\tilde{c}_{t+1}^2 = \frac{1+r_{t+1}}{1+\gamma} s_t,$$

and the intertemporal budget constraint

$$w_t = \tilde{c}_t^1 + \frac{1+\gamma}{1+r_{t+1}} \tilde{c}_{t+1}^2. \tag{3.49}$$

The household maximizes its Lagrangean function

$$\mathscr{L} = \frac{\left(\tilde{c}_t^1\right)^{1-\sigma} - 1}{1-\sigma} + \beta \frac{\left(\tilde{c}_{t+1}^2\right)^{1-\sigma} - 1}{1-\sigma} + \lambda_t \left[w_t - \tilde{c}_t^1 - \frac{1+\gamma}{1+r_{t+1}} \tilde{c}_{t+1}^2 \right],$$

with respect to $\tilde{c}_t^1$ and $\tilde{c}_{t+1}^2$ resulting in the first-order conditions:

$$\lambda_t = \left(\tilde{c}_t^1\right)^{-\sigma}, \tag{3.50a}$$

$$\lambda_t \frac{1+\gamma}{1+r_{t+1}} = \beta \left(\tilde{c}_{t+1}^2\right)^{-\sigma}, \tag{3.50b}$$

and, therefore,

$$\left(\frac{\tilde{c}_{t+1}^2}{\tilde{c}_t^1}\right)^{\sigma} = \beta \frac{1+r_{t+1}}{1+\gamma}. \tag{3.51}$$

Inserting this first-order condition in the intertemporal budget constraint (3.49), we derive optimal individual consumption

$$\tilde{c}_t^1 = \frac{w_t}{1+\beta^{\frac{1}{\sigma}} \left(\frac{1+r_{t+1}}{1+\gamma}\right)^{\frac{1}{\sigma}-1}}, \tag{3.52}$$

and, hence, savings

$$\begin{aligned} s_t &\equiv \frac{S_t}{A_t N_t} \\ &= w_t - \tilde{c}_t^1 \\ &= \left[1 - \frac{1}{1+\beta^{\frac{1}{\sigma}} \left(\frac{1+r_{t+1}}{1+\gamma}\right)^{\frac{1}{\sigma}-1}} \right] w_t. \end{aligned}$$

Evidently, the two specifications of the lifetime utility (3.43) and (3.48) imply different amounts of savings that are only equal to each other in the case of either no growth with $\gamma = 0$ or for $\sigma = 1$. Notice that the effect of the utility choice on savings depends on the intertemporal elasticity of substitution $1/\sigma$. Empirical evidence supports the hypothesis that this elasticity is below one, $1/\sigma \leq 1.0$. In this case, specification (3.43) results in lower savings than the alternative (3.48) for positive growth $\gamma > 0$.

Problems

3.1. Show that (3.12a) and (3.12b) hold in the case of a production function with constant returns to scale.

3.2. Show that the optimal savings function in the Numerical Example in Sect. 3.2.6 with log-linear utility and Cobb-Douglas production is given by (3.20).

3.3. Consider the Numerical Example in Sect. 3.2.6 with log-linear utility and Cobb-Douglas production.[23] Analyze whether the allocation in the market economy is efficient.

Next, analyze the effects of a transfer from the young to the old that is administered by a government authority that maintains a balanced budget. Compute the optimal (possibly negative) transfer. How does the transfer depend on the population growth rate? Consider different values $n \in \{0, 0.1, 0.2, \ldots, 2.0\}$.

3.4. Derive (3.35).

3.5. Compute the solution for the transition in the 20-period OLG example model by *backward shooting*, i.e., starting by providing a guess for k_T and finding the solution for k_{T-1} and so forth.

3.6. Compute the transition in the following 60-period finite-horizon Ramsey model:

Let U be given by a constant elasticity of substitution function

$$U(C_0, \ldots, C_T) := \left\{ \sum_{t=0}^{T} C_t^{\varrho} \right\}^{1/\varrho}, \qquad \varrho \in (-\infty, 1],$$

and define $f(K_t) := K_t^{\alpha}$, $\alpha \in (0, 1)$. Let the household maximize $U(C_0, \ldots, C_t)$ subject to the budget constraint

$$K_{t+1} = K_t^{\alpha} - C_t, \qquad t = 0, 1, \ldots, T, \tag{3.53}$$

[23] The problem is inspired by the example in Section 3.1 of de la Croix and Michel (2002).

implying the first-order conditions

$$\left[\frac{C_t}{C_{t+1}}\right]^{1-\varrho} \alpha K_{t+1}^{\alpha-1} = 1, \qquad t = 0, 1, \ldots, T-1. \tag{3.54}$$

If we eliminate consumption from the second set of equations using the first $T+1$ equations, we arrive at a set of T non-linear equations in the T unknowns $(K_1, K_2, \ldots, K_T)$:

$$\begin{aligned} 0 &= \left(\frac{K_1^\alpha - K_2}{K_0^\alpha - K_1}\right)^{1-\varrho} - \alpha K_1^{\alpha-1}, \\ 0 &= \left(\frac{K_2^\alpha - K_3}{K_1^\alpha - K_2}\right)^{1-\varrho} - \alpha K_2^{\alpha-1}, \\ &\vdots \\ 0 &= \left(\frac{K_T^\alpha}{K_{T-1}^\alpha - K_T}\right)^{1-\varrho} - \alpha K_T^{\alpha-1}. \end{aligned} \tag{3.55}$$

Solve the problem for $T = 59$, $\alpha = 0.35$, $\varrho = 0.5$, $k_0 = 0.1$ and $K_{60} = 0$.

References

Altonij, J. G., Hayashi, F., & Kotlikoff, L. (1997). Parental altruism and intervivos transfers: Theory and evidence. *Journal of Political Economy, 105*, 1121–1166.

Blanchard, O. J., & Fischer, S. (1989). *Lectures on macroeconomics*. Cambridge: MIT Press.

Clark, A. E., Frijters, P., & Shields, M. A. (2008). Relative income, happiness, and utility: An explanation for the Easterlin paradox and other puzzles. *Journal of Economic Literature, 46*(1), 95–144.

de la Croix, D., & Michel, P. (2002). *A theory of economic growth: Dynamics and policy in overlapping generations*. Cambridge: Cambridge University Press.

Diamond, P. (1965). National debt in a neoclassical growth model. *American Economic Review, 55*, 1126–1150.

Dynan, K. E., Skinner, J., & Zeldes, S. P. (2004). Do the rich save more? *Journal of Political Economy, 112*(2), 397–444.

Gale, W. G., & Scholz, J. K. (1994). Intergenerational transfers and the accumulation of wealth. *Journal of Economic Perspectives, 8*(4), 145–160.

Hendricks, L. (2007). Retirement wealth and lifetime earnings. *International Economic Review, 48*, 421–456.

Huggett, M. (1996). Wealth distribution in life-cycle economies. *Journal of Monetary Economics, 38*, 373–396.

Hurd, M. D. (1989). Mortality risk and bequests. *Econometrica, 57*, 799–813.

Jagannathan, R., McGrattan, E. R., & Scherbina, A. (2000). The declining U.S. equity premium. *Quarterly Review, Federal Reserve Bank of Minneapolis, 24*(4), 3–19.

Kotlikoff, L., & Summers, L. H. (1981). The role of intergenerational transfers in aggregate capital accumulation. *Journal of Political Economy, 89*(4), 706–732.

Mehra, R., & Prescott, E. C. (2003). The equity premium in retrospect. In G. Constantinides, M. Harris, & R. M. Stulz (Eds.), *Handbook of the economics of finance*. Amsterdam: Elsevier.
Samuelson, P. A. (1958). An exact consumption-loan model of interest, with or without the social contrivance of money. *Journal of Monetary Economics, 66*, 467–482.
Skinner, J., & Zeldes, S. P. (2002). The importance of bequests and life-cycle saving in capital accumulation: A new answer. *American Economic Review, 22*(2), 274–278.
Wickens, M. R. (2011). *Macroeconomic theory — A dynamic general equilibrium approach* (2nd ed.). Princeton/Oxford: Princeton University Press.

Part II

Fiscal Policy

Government Consumption

4

4.1 Introduction

Empirically, government expenditures represent a large share of total demand and significantly affect output, employment, and welfare. In the subsequent Sect. 4.2 of this chapter, we document some selected empirical facts of government consumption. In particular, we find that government consumption is procyclical, and after an unexpected increase in consumption, output, employment, and (to a smaller extent) private consumption all increase.

In Sect. 4.3, we show that in the neoclassical model with a government sector, permanent changes in government consumption have a larger impact on output than temporary changes in government consumption. In Sect. 4.4, we extend our analysis by introducing uncertainty and presenting a real business cycle (RBC) model that helps to replicate the empirical government consumption multiplier. A crucial model element is the assumption of the substitutability of private and public consumption in individual utility.

Finally, in Sect. 4.5, we present a complex New Keynesian business cycle model with sticky prices and wages that can account for empirical evidence of the effects of higher government consumption on output, private consumption, investment, and real wages. In this model, we study a government consumption rule with a reaction coefficient on output. We find that a government policy intended to minimize output fluctuations is characterized by a reaction coefficient of -1.3, i.e., the government should decrease its expenditures by 1.3% in response to GDP increasing 1% above its trend level.

4.2 Empirical Regularities

Government spending has increased in both absolute value and relative to GDP in the postwar period in most modern industrialized countries. Figure 4.1 presents government expenditures relative to GDP during the period 1980–2018 for selected

B. Heer, *Public Economics*, Springer Texts in Business and Economics,
https://doi.org/10.1007/978-3-030-00989-2_4

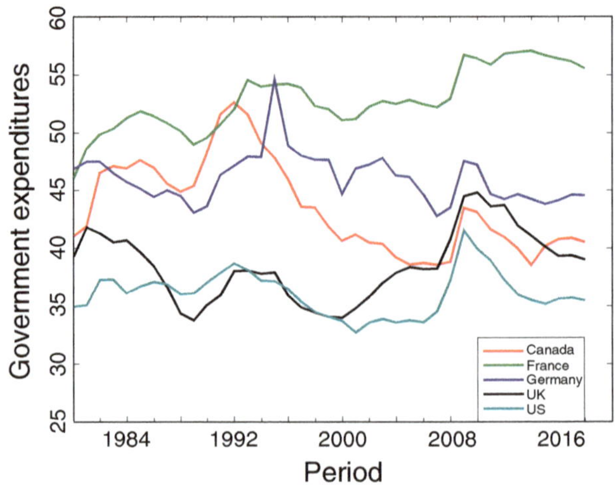

Fig. 4.1 Government expenditures relative to GDP, 1980–2018

major industrialized countries.[1] Clearly, the government share has increased during and in the aftermath of the Great Recession 2007–2009 in some countries, e.g., in France from 46% in 1980 to 56% in 2016, while there has not been any upward trend in countries such as Canada or Germany. Only during the recent financial crisis of 2009–2012 did governments unanimously increase their spending.[2] While government expenditures remain below 50% of GDP in most countries, the government share amounts to 55% in France today. From a historical perspective, all spending levels are extremely high. According to Hindriks and Myles (2006), government spending in these countries was approximately 10% of GDP in 1870 and only increased significantly during the years prior to and during World War I.

Figure 4.2 presents data on all OECD countries (as of this writing).[3] Chile and Mexico had the lowest government expenditures during the period 2013–2015, amounting to 24.0% and 24.5% of GDP. There are nine countries with a government share in excess of 50% of GDP, with the Finnish and French public sectors being the biggest spenders. Notice that the GIIPS countries (Greece, Italy, Ireland, Portugal and Spain) with the exception of Ireland and Spain are characterized by government shares of 50% of GDP or above. In addition, almost all OECD countries increased their spending during the recent financial crisis of 2007–2009,

[1]The data are taken from IMF, OECD, and Bundesbank statistics. Please see Appendix 4.6 for the documentation of the data. The numbers for the years 2016–2018 represent estimates. The statistics are loaded and graphed with the help of the Gauss program `Ch4_data.g`.

[2]The spike in government spending in Germany during the period 1990–1996 was caused by German reunification and the higher public spending in East Germany.

[3]In April 2015, the OECD invited Costa Rica and Lithuania to open formal OECD accession talks.

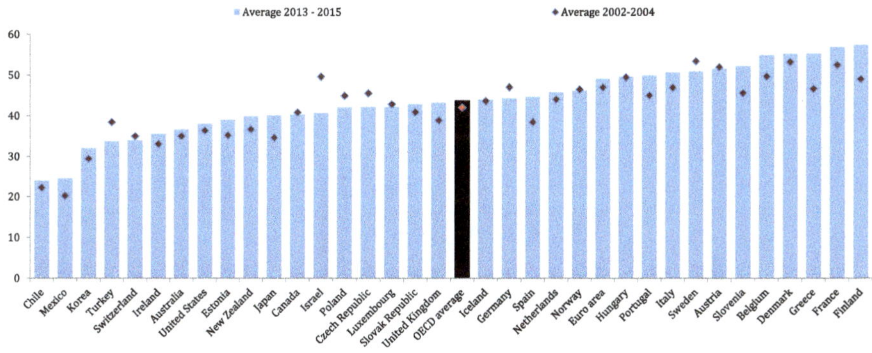

Fig. 4.2 Government expenditures relative to GDP, OECD countries, Average 2013–2015

with notable exceptions being Israel, Switzerland, Turkey, and Norway, among others.

In the following, we first present some evidence of the components of government spending. In the second part of this section, we examine the time series behavior of government consumption.

4.2.1 Composition of Government Spending

Government expenditures arise at various levels of the government, i.e., the local, state, or federal level. Hereinafter, we consolidate these numbers and present total government expenditures. Comparing the composition of government expenditures across countries, we notice important differences. For example, the US and, in particular, Israel spend much more on defense than Germany or France, while the latter countries have a much higher share of spending on social security and welfare. We will examine the most important components in turn.

The largest component of social spending in every major industrialized country is on public pensions and social security. Figure 4.3 displays the spending of OECD countries on public and private pensions in 2011 as a percentage of GDP. The OECD average of public spending on pensions amounted to 7.9% of GDP in 2011.[4] Pension spending varies significantly across countries. For example, Italy spent almost twice as much as the OECD average on pensions, amounting to 15.8% of GDP spent on public social security in 2011. Italy formerly had one of the most generous social security systems until it was revised in 1996 under the Dini reform. However, it will take many years until the reforms will manifest themselves in lower pension payments. Because of its important role in total government spending, we will separately examine this individual component in Chap. 6, which covers pensions.

[4]OECD average public and private pension spending amounted to 9.5%.

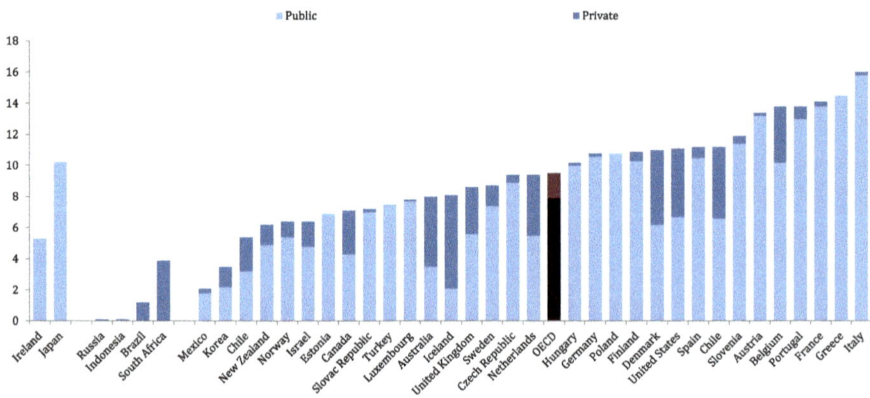

Fig. 4.3 Public and private pension expenditures, % of GDP, OECD, 2011

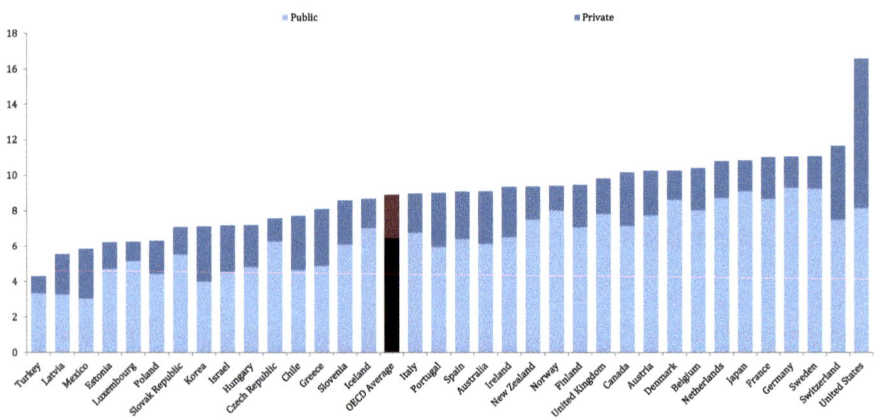

Fig. 4.4 Health expenditures, % of GDP, OECD, Average 2013–2015

The second largest component of public spending in most OECD countries is health expenditures. Figure 4.4 displays the average private and public health expenditures of OECD countries during 2013–2015. The OECD average of public health expenditures amounts to 6.5% of GDP. Notice that in many countries, private health expenditures are a large fraction of total health expenditures. For example, in the US economy, private health expenditures (8.4% of GDP) exceeded public health expenditures (8.2% of GDP) during 2013–2015.[5]

The third large component of government expenditures is education expenditures, as displayed in Fig. 4.5. The OECD average public spending on education

[5]Please take care not to equate (private and public) health expenditures with health. For example, Italy spends only half as much on health as the United States; however, in 2014, the average life expectancy in Italy was approximately 4 years longer than in the US.

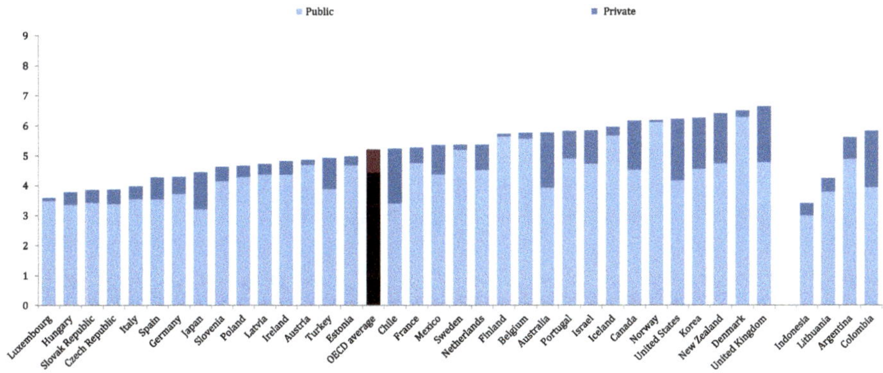

Fig. 4.5 Public and private education expenditures, % of GDP, OECD, 2014

amounted to 4.4% of GDP in 2014. Public expenditures on education varied between 3.2% in Japan and 6.3% in Denmark, while total education expenditures (private plus public) were highest in the UK.[6]

In later chapters, we will highlight the effect of taxes on growth. Therefore, a closer examination of how government expenditures affect growth rates may be fruitful. When addressing education, regressions often include years of schooling, rather than expenditure levels, as a proxy.[7] Barro and Sala-i-Martin (2003) review cross-country empirical evidence in Chapter 12 of their textbook and find a significant effect of years of schooling on growth, while health expenditure variables beyond life expectancy do not add explanatory power to the regression.

Figure 4.6 presents government spending on defense and public order and safety in 2012. Strikingly, the US and Israel use a larger share of their tax revenues for national and international defense than other countries. In these two countries, spending on defense amounted to 4.2% and 6.0% of GDP, respectively, while in countries such as Germany or Belgium, defense spending only accounts for approximately 1% of GDP.[8] Military spending by governments is often a very useful variable for economists in their empirical research because shocks to this spending component, such as the war on terror after 9/11, are usually completely

[6]Again, take care to not equate higher education spending with better education. In their article on "The Economics of International Differences in Educational Achievement" in the *Handbook of Economics of Education*, Hanushek and Woessmann (2011) review the literature on the determinants of educational attainment. In particular, they find that input measures such as class size or educational expenditures show little impact, while several measures of institutional structures such as school autonomy, later tracking, and the quality of the teaching force explain a significant portion of the international differences in student achievements.

[7]See also Footnote 6 in this section for an explanation of why years of schooling might represent a better measure of educational attainment.

[8]NATO members such as Germany and Belgium agreed to spend 2% of GDP on defense. Evidently, many NATO countries interpret this official target as a guideline.

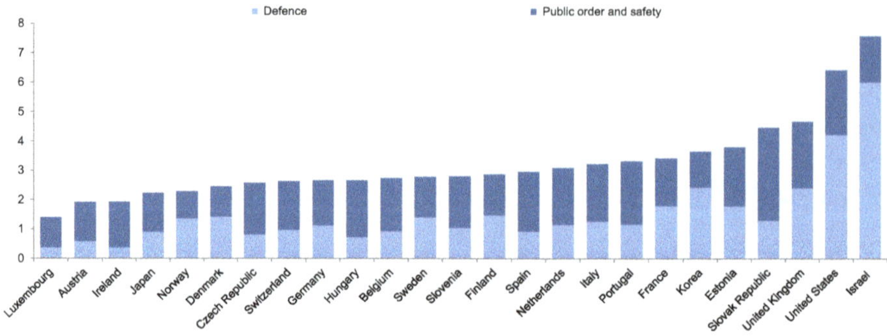

Fig. 4.6 Defense and public order and safety, % of GDP, OECD, 2012

exogenous. Therefore, this component often helps to identify parameters in vector autoregressions (VARs) or is used as an instrumental variable in econometric studies.

4.2.2 Time Series Evidence

Figure 4.7 presents US government consumption (in logarithms) during the period 1948–2014.[9] The original series is presented by the solid red line, while the broken green line presents the trend of the series after the Hodrick-Prescott (HP) filter has removed the cyclical component.[10] To compute the cyclical component of the time series, we apply the HP filter.[11] The HP filter removes all cycles longer than 32 quarters from the data while leaving shorter cycles unchanged.[12] The difference between the original series and the trend line is the cyclical component that is illustrated in Fig. 4.8.

Two periods in the cyclical government spending pattern are striking. First, cyclical government consumption increased during the Korean War in 1950–1953 and during the Vietnam War, both when US involvement increased significantly in 1961–1962 under president John F. Kennedy and also later in 1968–1969. For example, government consumption increased by 12% above its trend level in 1952. Second, government consumption increased again during and in the aftermath of the recent financial crisis of 2009–2010.

[9]Please see Appendix 4.6 for a description of the data on government consumption. The source of the data on real GDP, private consumption, and labor supply which is used in the computation of the results displayed in Table 4.1 is presented in Appendix 2.4.

[10]The GAUSS computer program *Ch4_data.g* together with the data file *Fred_data1a.txt* that computes Figs. 4.7 and 4.8 is available as a download from my web page.

[11]The HP Filter is described in Chap. 2 above. We use the parameter $\lambda = 1600$ for quarterly data.

[12]See Brandner and Neusser (1992).

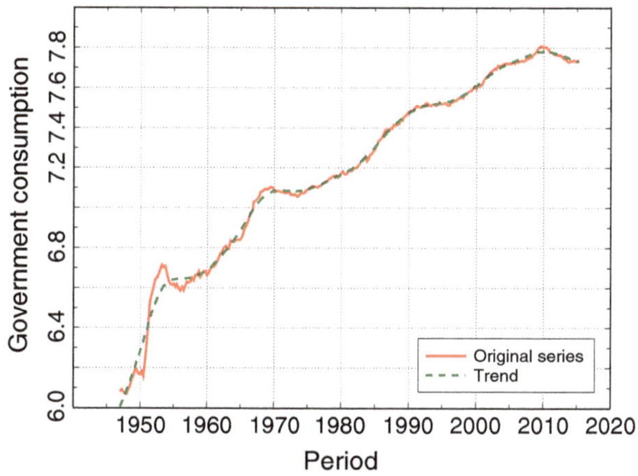

Fig. 4.7 US government consumption, 1948–2014

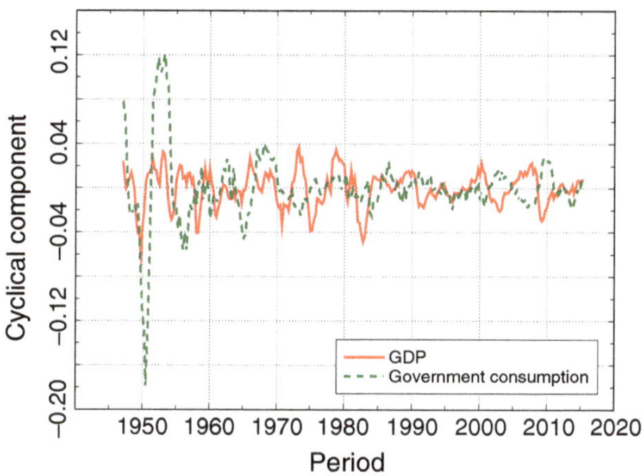

Fig. 4.8 Cyclical component of US GDP and government consumption, 1948–2014

In Fig. 4.8, we can also discern another important phenomenon. The years between the mid-1980s and the recent financial crisis were characterized by a time of low volatility in output (and, hence, consumption), the so-called *Great Moderation*. The explanations for the decline in output volatility are manifold. Clarida, Gali, and Gertler (2000) attribute the reduction in aggregate volatility to more effective monetary policy, whereas the good luck hypothesis proposed by Stock and Watson (2003) emphasizes the contribution of a reduction in the variance of business cycle shocks. Other studies, for example Davis and Kahn (2008) or Dynan, Elmendorf, and Sichel (2006), identify changes in inventory behavior or financial innovations as possible causes. Furthermore, Jaimovich and Siu (2009) provide empirical evidence in a cohort-based panel of the G7 countries

Table 4.1 Cyclical behavior of US government consumption

Variable	s_x	r_{xY}	r_{xG}
1948–2014			
Output Y	1.67	1.000	0.146
Public consumption G	3.25	0.146	1.000
Private consumption C^p	1.95	0.874	0.018
Hours L	1.29	0.773	−0.243
1956–2014			
Output Y	1.54	1.000	−0.108
Public consumption G	1.41	−0.108	1.000
Private consumption C^p	1.90	0.871	−0.297
Hours L	1.23	0.881	−0.169

Notes: $s_x := $ Standard deviation of time series x in percentages, where $x \in \{Y, G, C^p, L\}$. Empirical time series were HP filtered with weight 1600. $s_x :=$ standard deviation of the variable x. $r_{xY} :=$ Cross-correlation of the variable hours with output, $r_{xG} :=$ Cross-correlation of the variable with government consumption

that a demographic transition is closely linked to the volatility of cyclical output. Their analysis is supported by Lugauer (2012) and Lugauer and Redmond (2012), while Heer, Rohrbacher, and Scharrer (2017) emphasize the decline in labor supply elasticities across cohorts. In light of the empirical evidence presented in Fig. 4.8, it seems reasonable to add the decline in government consumption volatility as a likely candidate explanation for the Great Moderation.

Table 4.1 presents summary statistics for the HP-filtered (log) time series of US GDP, government consumption, private consumption, and labor. During the whole period 1948–2014, all variables were more volatile as measured by the standard deviation than during the subperiod 1956–2014. When we consider the data from the period after the Korean War (1956–2014), we find that output Y, with a standard deviation of 1.54%, is more volatile than government expenditures G (1.41%) and hours L (1.23%) but less volatile than private consumption C^p (1.90%).[13] During the period 1956–2014, government consumption is negatively correlated with output Y, labor L, and private consumption C^p. Apparently, government consumption is (weakly) countercyclical with a contemporaneous correlation coefficient of -0.108.

In addition, we find that the correlation of government consumption and output increases monotonically if we introduce more lags on output. As presented in

[13]The observation that private consumption is more volatile than output does not hold for all subperiods. For example, Cooley and Prescott (1995) find that the relative volatility of personal consumption with respect to output is only 74% in the US during the period 1954–1991. In addition, these authors document that durable consumption expenditures are much more volatile than the consumption of non-durables and services. Similarly, Heer and Maußner (2009) present empirical evidence for West Germany prior to German reunification over the period 1975–1989, when consumption is only approximately half as volatile as output.

Table 4.2 Correlation of government consumption and output

Lag i	0	1	2	3	4	5	6	7	8
$\text{Corr}(G_t, Y_{t-i})$	−0.108	−0.099	−0.076	−0.015	0.075	0.143	0.223	0.305	0.350

Table 4.2, the highest correlation of government consumption with output occurs with a lag on output of 8 quarters and amounts to 0.350 for the period 1956–2014. This observation may be interpreted as evidence that the government adjusts its budget to tax revenues with a decision and implementation lag of approximately two years. Alternatively, one could explain the observed lagged government consumption changes with the help of the additional time required to collect (income) taxes.

Our data on the cyclical behavior of government expenditures are related to the study of Ambler and Paquet (1996), who find that aggregate government spending in the US during the period 1959:1–1992:3 was weakly procyclical, with a contemporaneous correlation of 0.231.[14] In their study, the largest correlation between output and government consumption occurs with a lag of 5 quarters for government consumption. This observation lends further support to the hypothesis that the government decision occurs with a lag due to the lengthy decision-making process. Amber and Paquet also distinguish among various components of government spending, including military government spending (0.0951), non-military government expenditures (0.1737), and public investment (0.2733), where the contemporaneous correlations with output are given in parentheses. Total government consumption is as volatile as output, while the component of non-military government expenditures is less volatile, and the components of both military expenditures and public investment are more volatile than output.

In more elaborate empirical studies, researchers have applied VARs and analyzed the effects of a shock to government spending on economic variables.[15] The results can be summarized as follows:

- **GDP** (+) A government consumption shock increases GDP, e.g., in Blanchard and Perotti (2002). Using structural panel VAR analysis of four industrialized countries, Ravn, Schmitt-Grohé, and Uribe (2012) also provide cross-country evidence for this hypothesis.
- **Private Consumption** (+) There is mixed evidence and no clear consensus regarding the effect of government consumption on private consumption, but most studies find a positive effect, e.g., Blanchard and Perotti (2002), Galí and Lopez-Salido (2007), and Ravn, Schmitt-Grohé, and Uribe (2012).

[14] Compare this with Table 3 in Ambler and Paquet (1996). The time series are measured in logs and passed through the HP filter.

[15] In most vector autoregression studies, the assumption that no innovation other than government spending shocks can affect government spending within a given quarter is used for identification.

- **Labor (+)** Employment (total hours) increases after a shock to government consumption, e.g., in Blanchard and Perotti (2002).
- **Wages (+)** Rotemberg and Woodford (1992) find evidence that real wages also increase after a government spending shock, while Monacelli, Perotti, and Trigari (2010) only find a statistically insignificant increase of the real wage for men.
- **Investment (-)** Investment declines strongly after a government spending shock, e.g., in Blanchard and Perotti (2002).
- **Mark-up (-)** The mark-up declines, e.g., as in Monacelli and Perotti (2008). At the 5% confidence level, however, the decline in the mark-up is no longer statistically significant.
- **Interest Rates (- or 0)** With respect to the effect of government spending on real interest rates, the empirical evidence is more mixed. Murphy and Walsh (2016) survey the literature on this topic and come to the conclusion that real interest rates increase in response to higher government consumption in the US.[16] For example, Ramey and Shapiro (1998) and Fatás and Mihov (2001) provide empirical evidence that real T-bill rates increase after a positive government spending shock, while more recent studies such as Fisher and Peters (2010) and even Ramey (2011) find a negative (transient) effect of government spending on real interest rates.

Some qualifying remarks are warranted regarding the results from VAR studies. First, the cited VAR results are not completely robust with respect to different empirical methodologies. Standard regression analyses that simply include a dummy variable for a surge in government spending find that periods of higher fiscal spending result in employment increases, but real wages fall.

Second, Ramey (2011) criticizes standard VAR analyses because they assume that the effects of a shock to government consumption are manifest when the change can be observed in the data. Instead, she considers several variables, such as news in the journal *Business Week*, that can be used to estimate changes in the expected present value of government expenditures. For example, most military government spending is predetermined many quarters or even years ahead and available as public information, and economic agents react ahead of any actual change. After she corrects for these news shocks, she finds that most measures of consumption and wages fall.

Third, if one excludes the data from World War II and the Korean War (as suggested by a visual inspection of our Fig. 4.8), Ramey (2011) finds that shocks to temporary government spending actually lead to declines in output, hours, consumption, and investment. However, this view is challenged by Monacelli and Perotti (2008). These authors also use both narrative evidence and structural VARs and find that variations in government purchases generate an increase in consumption, while the real wage declines. Moreover, Monacelli, Perotti, and

[16]On p. 2 and in Table 1 on p. 8, Murphy and Walsh (2016) summarizes the studies on the relationship between interest rates and government spending shocks.

Trigari (2010) explain, in Section 2.2 of their article (pages 534–36), why they are skeptical that one can learn much from the results of Ramey (2011). In particular, the episodes that are crucial for the identification of news about military spending (World War II and the Korean War) are not typical of US economic history.

In summary, although empirical analysis has yielded some mixed evidence, most of the evidence suggests a positive effect of government consumption on output, private consumption, labor and real wages, while investment and, to a less significant extent, mark-ups and real interest rates fall. Why is this result important? As is often argued in the literature, this empirical evidence may help to discriminate between two different model types that we will study, the neoclassical model and the New Keynesian model. In the former, increases in government consumption imply a negative wealth shock, meaning that households increase labor and reduce private consumption in general (we will show that this need not be true if public and private consumption both enter utility). Moreover, real wages will fall. In New Keynesian models, however, private consumption (and wages) increase. Both models in their basic form also imply a rise in the interest rate.[17] We will successively discuss the two model types and their implications for the effects of government consumption in greater detail in the following.

4.3 Permanent and Temporary Changes in Government Consumption

In the following, we introduce government consumption into the deterministic Ramsey model with endogenous labor supply from Chap. 2. To study the role of government consumption, we assume that the individual derives utility from public consumption. For example, if you use your private car to ride on a public highway, utility from private consumption is enhanced.

To derive some analytical results, we will introduce public consumption following Barro (1981), Aschauer (1985), and Christiano and Eichenbaum (1992). Therefore, we define effective consumption as the sum of private and public consumption:

$$C_t = C_t^p + \mu G_t, \tag{4.1}$$

where the magnitude and sign of μ govern the effect of an increase in government consumption on marginal utility.

We will show in the following that if $\mu = 1$, an increase in public consumption perfectly crowds out private consumption; in other words, nothing happens following the increase. The increase in public consumption G_t is exactly balanced

[17] This need not hold in all specifications of New Keynesian models. For example, Heer and Scharrer (2018) introduce a variable price of capital into these types of models, and consequently, real interest rates decline in response to higher unanticipated government consumption.

by an equal reduction in private consumption C_t^p.[18] For $\mu < 1$, however, there is a change in total consumption C_t if government consumption G_t increases, and optimal government spending would be zero.[19]

We assume that the population in the model is constant and normalized to one. The representative household maximizes his intertemporal utility in period $t = 0$:

$$U = \sum_{t=0}^{\infty} \beta^t u(C_t, 1 - L_t), \tag{4.2}$$

where β, again, denotes the discount factor. The household supplies L_t units of labor, and its total endowment is normalized to one, and thus, $1 - L_t$ denotes leisure.

Instantaneous utility is specified as follows:

$$u(C, 1 - L) = \frac{\left(C^{\iota}(1 - L)^{1-\iota}\right)^{1-\sigma} - 1}{1 - \sigma}, \tag{4.3}$$

where $1/\sigma$ denotes the intertemporal elasticity of substitution, and ι and $1 - \iota$ are the relative weights of consumption and leisure in utility.

The household owns the capital stock K_t in period t, which evolves according to

$$K_{t+1} = (1 - \delta)K_t + I_t. \tag{4.4}$$

Capital K_t depreciates at rate δ. The household lends the capital stock to the firms, which pay real interest rate r_t. The household faces wage rate w_t, and thus, its labor income is equal to $w_t L_t$. In addition, it has to pay lump-sum taxes T_t.[20] Net household income is spent on private consumption C_t^p and savings, which are equal to the increase in capital holdings, $K_{t+1} - K_t$. Consequently, the household's budget constraint is represented by

$$w_t L_t + r_t K_t - T_t = C_t^p + K_{t+1} - (1 - \delta)K_t. \tag{4.5}$$

The first-order condition follows from the derivation of the Lagrangian

$$\mathscr{L} = \sum_{t=0}^{\infty} \beta^t \left[\frac{\left(C_t^{\iota}(1 - L_t)^{1-\iota}\right)^{1-\sigma} - 1}{1 - \sigma} + \lambda_t \left(w_t L_t + (1 + r_t - \delta)K_t - T_t - C_t^p - K_{t+1}\right) \right] \tag{4.6}$$

[18] For this observation to hold, we need to assume that the change in government consumption is fully anticipated.

[19] The special case of $\mu = 0$ is considered by Baxter and King (1993) and Aiyagari, Christiano, and Eichenbaum (1994) and is often adapted in business cycle research.

[20] Proportional income taxes will be introduced into the model in Chap. 5.

with respect to C_t^p, L_t, and K_{t+1}, taking government consumption G_t and taxes T_t to be exogenous. In particular, the first-order conditions of the household's utility maximization problem are given by:

$$\lambda_t = \iota C_t^{\iota(1-\sigma)-1}(1 - L_t)^{(1-\iota)(1-\sigma)}, \tag{4.7a}$$

$$\lambda_t w_t = (1 - \iota)C_t^{\iota(1-\sigma)}(1 - L_t)^{(1-\iota)(1-\sigma)-1}, \tag{4.7b}$$

$$\lambda_t = \lambda_{t+1}\beta(1 + r_{t+1} - \delta). \tag{4.7c}$$

According to (4.7a), the Lagrangian multiplier λ_t is equal to the marginal utility of (aggregate) consumption C_t. In (4.7b), the marginal utility from working an extra hour $\lambda_t w_t$ is equated to the marginal utility from an extra hour of leisure $(1 - \iota)C_t^{\iota(1-\sigma)}(1 - L_t)^{(1-\iota)(1-\sigma)-1}$. The Euler equation (4.7c) describes the optimal intertemporal consumption allocation.

Goods and factor markets are characterized by perfect competition. We, again, assume that production is described by a Cobb-Douglas technology:

$$Y_t = K_t^{\alpha} L_t^{1-\alpha}. \tag{4.8}$$

Firms rent capital from households. Therefore, wages and the real interest rate are given by:

$$w_t = (1 - \alpha)K_t^{\alpha} L_t^{-\alpha}, \tag{4.9a}$$

$$r_t = \alpha K_t^{\alpha-1} L_t^{1-\alpha}. \tag{4.9b}$$

Finally, the government budget is assumed to balance, meaning that

$$G_t = T_t. \tag{4.10}$$

In equilibrium, the resource constraint of the economy is presented by[21]

$$Y_t = C_t^p + G_t + I_t. \tag{4.11}$$

In the following, we study the long-run effects of a permanent change in government consumption G_t. Therefore, we analyze comparative steady states. In the steady state, the variables of the model are constant, $C_t = C$, $G_t = G$, $C_t^p = C^p$, $L_t = L$, $K_t = K$, $r_t = r$, and $w_t = w$. Consequently, $\lambda_t = \lambda$ and (4.7c) implies

$$1 + r - \delta = \frac{1}{\beta}.$$

[21]To derive the aggregate resource constraint (4.11), substitute (4.4), (4.8), (4.9), and (4.10) into (4.5).

As a consequence, the capital-labor coefficient K/L is given by

$$\frac{K}{L} = \left(\frac{\alpha}{\frac{1}{\beta} - 1 + \delta}\right)^{\frac{1}{1-\alpha}}. \tag{4.12}$$

Next, we derive the effect of higher government consumption on labor, capital, and output in steady state by performing a comparative steady-state analysis. Since r is given by (4.9b), the capital-output coefficient K/L is constant in all steady states, meaning that

$$dK = \frac{K}{L} dL. \tag{4.13}$$

Accordingly, the resource constraint

$$C^p + G + I = C^p + G + \delta K = K^\alpha L^{1-\alpha} \tag{4.14}$$

implies

$$dC^p + dG + \delta dK = \alpha \left(\frac{L}{K}\right)^{1-\alpha} dK + (1-\alpha)\left(\frac{K}{L}\right)^{\alpha} dL. \tag{4.15}$$

Substitution of (4.13) into (4.15) results in

$$dC^p + dG = \left[\alpha\left(\frac{K}{L}\right)^{\alpha} - \delta\frac{K}{L}\right] dL + w\, dL = (r-\delta)\frac{K}{L} dL + w\, dL. \tag{4.16}$$

From the household's first-order condition with respect to labor, we derive the steady-state condition

$$w = \frac{1-\iota}{\iota}\frac{C^p + \mu G}{1-L} \tag{4.17}$$

implying (remember that $dw = 0$ in steady state because K/L is constant):

$$-wdL = \frac{1-\iota}{\iota} dC^P + \frac{1-\iota}{\iota}\mu dG. \tag{4.18}$$

Inserting (4.18) in (4.16), we get

$$dC^p = -\frac{1+\mu\zeta}{1+\zeta} dG \tag{4.19}$$

with

$$\zeta = \frac{1-\iota}{\iota}\frac{C^p+G}{wL} > 0.$$

Clearly, $dC^p/dG = -1$ if and only if $\mu = 1$. In this case, an increase in public consumption is exactly matched by an equal decrease in private consumption. For $\mu < 1$, private consumption is reduced by less than the change in government consumption (or is increased if μ is sufficiently small and below $-1/\zeta$; most authors, e.g. Barro (1981), however, restrict the parameter space of μ to the unit interval and exclude the consideration of negative values for μ meaning that government consumption is not a good, but a bad which reduces utility).

Substituting dC^p from (4.19) into (4.16), we find that

$$\frac{dL}{dG} = \frac{\frac{\zeta(1-\mu)}{1+\zeta}}{\frac{C^p+G}{L}}, \tag{4.20}$$

and, therefore, $\frac{dL}{dG} > 0$ if $\mu < 1$.

Consequently, higher government purchases that are financed lump-sum increase the labor supply and, hence, the capital stock (because of (4.13)) and, therefore, output. In this case, government consumption actually *crowds in* investment in the long run.

How does this mechanism operate? To see this, assume that output and investment are constant. As a consequence, higher public consumption reduces private consumption one-to-one according to the aggregate resource constraint (4.11). Since, however, the utility weight of public consumption is smaller than that of private consumption for $\mu < 1$, C_t decreases, and the marginal utility of income increases according to (4.7a). Equivalently, the Lagrangian multiplier λ rises. Therefore, the incentives to work longer increase, and labor supply L is augmented (if leisure is a superior good).

Another way of thinking about this is that a decline in total consumption, $C = C^P + \mu G$, and the subsequent decline in utility results in a change in the household optimization problem. Since consumption and leisure are assumed to be substitutes, some of the decrease in total consumption is absorbed by a decrease in leisure or, equivalently, an increase in labor supply. From this reasoning, it is clear that the response of labor, capital, and output crucially depends on the elasticity of substitution between consumption and leisure, which we will study next by means of a computational example.

4.3.1 Numerical Example: Government Spending and the Substitutability of Private and Public Consumption

To derive a better understanding of the effects of government spending on equilibrium values of our variables, we assume now that effective consumption is represented by a constant elasticity of substitution (CES) aggregator:

$$C = \left[\phi \left(C^{p}\right)^{1-1/\rho_c} + (1-\phi)G^{1-1/\rho_c}\right]^{\frac{1}{1-1/\rho_c}}, \tag{4.21}$$

where the CES between private and public consumption is equal to ρ_c, and the relative weights of private and public consumption are represented by ϕ and $1-\phi$, respectively.[22] Notice that our specification (4.21) includes (4.1) as a special case for $\rho_c \to \infty$ and $\mu = (1-\phi)/\phi$.

Instantaneous utility is also a CES aggregator of effective consumption C and leisure $1-L$:

$$u(C, 1-L) = \frac{1}{1-\sigma}\left[C^{1-1/\rho} + \kappa\,(1-L)^{1-1/\rho}\right]^{\frac{1-\sigma}{1-1/\rho}}, \tag{4.22}$$

where ρ denotes the intratemporal elasticity of substitution between effective consumption C and leisure $1-L$, and the intertemporal elasticity of substitution is given by $1/\sigma$.

Inserting (4.21) and (4.22) into the Lagrangian,

$$\mathscr{L} = \sum_{t=0}^{\infty} \beta^t \left[u(C_t, 1-L_t) + \lambda_t \left(w_t L_t + (1+r_t-\delta)K_t - T_t - C_t^p - K_{t+1}\right)\right], \tag{4.23}$$

we can derive the following first-order condition for the labor supply L_t:

$$w_t = \frac{\kappa}{\phi}\left(\frac{C_t}{1-L_t}\right)^{\frac{1}{\rho}} (\Xi_t)^{1-\frac{1}{1-1/\rho_c}} \left(C_t^p\right)^{\frac{1}{\rho_c}}, \tag{4.24}$$

where the auxiliary variable Ξ_t is defined as a function of the variables C_t^p and G_t as follows:

$$\Xi_t \equiv \phi\left(C_t^p\right)^{1-1/\rho_c} + (1-\phi)G_t^{1-1/\rho_c}. \tag{4.25}$$

[22] Our notational convention in this book is that we use the variable x (X) as a subscript (superscript) of the parameter ρ, i.e. ρ_x (ρ^X), in case of a utility parameter (autoregressive parameter or parameter of a policy rule). Accordingly, ρ_c denotes the substitution elasticity of private and public consumption, while ρ^C denoted the autoregressive parameter of the technology shock in the consumption sector in Chap. 2.

The intertemporal first-order condition, again, is represented by (4.7c), meaning that in steady state, the capital intensity K/L is equal to (4.12). The steady state, therefore, is characterized by the following three equations in the endogenous variables K, L, and C^p for a given value of exogenous government expenditures G:

$$K = \left(\frac{\alpha}{\frac{1}{\beta} - 1 + \delta} \right)^{\frac{1}{1-\alpha}} L, \tag{4.26a}$$

$$K^\alpha L^{1-\alpha} = C^p + G + \delta K, \tag{4.26b}$$

$$(1-\alpha) K^\alpha L^{-\alpha} = \frac{\kappa}{\phi} \left(\frac{C}{1-L} \right)^{\frac{1}{\rho}} (\Xi)^{1-\frac{1}{1-1/\rho_c}} \left(C^p\right)^{\frac{1}{\rho_c}}. \tag{4.26c}$$

We can solve this system of non-linear equations in three endogenous variables with the help of a non-linear equation solver. The solution is implemented in the GAUSS program *Ch4_subs_private_pub.g*. To compute the steady state, we need to choose specific values for the parameters $\{\alpha, \beta, \delta, G, \sigma, \rho, \rho_c, \phi, \kappa\}$. The production elasticity of capital is set as $\alpha = 0.36$. If we consider a period length equal to one year, a discount factor of $\beta = 0.96$ (implying a real interest rate r equal to 4%) is appropriate. The annual depreciation rate is set equal to 10%. The government share G/Y is set equal to 20%, which is close to the values observed in the US economy. The intertemporal elasticity of substitution $1/\sigma$ is set to 1/2. The weight of private consumption and public consumption are set equal to 3/4 and 1/4, approximately representing the relative shares of private consumption (approximately 60% of GDP) and public consumption in GDP. The elasticity of substitution between leisure and consumption is set equal to $\rho = 0.6$ following Fehr, Kallweit, and Kindermann (2013).

The empirical evidence for the elasticity of substitution between private and public consumption ρ_c is less clear. For example, Ni (1995) provides a summary of studies (in addition to his own GMM study) on the empirical values of ρ_c for the US economy in the range of -1.8 to 0.8 with a median of approximately 0.3.[23] For a cross-country sample of Asian economies, Kwan (2006) finds that the elasticity of most countries is contained in the interval [0, 1], but some countries display values below or above this range. We will therefore test the sensitivity of our results for $\rho_c \in [0, 1]$.

The final parameter κ is separately calibrated for each value of ρ_c, such that the equilibrium labor supply is always equal to $L = 0.30$ in the case with $G/Y = 0.2$. As a consequence of our calibration, K, L, Y, and therefore G are constant for all values of the parameter ρ_c; in particular, government consumption amounts to

[23] Among others, the empirical estimates of the elasticities depend on the classification of the government expenditures, e.g., whether military spending is included.

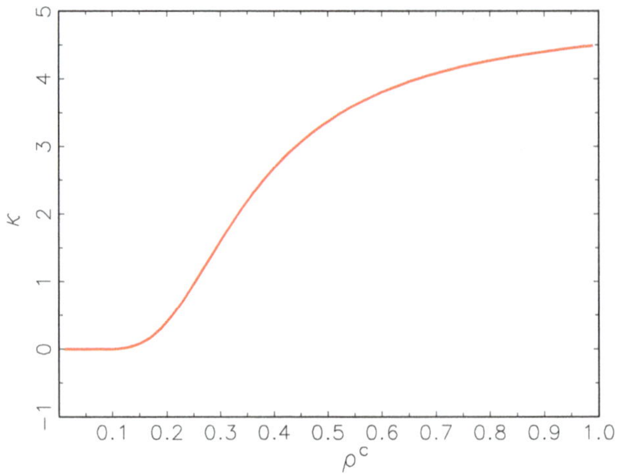

Fig. 4.9 Calibration of κ

$\bar{G} = 0.1014$. We simply use the non-linear system of equations (4.26) and solve it endogenously for the values of $\{\kappa, K, C^p\}$ given $L = 0.3$.[24] Our values for $\kappa(\rho_c)$ are illustrated in Fig. 4.9.

With the help of Gauss computer program *Ch4_subs_private_pub.g*, we compute the new steady-state values of $\{K, L, C^p\}$ for a 1% increase in government consumption to $\tilde{G} = 0.1024$ for each ρ_c (and the corresponding $\kappa(\rho_c)$). The effect on labor supply and output (presented in percentage changes) is illustrated in Fig. 4.10. In general, labor supply increases via the mechanism explained above. Marginal utility falls, ceteris paribus, for higher public consumption (and an equivalent decrease in private consumption), and the household increases labor supply because effective consumption C and leisure $1 - L$ are substitutes, $\rho > 0$. If private and public consumption are poorer substitutes (lower ρ_c), the effect becomes more pronounced. For example, for $\rho_c = 0.1$, the labor supply increases by 1%, from $L = 0.300$ to $L = 0.303$. The percentage change in production Y is exactly the same (because K/L remains fixed, such that K also increases by 1%, and since production is subject to constant returns to scale, output also increases by 1%). Accordingly, the long-run output government multiplier dY/dG amounts to nearly 5.0 (recall that $G/Y = 20\%$) for $\rho_c = 0.1$ and decreases to approximately 0.5 for $\rho_c \to 1.0$.

[24]In the Gauss computer program *Ch4_subs_private_pub.g*, the procedure *steadystate1(.)* is used in the non-linear equation problem solver to compute the solution to this problem.

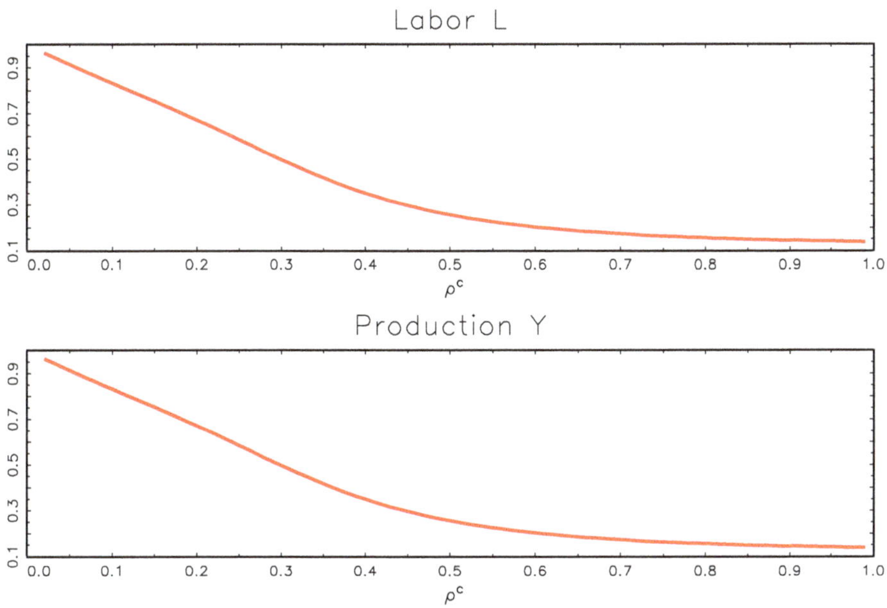

Fig. 4.10 Steady-state effect of higher permanent government expenditures. Labor L and production Y are presented as percentage changes relative to their steady-state values

4.3.2 Transition Dynamics After a Permanent and a Temporary Government Consumption Shock

Above, we have shown that a permanent increase in government consumption increases labor supply and output if leisure and consumption are substitutes. Aiyagari, Christiano, and Eichenbaum (1994) argue that a permanent increase in government consumption has a larger impact on employment and output than does a transitory shock. In the case of a transitory increase in government consumption, the household also reduces investment in an effort to smooth consumption over time.

We will show that the quantitative magnitude of this effect depends critically on the elasticity of substitution between private and public consumption ρ_C. While Aiyagari, Christiano, and Eichenbaum (1994) analyze the stochastic neoclassical growth model, we will study the dynamics in the simpler deterministic model.[25] In particular, we will illustrate the behavior of employment, consumption, and output

[25]The business cycle properties of the stochastic neoclassical model with government consumption will be studied subsequently. At this point, however, it may be instructive for the reader to see that one can show this result in the deterministic neoclassical growth model with the help of a simple solution technique. We will just use the non-linear equation solver that is applied abundantly in this book.

following an unexpected increase in government consumption with (1) a permanent duration and (2) a duration of four quarters (equal to one year) only.

We conduct our analysis of the transition dynamics in the Numerical Example. In contrast to Aiyagari, Christiano, and Eichenbaum (1994), we assume that government consumption increases the marginal utility of consumption, i.e., $\phi < 1$ in (4.21).[26] As our value of the elasticity of substitution between private and public consumption, we choose a conservative value of $\rho_c = 0.3$, which emerges as the median value of the empirical studies reviewed by Ni (1995). Furthermore, leisure is a superior good, and we assume an empirically plausible elasticity of substitution equal to $\rho = 0.6$. The rest of the parameters are chosen as above, with $\alpha = 0.36$, $\beta = 0.96$, $\delta = 0.10$, $\phi = 3/4$, and $\sigma = 2.0$. For this parameterization, we choose $\kappa = 0.08892$, such that the steady-state labor supply is equal to $L = 0.30$.

We assume that the economy is in steady state in period $t = 0$ with a government share equal to 20% of GDP. In this case, absolute government expenditures amount to $G = 0.1013$. In period $t = 0$, we will denote the endogenous variables with a bar, e.g., $\bar{L} = 0.3$, $\bar{K} = 1.288$, $\bar{C}^p = 0.2767$, and $\bar{Y} = 0.5069$. At the beginning of period $t = 1$, the government unexpectedly increases its expenditures to $\tilde{G} = 0.1024$, a 1% increase. In the first case, we consider this change to be permanent. In this case, the economy reaches its new steady state with $\tilde{L} = 0.3029$, $\tilde{K} = 1.3008$, $\tilde{C}^p = 0.2794$, and $\tilde{Y} = 0.5119$. Notice that we have denoted the new steady state variables by inserting a tilde "~" over the variable name.

In the case of a transitory increase in G_t, we assume a temporary increase in government consumption over four periods, such that

$$G_t = \begin{cases} \tilde{G} & \text{for t=1, 2, 3, 4,} \\ \bar{G} & \text{else.} \end{cases}$$

Therefore, in the long run, the endogenous variables approach the initial steady-state values $\{\bar{L}, \bar{C}^p, \bar{K}\}$ in case 2.

To compute the dynamics, we assume that the transition takes 40 periods (equal to 40 years). We can reduce the problem to a one-dimensional non-linear-equation problem by exploiting the recursive structure of the model. To find the solution, we apply the method of *reverse shooting*. The computational details are described in Appendix 4.1. The computation is implemented in the GAUSS program *Ch4_subs_private_pub_dyn.g*.

The results are illustrated in Fig. 4.11 for the case of an elasticity of substitution between private and public consumption equal to $\rho_c = 0.3$ and an intertemporal elasticity of substitution of $1/\sigma = 1/2$. The temporary and permanent government spending shocks are represented by the solid and broken lines, respectively. Clearly, output, capital, and labor supply all increase to a larger extent in the case of a

[26]In addition to $\phi = 1$, Aiyagari, Christiano, and Eichenbaum (1994) also assume that utility is additively logarithmic in total consumption and leisure.

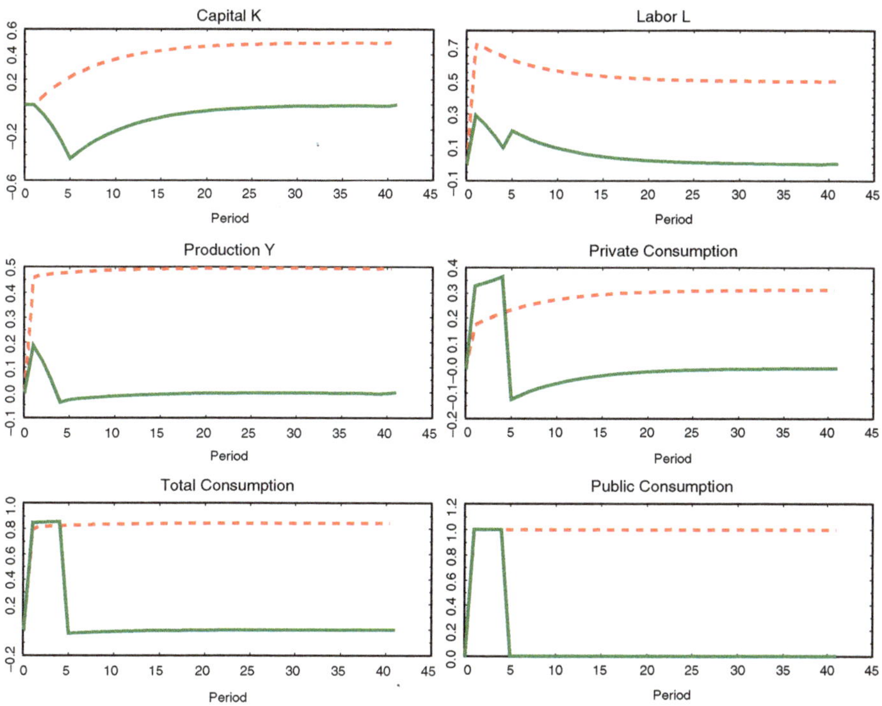

Fig. 4.11 Effect of temporary (solid line) and permanent shock (broken line), $\rho_c = 0.3$, $1/\sigma = 1/2$. Values are presented in percentage deviations from the initial steady state

permanent shock than in the case of a temporary shock to government consumption. Our result emphasizes that the conclusions of Aiyagari, Christiano, and Eichenbaum (1994) are valid and extend to the case $\phi < 1$.[27] Notice also that private consumption C^p initially increases after a temporary increase in government consumption. This is in accordance with the empirical evidence summarized in Sect. 4.2. Finally, notice that, in the case of a temporary increase in government consumption, households also smooth intertemporal consumption by spending down their capital stock during the transition. Accordingly, we conclude that government consumption crowds out (in) private investment if higher government spending is expected to be temporary (permanent).

[27]You can also verify in the program *Ch4_subs_private_pub_dyn.g* that this results holds for other values of $\rho_c \in [0, 1]$.

4.4 The RBC Model with Stochastic Government Consumption

In Sect. 4.3, government consumption was assumed to be deterministic, and all changes in government spending were preannounced in period 0. We found that private consumption increased in response to higher government spending and that the effect on private investment depended on the nature of the change, i.e., whether it was permanent or transitory. The government multiplier was found to be in the range of 1.0–5.0, depending on the elasticity of substitution between private and public goods. In this section, we study whether our previous results also hold under the more realistic assumption that, at least, some part of the changes in government consumption is not anticipated in period 0 or stochastic in nature.

4.4.1 The Model

The model is based on the stochastic neoclassical growth model presented in Chap. 2. In the following, we will augment it with a government sector in which stochastic government spending is subject to an exogenous shock.

4.4.1.1 Households

We study the behavior of a representative household. For this reason, we assume that households are identical and of measure one. Households are infinitely lived and maximize the expected value of intertemporal utility

$$\mathbb{E}_0 \sum_{t=0}^{\infty} \beta^t u(C_t, 1-L_t), \quad 0 < \beta < 1, \tag{4.27}$$

where instantaneous utility $u(.,.)$ is discounted by the factor β and described by a function of effective consumption C_t and leisure $1 - L_t$ according to:

$$u(C_t, 1-L_t) = \frac{\left(C_t^{\iota}(1-L_t)^{1-\iota}\right)^{1-\sigma}}{1-\sigma}. \tag{4.28}$$

The consumption bundle C_t is presented by the CES aggregator (4.21) of private consumption C_t^p and government consumption G_t. We will also consider the special case with $\phi = 1$ in which government consumption does not affect utility, meaning that $C_t = C_t^p$.

The household owns the capital stock, which evolves according to

$$K_{t+1} = (1-\delta)K_t + I_t. \tag{4.29}$$

Capital depreciates at rate δ. The household receives income from labor $w_t L_t$ and capital $r_t K_t$. It spends its income on private consumption C_t^P, investment I_t, and lump-sum taxes T_t. The individual budget constraint is represented by

$$C_t^P + I_t = w_t L_t + r_t K_t - T_t. \tag{4.30}$$

The household maximizes intertemporal utility (4.27) subject to (4.30) resulting in the first-order conditions:

$$\lambda_t = \iota\phi C_t^{\iota(1-\sigma)-1}(1-L_t)^{(1-\iota)(1-\sigma)}\left(\Xi_t\right)^{\frac{1}{1-1/\rho_C}-1}\left(C_t^P\right)^{-\frac{1}{\rho_C}}, \tag{4.31a}$$

$$\lambda_t w_t = (1-\iota)C_t^{\iota(1-\sigma)}(1-L_t)^{(1-\iota)(1-\sigma)-1}, \tag{4.31b}$$

$$\lambda_t = \mathbb{E}_t \lambda_{t+1}\beta(1 + r_{t+1} - \delta), \tag{4.31c}$$

with Ξ as defined in (4.25).

4.4.1.2 Production

Firms are owned by the households and maximize profits with respect to their labor and capital demand. Production Y_t is characterized by constant returns to scale in labor L_t and capital K_t:

$$Y_t = Z_t K_t^{\alpha} L_t^{1-\alpha}. \tag{4.32}$$

Production is also subject to a technology shock Z_t that is governed by the following AR(1) process:

$$\ln Z_t = \rho^Z \ln Z_{t-1} + \epsilon_t^Z, \quad \epsilon_t^Z \sim N\left(0, \sigma^Z\right), \tag{4.33}$$

The individual firm takes Z_t as exogenous.

In factor market equilibrium, factors are compensated by their marginal products:

$$w_t = (1-\alpha)Z_t K_t^{\alpha} L_t^{-\alpha}, \tag{4.34a}$$

$$r_t = \alpha Z_t K_t^{\alpha-1} L_t^{1-\alpha}. \tag{4.34b}$$

4.4.1.3 Government

The government purchases an amount of G_t of the final good in each period. G_t follows a first-order autoregressive process:

$$\ln G_t = (1-\rho^G)\ln G + \rho^G G_{t-1} + \epsilon_t^G, \quad \epsilon_t^G \sim N(0, \sigma^G), \tag{4.35}$$

where G denotes steady-state government consumption. Government consumption will be financed with a lump-sum tax T_t and the government budget is assumed to balance in each period t:

$$G_t = T_t. \tag{4.36}$$

4.4.1.4 Competitive Equilibrium

In a competitive equilibrium, (1) households maximize their intertemporal utility, (2) firms maximize profits, (3) the government balances its budget, and (4) the goods market clears:

$$Y_t = C_t^p + G_t + I_t. \tag{4.37}$$

The last equation can be derived by inserting (4.34) and (4.36) into the individual budget constraint (4.30) and noticing that production is characterized by constant returns to scale such that $Y_t = w_t L_t + r_t K_t$.

To summarize, the equilibrium of the economy can be characterized by the following 8 equations in the 8 variables Y_t, C_t^p, C_t, I_t, L_t, w_t, r_t, λ_t:

$$\lambda_t = \iota\phi C_t^{\iota(1-\sigma)-1}(1-L_t)^{(1-\iota)(1-\sigma)}\left(\Xi_t\right)^{\frac{1}{1-1/\rho_C}-1}\left(C_t^p\right)^{-\frac{1}{\rho_C}}, \tag{4.38a}$$

$$\lambda_t w_t = (1-\iota)C_t^{\iota(1-\sigma)}(1-L_t)^{(1-\iota)(1-\sigma)-1}, \tag{4.38b}$$

$$\lambda_t = \mathbb{E}_t \lambda_{t+1}\beta(1+r_{t+1}-\delta), \tag{4.38c}$$

$$C_t = \left[\phi\left(C_t^p\right)^{1-1/\rho_C} + (1-\phi)G_t^{1-1/\rho_C}\right]^{\frac{1}{1-1/\rho_C}}, \tag{4.38d}$$

$$Y_t = C_t^p + I_t + G_t, \tag{4.38e}$$

$$K_{t+1} = (1-\delta)K_t + I_t, \tag{4.38f}$$

$$w_t = (1-\alpha)Z_t K_t^{\alpha} L_t^{-\alpha}, \tag{4.38g}$$

$$r_t = \alpha Z_t K_t^{\alpha-1} L_t^{1-\alpha}, \tag{4.38h}$$

where

$$\Xi_t \equiv \phi\left(C_t^p\right)^{1-1/\rho_C} + (1-\phi)G_t^{1-1/\rho_C}.$$

The two exogenous variables $\{Z_t, G_t\}$ follow (4.33) and (4.35).

4.4.1.5 Calibration

To compute the model, we need to calibrate the parameters α, δ, β, ι, ϕ, σ, ρ_C, ρ^Z, σ^Z, ρ^G, and σ^G. The period length in the business cycle model is equal to one quarter. A description of these parameters is contained in Table 4.3.

Table 4.3 Calibration of the model with stochastic government consumption

Parameter	Value	Description
β	0.99	Subjective discount factor
$1/\sigma$	1/2	Intertemporal elasticity of substitution
ϕ	3/4	Relative weight of private consumption in effective consumption
ρ_c	0.50 (1.10)	Elasticity of substitution between private and public consumption
ι	0.4096 (0.3219)	Relative weight of consumption in utility
L	0.3	Steady-state labor supply
α	0.36	Share of capital in value added
δ	0.025	Rate of capital depreciation
ρ^Z	0.95	Autocorrelation of TFP shock
σ^Z	0.0072	Standard deviation of innovations in a TFP shock
G/Y	0.20	Share of government spending in steady-state production
ρ^G	0.90	Autocorrelation parameter in a government spending shock
σ^G	0.01	Standard deviation of innovations in a government spending shock

Notes: Values in parentheses are reported for the sensitivity analysis with respect to $\rho_c = 1.10$

We assume that government consumption is equal to 20.0% of output, which is approximately the value for the US economy during the period 1980–2010. The production parameters are set equal to standard parameter values in the RBC literature for the US economy. We choose a production elasticity of capital equal to $\alpha = 0.36$ and an annual depreciation rate of 10%, implying $\delta = 2.5\%$. To have a quarterly real interest rate (net of depreciation) equal to 1%, we choose a discount factor of $\beta = 0.99$. With respect to our preferences, we choose the relative weight of private consumption in effective consumption $\phi = 3/4$ as in the previous section. We will also report results for the case in which $\phi = 1.0$ hereinafter. Similarly, we set the intertemporal elasticity of substitution equal to $1/\sigma = 1/2$ and the intratemporal elasticity of substitution between private and public consumption equal to $\rho_c = 0.50$. We will also present a sensitivity analysis with respect to $\rho_c = 1.1$.

The utility parameter for the weight of leisure in utility, $\iota = 0.4096$, is set such that steady-state labor supply is equal to $L = 0.3$. For this reason, notice that in steady state, $Z_t = Z_{t+1} = Z = 1.0$ and

$$r = 1/\beta - 1 + \delta = 3.51\%.$$

From (4.34b), we can compute the steady-state capital intensity K/L:

$$\frac{K}{L} = \left(\frac{\alpha}{r}\right)^{\frac{1}{1-\alpha}} = 37.99,$$

which implies $K = \frac{K}{L}L = 11.40$ for $L = 0.30$. Therefore, production amounts to $Y = ZK^{\alpha}L^{1-\alpha} = 1.11$. Government consumption is equal to 20% of production, or $G = 0.222$. Steady-state investment follows from (4.29) for $K_{t+1} = K_t = K$ such that $I = \delta K = 0.285$. From the goods market equilibrium (4.37), we obtain $C^p = Y - G - I = 0.604$. Given the values of ρ_c and ϕ, we can compute $C = 0.423$ with the help of (4.21). The equilibrium wage follows with the help of (4.34a), $w = 2.37$. Dividing (4.31b) by (4.31a), we find that

$$w = \frac{1-\iota}{\iota}\frac{1}{\phi}\frac{C}{1-L}\frac{1}{\Xi^{\frac{1}{1-1/\rho_c}-1}(C^p)^{-\frac{1}{\rho_c}}},$$

which we can use to solve for $\iota = 0.4096$. The solution is implemented in the GAUSS program *Ch4_RBC1.g*.

In accordance with prominent articles on RBC models, we choose the parameters $\rho^Z = 0.95$ and $\sigma^Z = 0.0072$ for the autoregressive process of (log) technology as estimated by Cooley and Prescott (1995) and $\rho^G = 0.90$ and $\sigma^G = 0.01$ for the process of (log) government consumption.[28] The parameters are summarized in Table 4.3.

The GAUSS program *Ch4_RBC1.g* computes the solution in the form of policy functions for the next-period capital stock, $K'(K,Z,G)$, consumption $C^p(K,Z,G)$, labor supply $L(K,Z,G)$, and investment $I(K,Z,G)$ as functions of the state variables K, Z, and G. The numerical solution procedure is sketched in Appendix 2.2 and explained in greater detail in Chapters 1 and 2 of Heer and Maußner (2009). We will also use the numerical routines that accompany that book.

4.4.2 Effects of an Unanticipated Increase in Government Consumption

To explain the empirical time series behavior, we will simulate the model for random variables $\{\epsilon_t^Z, \epsilon_t^G\}_{t_0}^{t_1}$. Therefore, we have to make a decision about the value of the endogenous and exogenous state variables $\{K_t\}$ and $\{Z_t, G_t\}$ in the first period t_0. As is common, we choose to have the variables in steady state, $K_{t_0} = K = 11.40$, $Z_{t_0} = Z = 1$, and $G_{t_0} = G = 0.222$. We then repeatedly generate the normally distributed random variables ϵ_t^Z and ϵ_t^G for a time horizon of 80 quarters (20 years). In particular, we use 800 simulations to ensure that the law of large number applies and the averages of the second moments for the variables stabilize. The time series of the model are filtered with the HP-filter of Hodrick and Prescott with weight 1600 for quarterly data.

To understand the behavior of the simulated time series moments, we first study the impulse response functions. These functions show how the variables react to an

[28]For example, Schmitt-Grohé and Uribe (2007) use $\rho^G = 0.87$ and $\sigma^G = 0.016$, while Christiano and Eichenbaum (1992) apply the estimates $\rho^G = 0.96$ and $\sigma^G = 0.020$.

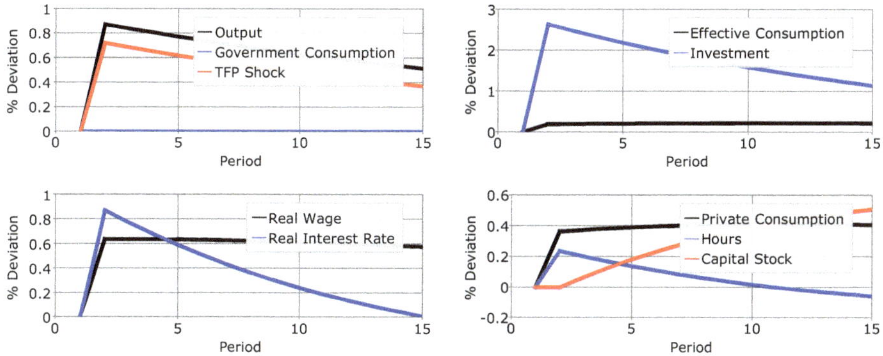

Fig. 4.12 Impulse responses: technology shock

exogenous shock of one standard deviation in the period $t = 2$, i.e., $\epsilon_2^Z = 0.0072$ or $\epsilon_2^G = 0.010$, while the shock is zero in the remaining periods $t = 3, \ldots$.[29] We will first study the effects of a technology shock before we consider the effects of a positive government consumption shock.

4.4.2.1 Impulse Responses to a Technology Shock

Figure 4.12 plots the percentage changes in output Y, technology Z, government consumption G, effective consumption C, investment I, the real wage w, the real interest rate r, hours L, private consumption C^p, and the capital stock K (from the upper left to the bottom right). In period $t = 1$, the economy is in steady state. In period $t = 2$, the technology level Z_t increases by one standard deviation, $\epsilon_2^Z = 0.0072$, and gradually declines to zero thereafter with an autoregressive parameter $\rho^Z = 0.95$. As a consequence of the increase in productivity, the marginal products of both labor and capital increase, and hence, so do the factor prices w and r. Because of the higher wage, the household increases its labor supply (the substitution effect dominates the income effect of higher wages), and hours increase by 0.24% on impact in period 2. The higher wage income of the household generates additional income that it spends on both higher private consumption C^p and higher investment I. Notice that investment I reacts much more strongly (+2.7%) than private consumption C^p (+0.36%), as observed empirically. Because of the increase in the real interest rate r, the households intertemporally substitutes consumption and increases consumption in later periods. For this reason, savings (and hence investment) react even more strongly.

The response of capital is sluggish, and capital increases gradually over the time horizon of 15 periods, as displayed in the bottom-right panel of Fig. 4.12. If we increase the time horizon for the impulse responses (not illustrated), we find that

[29] Some studies prefer to display impulses responses for a shock of 1% rather than one standard deviation.

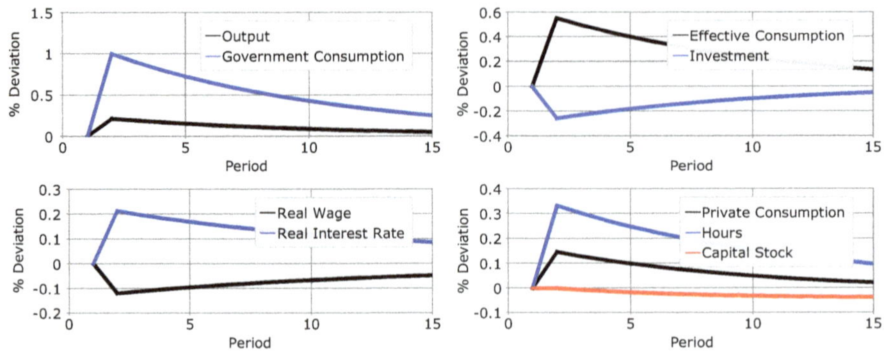

Fig. 4.13 Impulse responses: Government consumption shock, $\rho_c = 0.5$ and $\phi = 3/4$

the maximum response of capital takes place after 23 periods (or 5.75 years). Since both productivity Z and inputs K and L increase, production Y also increases by 1.05%. Notice that the increase in output is higher than the increase in technology (0.72%) due to the higher amount of inputs used in production. In addition, output reacts more strongly than labor and consumption, as observed empirically.

4.4.2.2 Impulse Responses to a Government Consumption Shock

Figure 4.13 plots the impulse responses of the variables to a 1% shock to government consumption. The government consumption shock, $\epsilon_2^G = 0.01$, results in a buildup of government consumption G_t on impact that gradually declines according to the autoregressive process (4.35) as displayed in the top-left panel of Fig. 4.13.

Since additional government consumption is financed by lump-sum taxes, the household is subject to a negative wealth effect. In addition, for $\phi \neq 1.0$, there is also an effect of higher government consumption on the marginal utility of consumption, which decreases due to a higher G_t and, therefore, a higher C_t. For $\rho_c = 0.50$ and $\phi = 3/4$, the incentive effect of lower marginal utility from consumption on the labor supply is smaller than the wealth effect, and thus, labor increases by 0.33% on impact (bottom-right panel). Due to higher income, the household can also afford higher private consumption, which increases by 0.13% on impact (bottom-right panel). Since the household increases labor supply, the marginal product of labor declines and real wages fall, while the real interest rate increases. In addition, higher government consumption crowds out investment I_t, which falls by 0.24%. Although the capital stock decreases, the effect of higher labor supply dominates, meaning that output increases by 0.22%.

All these responses are in accordance with the empirical observations presented in Sect. 4.2.2 of this chapter, with the possible exception of wages for which there is mixed empirical evidence. Consequently, the neoclassical growth model with stochastic government is in good accordance with the qualitative behavior observed in empirical VARs when private and public consumption are substitutes (with an elasticity of substitution equal to 0.5).

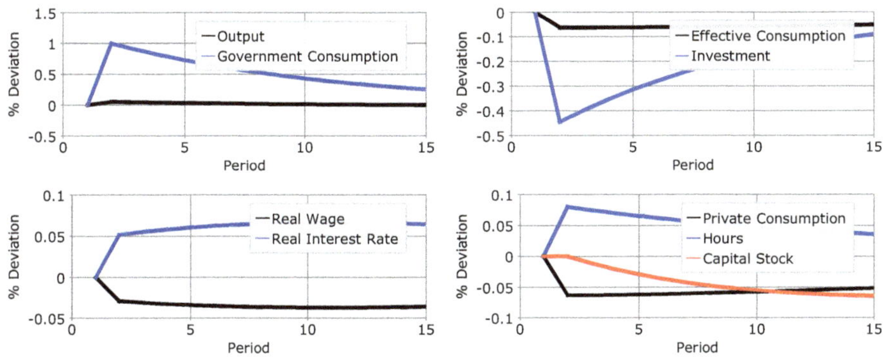

Fig. 4.14 Impulse responses: Government consumption shock, $\phi = 1.0$

How sensitive is this result with respect to the substitutability of private and public consumption? First, we consider the case of $\phi = 1.0$ in which public consumption does not affect utility. For this case, the impulse responses to a government consumption shock are presented in Fig. 4.14.[30] In this case, the marginal utility of private consumption is not increased by a higher G_t. Therefore, the household immediately decreases its private consumption (by 0.06%) in response to the higher G_t because its net income is reduced by the additional taxes. The household also increases its labor supply by 0.08% in period $t = 2$ due to the wealth effect. As a consequence, output increases by 0.04% on impact. Notice that in this case, the response of private consumption is not in accordance with the empirical observations.

Next, consider the case in which the elasticity of substitution between private and public consumption increases above one such that the two goods are gross substitutes. For $\phi = 3/4$ and $\rho_c = 1.1$, the impulse responses to a government spending shock are graphed in Fig. 4.15. Notice that in this case (1) private consumption falls even more significantly than in the case with $\phi = 1.0$ (-0.11% compared with -0.06%) because private and public consumption are gross substitutes, (2) output and hours increase, and (3) real wages fall.

To summarize, we have shown that, in the present model, we can restore the basic empirical facts in the form of the qualitative impulse responses if (1) government consumption affects utility from effective consumption and (2) public and private consumption are not close substitutes. Otherwise, we find that private consumption declines on impact if public consumption increases.[31]

[30]In each case where we changed the value of ϕ or/and ρ_c, we re-calibrated the parameter ι so that steady-state labor supply is equal to $L = 0.30$.

[31]For our model and the calibration presented in Table 4.1, private consumption declines on impact for all $\rho_c \geq 0.67$.

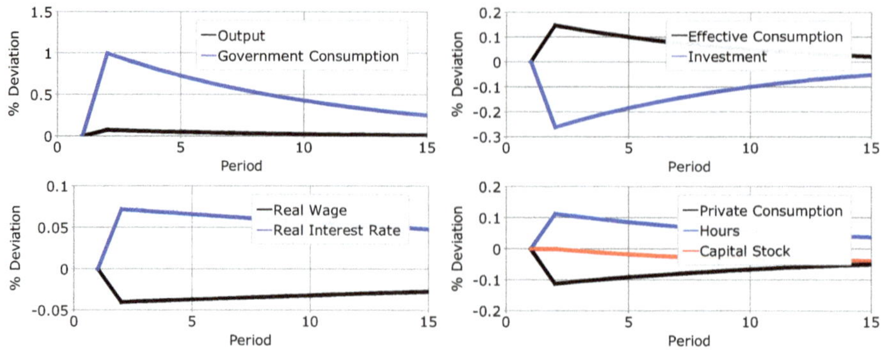

Fig. 4.15 Impulse responses: Government consumption shock, $\rho_c = 1.1$ and $\phi = 3/4$

If government consumption does not affect utility, there are also other ways to change the simple RBC model to make the model's implications in better accordance with the data. In this vein, Devereux, Head, and Lapham (1996) show that both the presence of monopolistic competition and increasing returns to scale in production help to reconcile the theoretical effects of a government spending shock in an RBC model with the empirical observations. As we have argued above, a government spending shock results in a negative wealth effect, meaning that the individual increases labor supply and decreases consumption in the standard model (with $\phi = 1.0$). As a consequence, wages also fall because the marginal product of labor declines. With monopolistic competition and an endogenous number of intermediate goods in the model of Devereux, Head, and Lapham (1996), higher government demand increases the number of goods and the productivity of final goods producers. As a consequence, a positive government spending shock increases private consumption and, in contrast to the above, real wages. However, since the empirical evidence regarding the qualitative response of wages following a government consumption shock is mixed, it is difficult to evaluate the relative performance of the simple RBC model with public consumption in utility and the model of Devereux, Head, and Lapham (1996).[32]

4.4.3 Government Spending Multiplier

With the help of the impulse responses, we can also compute the government spending multiplier. To do so, let us consider the case with $\phi = 3/4$ and $\rho_c = 0.5$, in which higher government consumption results in an increase of output and private

[32] In Sect. 4.5, we also consider a standard New Keynesian model with frictions in the form of sticky prices and wages. Galí and Lopez-Salido (2007) show that this model is also able to replicate the empirical fact that private consumption rises in response to an unexpected increase in government consumption if one allows for the presence of rule-of-thumb consumers who do not save.

consumption. The government multiplier is defined by $\Delta Y_t / \Delta G_t$, which, for our calibration, is approximately equal to

$$\frac{\Delta Y}{\Delta G} \approx \frac{dY}{dG} = \frac{dY}{Y} / \frac{dG}{G} \times Y/G.$$

In our simulated economy, $G/Y = 0.20$ in steady state. In addition, we can read off the percentage changes in government consumption and output from the impulse responses, with $\frac{dG}{G} = 0.01$ and $\frac{dY}{Y} = 0.0022$. Accordingly, the multiplier amounts to $\frac{\Delta Y_t}{\Delta G_t} \approx \frac{0.0022}{0.01} \times 5 = 1.1$ on impact. If we sum up the effects over the first year (the first four quarters), additional government consumption amounts to 3.44%, while output increases by 0.76%. Therefore, the one-year multiplier also amounts to approximately 1.1. By inspecting Figs. 4.13, 4.14, and 4.15, it becomes evident that the multiplier is very sensitive to the substitutability of private and public consumption and increases if public and private consumption are more complementary. For both cases, $\{\phi, \rho_c\} = \{1.0, 0.5\}$ and $\{\phi, \rho_c\} = \{0.75, 1.1\}$, the government multiplier drops below 0.2.

How does the multiplier from our simple RBC model compare with empirical values? Using standard regression methods, Barro (1981) finds a multiplier of approximately 0.8 in his empirical analysis. In more recent empirical work, researchers predominantly apply more elaborate VARs. Assuming that government consumption does not respond to contemporaneous shocks, Blanchard and Perotti (2002) find a small positive multiplier that varies across specifications (e.g., if a deterministic or stochastic time trend is used) and periods. Applying news variables to estimate the effects of government spending shocks when they are publicly known, Ramey (2011) estimates a multiplier of 0.6 to 0.8 in her VAR analysis. However, if she excludes the data from World War II and the Korean War, the multiplier becomes negative and the standard errors become very large. Monacelli, Perotti, and Trigari (2010) distinguish between employment at the intensive margin (hours) and extensive margin (number of employed) and find that increasing government consumption by 1% of GDP generates output multipliers of 1.2 (at one year) and 1.6 (at the peak after approximately two and a half years).

In a more recent contribution, Féve, Matherod, and Sahuc (2012) also consider the case in which the government uses a feedback rule for its spending policy such that (lagged) output enters the equation (4.35) as an additional variable on the right-hand side. The coefficient of output is negative, and thus, the endogenous government policy is countercyclical. In this case, they show that an econometrician would underestimate the complementarity of public and private consumption, and thus, the government spending multiplier would also be underestimated. They find a multiplier that exceeds unity.

Auerbach and Gorodnichenko (2011) show that for most OECD countries, the multiplier is large during recessions and low in normal times. The multiplier also seems to be larger in countries with high wealth inequality and a large fraction of the population facing binding credit constraints. For countries with higher wealth inequality or lower average wealth, Brinca, Holter, Krusell, and Malafry (2016) show in their VAR analysis that the fiscal multiplier is larger. In a sample of 15 OECD countries, they find that fiscal multipliers increase with the country's wealth Gini coefficient and decrease with the capital-output ratio. The regression coefficients in their structural VARs are quantitatively significant and a one-standard-deviation increase in the wealth Gini raises the multiplier by approximately 17% of the average multiplier value.

In light of these empirical studies, we find that our benchmark calibration with $\{\phi, \rho_c\} = \{0.75, 0.5\}$ generates a reasonable government multiplier of 1.1. Our model, however, is unable to replicate the asynchronous behavior of the multiplier over the cycle or its dependence on wealth inequality and the number of credit-constrained households.

4.4.4 Time Series Analysis

Table 4.4 reports our results on the second moments of the simulated series for the benchmark case with $\phi = 3/4$ and $\rho_c = 0.5$. We closely replicate the empirical facts on the relative volatility of output, consumption, and investment. Output is approximately as volatile as government consumption, while it is more volatile (less volatile) than private consumption, labor, and wages (investment). Government consumption, however, is excessively positively correlated with labor. While the empirical value presented in Table 4.1 is equal to −0.243 during the period 1956–2014, our model implies a value of 0.79.

Table 4.4 Second moments for the benchmark case with $\phi = 0.75$ and $\rho_c = 0.5$

Variable	s_x	s_x/s_Y	r_{xY}	r_{xL}	r_{xG}
Output Y	1.12	1.00	1.00	0.73	0.23
Effective consumption C	0.72	0.64	0.53	0.92	0.94
Private consumption C^p	0.50	0.44	0.97	0.74	0.35
Investment I	3.30	2.95	0.94	0.50	−0.09
Hours L	0.52	0.46	0.73	1.00	0.79
Real wage w	0.82	0.73	0.90	0.37	−0.18
Real interest rate r	1.15	1.03	0.97	0.76	0.23
Government consumption G	1.23	1.10	0.23	0.79	1.00

Notes: s_x:= Standard deviation of the time series x, where $x \in \{Y, C, C^p, I, L, w, r, G\}$. The model-generated time series were HP filtered with weight 1600. s_x/s_Y:= standard deviation of the variable x relative to the standard deviation of output Y. r_{xY}:= Cross-correlation of the variable hours with output, r_{xL}:= Cross-correlation of the variable with labor, r_{xG}:= Cross-correlation of the variable with government consumption

4.4.5 Sensitivity Analysis

Using our choice of preferences as presented by the Cobb-Douglas utility function (4.28), we have been able to analyze cases in which private consumption either falls (for $\phi = 1.0$) or increases (for $\phi = 3/4$) after an expansionary shock to public consumption. In the following, we perform a sensitivity analysis of the functional form of utility, in particular with respect to the so-called *Frisch labor supply elasticity*. This elasticity measures the percentage change in the labor supply in response to a 1% increase in the wage, given a constant marginal utility of wealth.

For the Cobb-Douglas case (4.28), the Frisch labor supply elasticity amounts to[33]:

$$\eta_{L,w} = \frac{1 - \iota(1-\sigma)}{\sigma} \frac{1-L}{L}. \tag{4.39}$$

For our benchmark calibration with $\iota = 0.4096$, this implies a Frisch labor supply elasticity for the Cobb-Douglas case equal to $\eta_{L,w} = 1.64$ in steady state (with $L = 0.30$). Notice further that if we choose ι to fix the steady-state labor supply $L = 0.30$, we simultaneously set $\eta_{L,w}$. We only have one free parameter, ι, that determines both the average labor supply and the labor supply elasticity.

Estimates of $\eta_{L,w}$ implied by microeconometric studies vary considerably. MaCurdy (1981) and Altonij (1986) both use PSID data to estimate values of 0.23 and 0.28, respectively, while Killingsworth (1983) finds a US labor supply elasticity equal to $\eta_{L,w} = 0.4$.[34] In macroeconomic studies such as Trabandt and Uhlig (2011), a value of unity is often chosen. Therefore, the Frisch labor supply elasticity $\eta_{L,w} = 1.64$ that we applied above in the case of Cobb-Douglas utility seems to be in the upper range of possible values; thus, we will use the more conservative estimate $\eta_{L,w} = 0.3$ in our sensitivity analysis.

To analyze the sensitivity of our results with respect to $\eta_{L,w}$, we consider a different utility function that is additively separable in consumption and leisure. In particular, let us consider

$$u(C_t, L_t) = \frac{C_t^{1-\sigma}}{1-\sigma} - \nu_0 \frac{L_t^{1+\frac{1}{\nu_1}}}{1+\frac{1}{\nu_1}}, \tag{4.40}$$

where ν_1 denotes the Frisch elasticity of labor supply.[35] In comparison with the Cobb-Douglas case (4.28), this utility function has the advantage that we can separately choose ν_0 (to fix $L = 0.30$ in steady state) and ν_1 (the labor supply

[33] The Frisch labor supply elasticity is derived in Appendix 4.2.

[34] Domeij and Floden (2006) argue that these estimates are biased downward due to the omission of borrowing constraints.

[35] The only equilibrium conditions that change in (4.38) are (4.38a) and (4.38b), which are replaced by

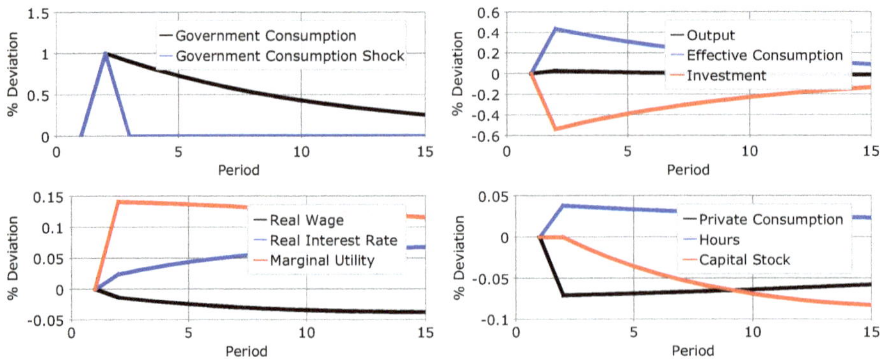

Fig. 4.16 Impulse responses: Government consumption shock, $\rho_c = 0.5$ and $\phi = 3/4$, additive utility

elasticity). For $\nu_1 = 0.30$, we need to set $\nu_0 = 269.58$, meaning that steady-state labor supply is again equal to $L = 0.3$.[36]

We find that the qualitative results are sensitive to the labor supply elasticity $\eta_{L,w}$. For the case with $\phi = 0.75$ and $\rho_c = 0.50$, we illustrate the impulse responses to a government shock in Fig. 4.16. Notice that private consumption decreases, while labor supply increases on impact. As a consequence, output increases as well, even though to a negligible quantitative extent (by approximately 0.02%). Therefore, the government multiplier is also reduced from 1.1 (benchmark case with $\eta_{L,w} = 1.64$) to 0.1 (with $\eta_{L,w} = 0.3$). The smaller Frisch labor supply elasticity only results in an increase in labor and, hence, labor income that is much smaller than the additional lump-sum taxes that are needed for the financing of the increase in government consumption. Consequently, net income falls and the household can only afford to spend less on both investment and private consumption. Evidently, additional government expenditures crowd out private investments, while the real interest rate increases. The impulse responses are therefore not in complete accordance to those in Fig. 4.13.[37]

$$\lambda_t = \phi C_t^{-\sigma} (\Xi_t)^{\frac{1}{1-1/\rho_c}-1} \left(C_t^p\right)^{-\frac{1}{\rho_c}},$$

$$\lambda_t w_t = \nu_0 L_t^{\frac{1}{\nu_1}}.$$

[36] You are asked to perform the numerical computation in Problem 4.3. The GAUSS program *Ch4_RBC2.g* that computes this problem can be downloaded from my homepage.

[37] The magnitude and the sign of the impulse responses also depend on the functional form of utility. If we employ utility function (4.40) with $\eta_{L,w} = 1.64$ rather than Cobb-Douglas utility (4.28), private consumption also decreases in response to higher government consumption (not illustrated). As expected, the response of labor supply is much stronger than for the case with $\eta_{L,w} = 0.30$ and amounts to $+0.11\%$.

4.5 Keynes' Argument for Countercyclical Government Spending

In this section, we pursue Keynes' original argument in favor of government spending. In times of high unemployment and in the presence of sticky prices, expansive government expenditures help to increase output. First, we briefly review the standard AS-AD model to demonstrate this effect. We assume that the reader is familiar with the AS-AD model. Next, we present a New Keynesian model that is able to replicate the main findings of this model.

Why do we need a New Keynesian model if the AS-AD model is able to replicate the same main mechanism? There are multiple reasons that the simple AS-AD model is not sufficient. Most important, if we want to conduct policy analysis, we want to be able to provide quantitative policy advice. For this reason, the AS-AD model was estimated using standard (multiple-equation) regression models. For example, private consumption was estimated as a linear function of income Y, $C^p = C_0^p + cY$. In addition, the total demand was given by $Y = C^p + I + G$. These equations, among others, were used as an input into the policy analysis, e.g., what happens if G increases by 1%. However, Lucas (1976) noted in his famous *Lucas critique* that coefficients such as c are themselves functions of the economic policy, and therefore, policy recommendations based upon such models were flawed. We, instead, will specify a general equilibrium model in which consumption, savings, and labor supply are all micro-founded and derived from individual optimization. Therefore, the policy functions only depend upon deep structural parameters, e.g., preference parameters such as the intertemporal elasticity of substitution or the policy parameters themselves. As a consequence, these models with deep parameters are not subject to the *Lucas critique*.[38]

4.5.1 AS-AD Revisited

Figure 4.17 presents the AS-AD model. The AS curve displays the aggregate supply curve, which is upward-sloping in the (Y, P) diagram. The reason for the positive relationship between the two variables, output Y and prices P, is as follows: A rise in production Y is associated with higher labor input. As a consequence, workers (or unions) demand higher wages, which increases marginal costs for given price expectations P^e. Since the firm uses mark-up pricing on marginal costs, prices increase.

[38]Even this statement is subject to restrictions. Preferences are not completely exogenous. For example, firms' advertising is used to influence consumer preferences or political institutions may have an effect on cultural development and, therefore, our deep utility parameters. One of the first modern economist to point out the endogeneity of preferences was Carl Christian von Weizsäcker (see, e.g., von Weizsäcker (1971)).

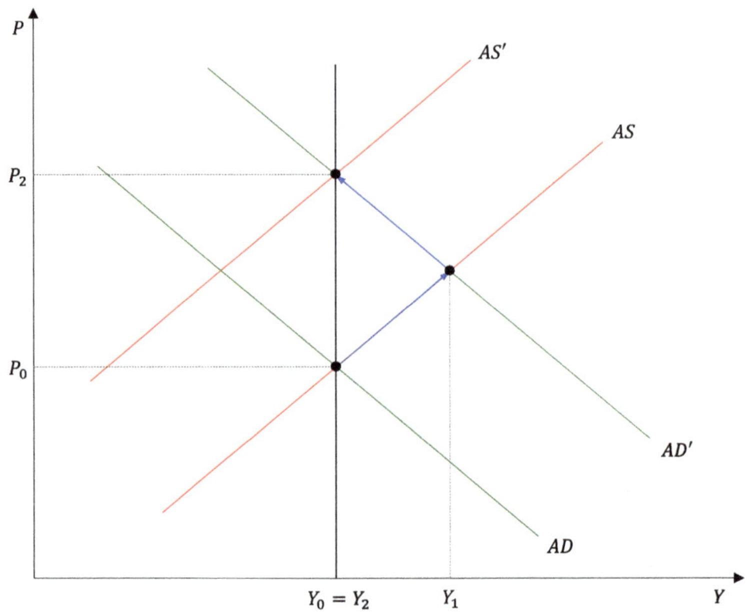

Fig. 4.17 Dynamic effects of higher government consumption in the AS-AD model

The AD (aggregate demand) curve characterizes the simultaneous equilibrium in the goods and money market (IS and LM curves). The curve is downward-sloping. The negative relationship between output Y and prices P can be explained as follows: A rise in the prices implies lower real money balances, M/P. As a consequence, interest rates increase and goods demand declines. In equilibrium, demand is equal to production, and thus, output Y also declines. The general equilibrium (Y_0, P_0) is the intersection of the AS and AD curves, where price expectations are equal to actual prices, $P_0 = P^e$.

An increase in government spending G results in an outward shift of the AD curve from AD to AD'. As a consequence, both prices P and output Y rise. For example, output increases from Y_0 to Y_1 in Fig. 4.17. Depending on the stickiness of prices, output may increase even more markedly. Figure 4.18 illustrates the case in which prices remain constant at $P = P_0$ for one period and output increases to a higher level. In this case, there is no general equilibrium (a simultaneous equilibrium in goods, money, and the labor market), but only the goods and the money market are in equilibrium, and thus, the new equilibrium is a point on the new AD curve, AD'. Only gradually do prices increase. Since the workers notice that their price expectations are too low, they increase wage demands, and the AS curve shifts upward until it has reached the new long-run equilibrium AS', in which price expectations are equal to actual prices, $P^e = P_2$. Therefore, an increase in government spending is associated with higher employment, output, prices, and wages during the transition. In addition, private consumption increases during the

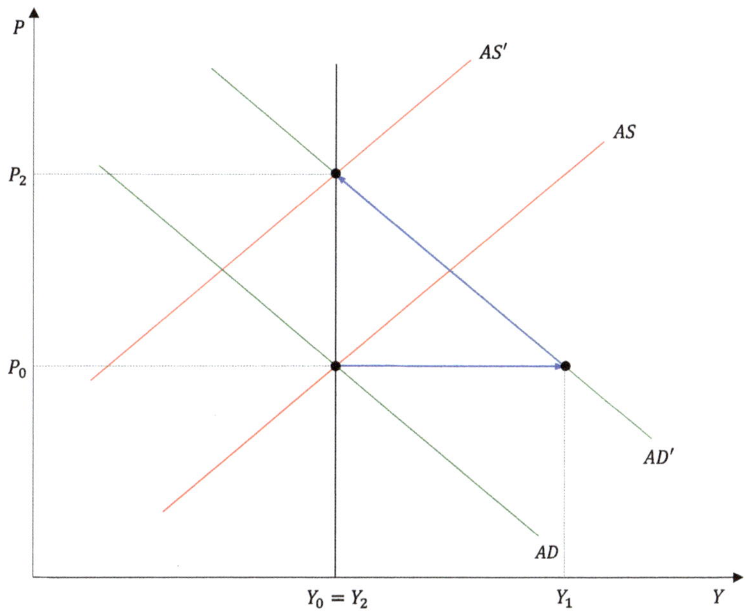

Fig. 4.18 Dynamic effects of higher government consumption in the AS-AD model, constant prices on impact

transition because private consumption is assumed to be a positive function of income.

Accordingly, the AS-AD model seems to be in accordance with empirical VAR evidence according to which output, private consumption, and real wages all increase after an increase in government spending. Two qualifying remarks with respect to the reaction of private consumption are necessary: (1) Private consumption is a function of disposable income. If additional government expenditures are financed by means of taxes rather than government debt (in the presence of non-Ricardian equivalence), the effect of additional government spending on private consumption is reduced. (2) We neglect any interest effect on consumption (or, respectively, savings). If prices increase and real money contracts, the interest rate will rise so that savings are augmented (if the substitution effect dominates the income effect). As a consequence, private consumption is reduced further.

A second qualifying remark is in order. Above, we discussed the effects of higher government consumption starting in a general equilibrium of the labor, goods, and money markets with output Y_0 and price expectations $P^e = P_0$. As a consequence, output increased above its general equilibrium level. If, however, goods markets are competitive, firms operate in their profit maximizing equilibrium at $Y = Y_0$. Therefore, they do not have any incentive to increase their production beyond Y_0. For firms to be interested in increasing their production beyond Y_0 in the case of sticky prices, we need to assume some form of monopolistic behavior such that profits increase with higher production. Therefore, a monopolistically competitive goods producer will be part of the micro-founded New Keynesian model, to which we turn next.

4.5.2 A New Keynesian Model

In the following, we describe a standard New Keynesian model with rigid prices and wages. The economy consists of households, a labor agency, a production sector with a monopolistically competitive wholesale sector and both intermediate goods firms and final goods firms, the government sector, and the central bank.

To introduce nominal frictions into our 'Keynesian' model, we assume sticky wages and prices. Sticky prices result from *Calvo price staggering,* as described in Calvo (1983), meaning that only a part of the producers (in the wholesale sector) can optimally adjust their prices. The rest of them can adjust their prices by the previous period's inflation rate. Similarly, only a part of the workers' wages can be adjusted optimally as in Erceg, Henderson, and Levin (2000). The wages of the other workers are only adjusted for the inflation rate prevailing in the previous period. To keep the model tractable, we will introduce an employment agency that bundles the labor services of the households.

As we consider a nominal model with goods prices and nominal wages, we need to introduce money into the model. There are various ways to do so, e.g., by assuming a cash-in-advance constraint for consumption, a shopping-time technology or so-called *'limited participation'* where firms have to pre-finance wages.[39] We will take a short-cut since our focus is not on monetary, but rather fiscal, policy and assume that households obtain utility from holding money, the so-called *'money-in-the-utility'* model. In addition, we need to introduce a central bank that controls the money supply. We follow the standard approach and assume that the central bank uses a Taylor rule and sets nominal interest rates with respect to the inflation rate and output. For this reason, we also have to introduce an additional asset in the form of nominal government bonds that are held by the households (we will set the equilibrium supply equal to zero and only study government debt in the final part of the book).

We also introduce various features into the model that help to improve the cyclical behavior of various variables. In particular, we will include capital adjustment costs and habits in consumption. These two assumptions help to improve the

[39]The expression 'limited participation,' as introduced by Christiano, Eichenbaum, and Evans (1997), results from the constraints that agents face in the financial market. Households can only lend funds to the firms with the help of a financial intermediary at the beginning of the period. The central bank injects money into the banking sector after the households have deposited their money at the bank. Hence, households can no longer participate in the financial market, i.e., they have limited participation. At the end of the period, the financial intermediary retrieves the loans from the firms that need to pre-finance labor costs. The different ways to introduce a motive for money demand in general equilibrium models are reviewed in Walsh (2010), among others.

dynamic behavior of the economic variables, e.g., output and consumption. As a consequence, output is more persistent, and the impulse response is more hump-shaped, as observed empirically.[40]

The present standard *New Keynesian* model such as presented in Christiano, Eichenbaum, and Evans (2005), therefore, includes the following features:

- Sticky prices
- Sticky wages
- Monopolistic competition
- Capital adjustment costs
- Habits in consumption
- Motive for money demand
- Various shocks (on government consumption, technology, and nominal interest rates).

4.5.2.1 Firms

In the production sector, we distinguish a final goods sector, a wholesale sector, and an intermediate goods production sector. Intermediate goods $Y_t(j)$ are produced by competitive firms that use capital and labor. They are sold to the wholesale sector at price P_{yt}. Firm j in the wholesale sector brands and sells these goods at price $P_t(j)$ to the final goods sector. The final goods sector produces the final good Y_t and sells it at price P_t. Notice that the only reason that we introduce a wholesale sector in the model is to simplify the solution of the model. In particular, we assume that the firms, denoted j, with $j \in [0, 1]$, in the wholesale factor cannot all adjust their prices in each period. A random fraction $1 - \varphi_y$ of them are able to do so, while the other fraction φ_y can only raise their prices by the inflation rate observed in the previous period. We thereby introduce sticky prices following Calvo (1983). To keep the solution tractable, we cannot assume that the intermediate goods producers set their price in a Calvo staggering way because this would greatly complicate integration over the individual demand functions of the intermediate goods producers to derive aggregate factor demands. Instead, we will assume that all intermediate goods producers are identical.

Let us start at the end of the production chain and work backward to the production of the intermediate good. Final output Y_t is produced from differentiated inputs $Y_t(j)$, $j \in [0, 1]$ according to the production function of Dixit and Stiglitz (1977):

$$Y_t = \left(\int_0^1 Y_t(j)^{\frac{\epsilon_y - 1}{\epsilon_y}} \, dj \right)^{\frac{\epsilon_y}{\epsilon_y - 1}}, \quad \epsilon_y > 1, \tag{4.41}$$

---

[40] Both features help to increase the cost of intertemporal substitution of consumption for the household. As a consequence, the premium on risky assets increases, and the model is also in better accordance with asset price implications. See, for example, Jermann (1998) and Uhlig (2007).

where ϵ_y denotes the price elasticity for the demand of the intermediate good $Y_t(j)$.[41] We assume that each producer j faces the same demand elasticity to facilitate computation of the equilibrium.

The intermediate good $Y_t(j)$ is bought at price $P_t(j)$, and the final good Y_t is sold at price P_t to the household as consumption good C_t^p, to the government as public good G_t, and to the production sector as investment good I_t.

Profits $\Pi_t(j)$ are defined as follows:

$$\Pi_t(j) = P_t Y_t - \int_0^1 P_t(j) Y_t(j)\, dj. \tag{4.42}$$

Inserting the production function of the final good (4.41) into the profit equation (4.42), we obtain

$$\Pi_t(j) = P_t \left(\int_0^1 Y_t(j)^{\frac{\epsilon_y - 1}{\epsilon_y}}\, dj \right)^{\frac{\epsilon_y}{\epsilon_y - 1}} - \int_0^1 P_t(j) Y_t(j)\, dj.$$

The first-order condition of the firm is derived from setting the derivative of this equation with respect to $Y_t(j)$ equal to zero[42]:

$$P_t \left(\int_0^1 Y_t(j)^{\frac{\epsilon_y - 1}{\epsilon_y}}\, dj \right)^{\frac{1}{\epsilon_y - 1}} Y_t(j)^{-\frac{1}{\epsilon_y}} - P_t(j) = 0$$

Noticing that the production function (4.41) implies

$$\left(\int_0^1 Y_t(j)^{\frac{\epsilon_y - 1}{\epsilon_y}}\, dj \right)^{\frac{1}{\epsilon_y - 1}} = Y_t^{\frac{1}{\epsilon_y}},$$

we can derive the demand function for the intermediate product $Y_t(j)$:

$$Y_t(j) = \left(\frac{P_t(j)}{P_t} \right)^{-\epsilon_y} Y_t. \tag{4.43}$$

Furthermore, we assume that the firms in the final goods sector make zero profits:

$$P_t Y_t = \int_0^1 P_t(j) Y_t(j)\, dj. \tag{4.44}$$

---

[41]To derive that ϵ_y is equal to the price elasticity, differentiate the demand equation (4.43) with respect to $P_t(j)$.

[42]We used the chain rule of differentiation and the *Leibniz integral rule* as presented in Footnote 42 of Chap. 2.

Inserting the demand function (4.43) into (4.44), we can also derive the price index of the final good P_t:

$$P_t = \left(\int_0^1 P_t(j)^{1-\epsilon_y} \, dj \right)^{\frac{1}{1-\epsilon_y}}. \tag{4.45}$$

4.5.2.2 Price Setting

A firm j in the wholesale sector buys goods at the nominal price P_{yt} from the production sector, brands them, and sells them at price $P_t(j)$ to the final goods sector. Its profits are equal to $(P_t(j) - P_{yt})Y_t(j)$ and, in the units of the final product, are given by

$$D_t(j) = \left(\frac{P_t(j)}{P_t} - g_t \right) Y_t(j), \quad g_t = \frac{P_{yt}}{P_t}. \tag{4.46}$$

Dividends $D_t(j)$ are distributed to the household sector. g_t denotes the inverse of the mark-up. Later in this chapter, we will also analyze the behavior of the mark-up, which according to empirical evidence presented in Sect. 4.2 falls after an increase in government spending.

As described above, monopolistic firms in the wholesale sector set prices in a Calvo staggering way. In each period, a randomly selected fraction $1 - \varphi_y$ of firms in this sector receive the signal to optimally choose their relative price $p_{At} = P_{At}/P_t$. The remaining fraction φ_y are allowed to raise their nominal price P_{Nt} according to the inflation rate observed in the previous period π_{t-1}[43]:

$$P_{Nt} = \pi_{t-1} P_{Nt-1}, \quad \pi_t = \frac{P_t}{P_{t-1}}. \tag{4.47}$$

To describe the solution for the price setting of the wholesale firm j, we define the variable s_t^y as follows:

$$s_t^y \equiv \int_0^1 \left(\frac{P_t(j)}{P_t} \right)^{-\epsilon_y} dj. \tag{4.48}$$

s_t^y is a measure of the price dispersion of individual prices $P_t(j)$. In steady state, where all firms in the wholesale sector demand the same price $P_t(j) = P_t$, the index is equal to one, $s^y = 1$.

[43] In Problem 4.6, you are asked to study the case in which the price P_{Nt} increases by the average inflation π rather than by the inflation in the last period. Also note that the inflation rate amounts to $\pi - 1$, while π denotes the inflation factor.

The equilibrium conditions are derived in Appendix 4.3 and are represented by:

$$s_t^y = (1-\varphi_y)p_{At}^{-\epsilon_y} + \varphi_y(\pi_{t-1}/\pi_t)^{-\epsilon_y} s_{t-1}^y, \tag{4.49a}$$

$$p_{At} = \frac{\epsilon_y}{\epsilon_y - 1}\frac{\Gamma_{1t}}{\Gamma_{2t}}, \tag{4.49b}$$

$$\Gamma_{1t} = g_t \Lambda_t Y_t + (\beta\varphi_y)\mathbb{E}_t\left(\frac{\pi_t}{\pi_{t+1}}\right)^{-\epsilon_y} \Gamma_{1t+1}, \tag{4.49c}$$

$$\Gamma_{2t} = \Lambda_t Y_t + (\beta\varphi_y)\mathbb{E}_t\left(\frac{\pi_t}{\pi_{t+1}}\right)^{1-\epsilon_y} \Gamma_{2t+1}, \tag{4.49d}$$

$$1 = (1-\varphi_y)p_{At}^{1-\epsilon_y} + \varphi_y(\pi_{t-1}/\pi_t)^{1-\epsilon_y}. \tag{4.49e}$$

In this system of equilibrium equations, Γ_{1t} and Γ_{2t} are simply auxiliary variables, while $p_{At} = P_{At}/P_t$, as noted above, is the optimal price relative to the price of the final good.

4.5.2.3 Production and Capital Accumulation

In this model, we assume for analytical convenience that capital is accumulated by the firms in the intermediate goods sector. Households own equity in the firms. Since all firms behave identically and their number is normalized to one, we can study their behavior by means of a representative firm.

The production function of a representative firm in the intermediate goods sector is given by

$$\tilde{Y}_t = Z_t L_t^{1-\alpha} K_t^{\alpha}, \quad \alpha \in (0,1), \tag{4.50}$$

where K_t and L_t denote the capital stock and labor.

The intermediate good is sold at the relative price g_t to the wholesale sector. The level of total factor productivity Z_t is governed by the AR(1) process

$$\ln Z_t = \rho^Z \ln Z_{t-1} + \epsilon_t^Z, \quad \epsilon_t^Z \sim N\left(0, \sigma^Z\right). \tag{4.51}$$

The firm accumulates its stock of capital and takes the funds for investment I_t from retained earnings RE_t and newly issued equity S_t at price v_t:

$$I_t = RE_t + v_t\,(S_{t+1} - S_t). \tag{4.52}$$

The producer distributes the remaining surplus profits over retained earnings as dividends to the households:

$$d_t S_t = g_t \tilde{Y}_t - w_t L_t - RE_t, \tag{4.53}$$

where d_t denotes the dividends per share S_t.

Capital accumulation is subject to adjustment costs such that

$$K_{t+1} = (1-\delta)K_t + \Phi\left(\frac{I_t}{K_t}\right)K_t, \quad \Phi'(.) > 0, \ \Phi''(.) \le 0. \tag{4.54}$$

In addition, we assume the adjustment cost function to have the following properties:

$$\Phi(\delta) = \delta, \ \Phi'(\delta) = 1.$$

Accordingly, in steady state with $I = \delta K$, adjustment costs are equal to $\Phi(I/K) = \Phi(\delta) = \delta$ and do not play any role in the steady state. As we will see shortly, the second assumption, $\Phi'(\delta) = 1$, implies that Tobin's q is equal to one in steady state.

For $\Phi(I/K) = I/K$, we are back to the standard capital accumulation equation. For our specification, capital accumulation is subject to frictions, and thus, investment I_t does not produce capital one-to-one. In particular, we parameterize the function Φ following Jermann (1998):

$$\Phi(I_t/K_t) := \frac{a_1}{1-\zeta}\left(\frac{I_t}{K_t}\right)^{1-\zeta} + a_2, \quad \zeta > 0, \tag{4.55}$$

where $1/\zeta$ denotes the elasticity of investment with respect to Tobin's q.

For the parameters a_1 and a_2, we select the values

$$a_1 = \delta^{\zeta}, \quad a_2 = \left(\frac{\zeta}{1-\zeta}\right)^{\delta},$$

such that $\Phi(\delta) = \delta$ and $\Phi'(\delta) = 1$ hold.

The firm chooses investment and labor demand to maximize its cum-dividend value, i.e., the value of the discounted dividends including the present period t:

$$V_t^{cd} = \mathbb{E}_t \sum_{s=0}^{\infty} \beta^s \frac{\Lambda_{t+s}}{\Lambda_t}\left[g_{t+s} Z_{t+s} L_{t+s}^{1-\alpha} K_{t+s}^{\alpha} - w_{t+s} L_{t+s} - I_{t+s}\right]$$

subject to (4.54) and a given initial stock of capital K_t. The term $\beta^s \Lambda_{t+s}/\Lambda_t$ is the stochastic discount factor of the stockholders (the households) that we will introduce shortly. The first-order conditions are given by:

$$w_t = (1-\alpha) g_t \frac{\tilde{Y}_t}{L_t}, \tag{4.56a}$$

$$q_t = \frac{1}{\Phi'(I_t/K_t)}, \tag{4.56b}$$

$$q_t = \beta \mathbb{E}_t \frac{\Lambda_{t+1}}{\Lambda_t}\left[\alpha g_{t+1} \frac{\tilde{Y}_{t+1}}{K_{t+1}} - \frac{I_{t+1}}{K_{t+1}} + q_{t+1}\left(1 - \delta + \Phi(I_{t+1}/K_{t+1})\right)\right], \tag{4.56c}$$

where q denotes Tobin's q, i.e., the price of capital.[44] Notice that $q = 1$ in steady state given our calibration of $\Phi(.)$.

Our main reason to introduce capital adjustment costs is to allow for more realistic behavior of the wealth effect from higher government consumption. When a shock (either a government or a technology shock) hits the economy, firms gradually adjust their capital due to adjustment costs. As a consequence, the price of capital and, as we will show in the following, the value of the shares respond more markedly.

4.5.2.4 Labor Demand

The household has a unit mass of members $h \in [0, 1]$ who sell their labor services $L_t(h)$ at wage rate $W_t(h)$. The household's labor services are only imperfect substitutes in the production process of the intermediate goods. To keep the model tractable, it is convenient to introduce an 'employment agency' that combines the households' labor hours in the same proportion as the firms. The agency bundles the individual labor services into a single service according to the form proposed by Dixit and Stiglitz (1977):

$$L_t = \left[\int_0^1 L_t(h)^{\frac{\epsilon_w - 1}{\epsilon_w}} dh \right]^{\frac{\epsilon_w}{\epsilon_w - 1}}, \quad \epsilon_w > 1. \tag{4.57}$$

and sells L_t at the nominal wage W_t to the intermediate goods producer. Maximizing profits, $W_t L_t - \int_0^1 W_t(h) L_t(h)\, dh$, subject to (4.57) yields the demand function for labor

$$L_t(h) = \left(\frac{W_t(h)}{W_t} \right)^{-\epsilon_w} L_t, \tag{4.58}$$

and the aggregate wage index

$$W_t = \left[\int_0^1 W_t(h)^{1-\epsilon_w} \right]^{\frac{1}{1-\epsilon_w}}. \tag{4.59}$$

Notice that ϵ_w denotes the wage elasticity of labor demand.

Profits of the labor agency are equal to zero, implying

$$W_t L_t = \int_0^1 W_t(h) L_t(h)\, dh.$$

[44] Heer and Maußner (2009) show that the (stock market) value of the firm at the beginning of period $t + 1$ is equal to the value of the capital at the end of period t (=beginning of period $t + 1$), $V_t^{cd} = q_t K_{t+1}$. Accordingly, q_t describes the (market) value of the firm relative to the replacement cost of capital and, therefore, amounts to the definition of Tobin's q.

The analytical derivation of the demand function is methodologically the same as that for the goods demand of the final goods' producers above as we consider simply two variants of the same *Dixit-Stiglitz model*. The derivation is left as an exercise for the reader.[45]

4.5.2.5 Household Preferences

The number of households is normalized to one. The household $h \in [0, 1]$ derives (dis)utility from effective consumption $C_t(h)$, labor $L_t(h)$, and real money $M_{t+1}(h)/P_t$:

$$u(C_t(h), \bar{C}_t, L_t(h), M_{t+1}(h)/P_t) = \frac{(C_t(h) - \chi^C \bar{C}_t)^{1-\sigma} - 1}{1-\sigma} - \frac{\nu_0}{1+\frac{1}{\nu_1}} L_t(h)^{1+\frac{1}{\nu_1}}$$
$$+ \gamma_0^M \frac{\left(\frac{M_{t+1}(h)}{P_t}\right)^{1-\gamma_1^M} - 1}{1-\gamma_1^M},$$
$$\sigma, \nu_0, \nu_1, \gamma_0^M, \gamma_1^M > 0, \ \chi^C \in [0, 1), \tag{4.60}$$

where effective consumption C_t is, again, given by the composite good of private and public consumption, as in (4.21). $\bar{C}_t$ denotes the consumption habits of the households.[46] The household takes its habits as exogenous.[47] In equilibrium, the habit $\bar{C}_t$ equals the aggregate consumption of the previous period, $\bar{C}_t = C_{t-1} = \int_0^1 C_{t-1}(h)\, dh$.

Household members hold three different types of assets: shares in the intermediate goods producer $S_t(h)$, nominal bonds $B_t(h)$, and nominal money as of the beginning of the next period (alternatively, the end of period t) $M_{t+1}(h)$. In terms of the final good, stocks have price v_t and pay dividend d_t. The real value of bonds is given by $B_t(h)/P_t$. Bonds pay the predetermined nominal interest rate Q_t. In addition to dividends and interest income, the households receive wage income $W_t(h)L_t(h)$, a share (equal to unity) of the profits distributed from the wholesale sector, and a share (equal to unity) of the government transfers in period t, $tr_t(h)$.

---

[45] The structure of the model seems to be very complicated. We distinguish three production sectors (final goods, wholesale, intermediate goods) and one employment agency. If you consider the alternative case where we only postulate one production sector without an employment agency, the benefits of this fragmentation of services in the production sector are evident. If we had only one production sector that is characterized by monopolistic competition and heterogeneous firms, each firm's labor demand and price-setting behavior would depend on its marginal costs and, hence, its capital stock. As a consequence, we would not be able to derive simple index functions in the form of (4.45) and (4.59) for the aggregate price and aggregate wage. Instead, these aggregate prices would depend on the distribution of capital in the production sector.

[46] This specification was introduced by Constandinides (1990). As an alternative, Abel (1990) uses the ratio of consumption and habits, $C_t/\bar{C}_t$, in the utility function.

[47] The results are not sensitive to this assumption.

The budget constraint, therefore, reads as:

$$\frac{W_t(h)}{P_t}L_t(h) + \int_0^1 D_t(j)\,dj + d_t S_t(h) + (Q_t - 1)\frac{B_t(h)}{P_t} - C_t^p(h) + tr_t(h) - T_t(h)$$
$$= v_t\left(S_{t+1}(h) - S_t(h)\right) + \frac{B_{t+1}(h) - B_t(h)}{P_t} + \frac{M_{t+1}(h) - M_t(h)}{P_t}, \tag{4.61}$$

where $T_t(h)$ denotes lump-sum taxes.

In each period, a random fraction $1 - \varphi_w$ of the household members receive a signal to optimally choose their nominal wage W_{At}. The remaining fraction φ_w are allowed to increase their wage W_{Nt} according to the price inflation observed in the previous period:

$$W_{Nt} = \pi_{t-1} W_{Nt-1}, \quad \pi_t = \frac{P_t}{P_{t-1}}. \tag{4.62}$$

Those who optimize set their real wage $\tilde{w}_t = W_{At}/P_t$ to maximize

$$\mathbb{E}_t \sum_{s=0}^{\infty} (\beta\varphi_w)^s u(C_{t+s}(h), L_{t+s}(h), \bar{C}_{t+s}, M_{t+s+1}(h)/P_{t+s})$$

subject to labor demand (4.58) and the budget constraint (4.61).

To derive the equilibrium conditions for the wage staggering, let us define the variable s_t^w, which is a measure of wage dispersion among households:

$$s_t^w \equiv \left(\int_0^1 \left(\frac{W_t(h)}{W_t} \right)^{-\epsilon_w (1 + \frac{1}{\nu_1})} dh \right)^{\frac{1}{1+\frac{1}{\nu_1}}}. \tag{4.63}$$

This measure will be helpful to describe the equilibrium dynamics of the economy and helps to relate average working hours $\tilde{L}_t$ and aggregate labor L_t. In particular, the average (or expected) working hours of a household member $\tilde{L}_t$ are derived in Appendix 4.4 as

$$\tilde{L}_t = \int_0^1 L_t(h)\,dh = s_t^w L_t.$$

In steady state, all households receive the same wage, and, hence, $s^w = 1$. Furthermore, average working hours are equal to aggregate labor in steady state because all workers supply the same amount of labor.

The equilibrium conditions are derived in Appendix 4.4 and are represented by:

$$\tilde{w}_t = \frac{\epsilon_w}{\epsilon_w - 1}\frac{\Delta_{1t}}{\Delta_{2t}}, \tag{4.64a}$$

$$\Delta_{1t} = \nu_0\left(\frac{\tilde{w}_t}{w_t}\right)^{-\epsilon_w(1+\frac{1}{\nu_1})} L_t^{1+\frac{1}{\nu_1}} + (\beta\varphi_w)\mathbb{E}_t\left(\frac{\pi_t\tilde{w}_t}{\pi_{t+1}\tilde{w}_{t+1}}\right)^{-\epsilon_w(1+\frac{1}{\nu_1})}\Delta_{1t+1}, \tag{4.64b}$$

$$\Delta_{2t} = \Lambda_t\left(\frac{\tilde{w}_t}{w_t}\right)^{-\epsilon_w} L_t + (\beta\varphi_w)\mathbb{E}_t\left(\frac{\tilde{w}_t}{\tilde{w}_{t+1}}\right)^{-\epsilon_w}\left(\frac{\pi_t}{\pi_{t+1}}\right)^{1-\epsilon_w}\Delta_{2t+1}, \tag{4.64c}$$

$$w_t^{1-\epsilon_w} = (1-\varphi_w)\tilde{w}_t^{1-\epsilon_w} + \varphi_w\left(\frac{\pi_{t-1}}{\pi_t}w_{t-1}\right)^{1-\epsilon_w}, \tag{4.64d}$$

$$(s_t^w)^{1+\frac{1}{\nu_1}} = (1-\varphi_w)\left(\frac{\tilde{w}_t}{w_t}\right)^{-\epsilon_w(1+\frac{1}{\nu_1})} + \varphi_w\left(\frac{\pi_{t-1}w_{t-1}}{\pi_t w_t}\right)^{-\epsilon_w(1+\frac{1}{\nu_1})}(s_{t-1}^w)^{1+\frac{1}{\nu_1}}. \tag{4.64e}$$

4.5.2.6 Consumption and Portfolio Choice

The pooling assumption allows us to derive the demand for consumption, bonds, stocks, and money from maximizing

$$\mathbb{E}_t\sum_{s=0}^{\infty}\beta^s u(C_{t+s}, \bar{C}_{t+s}, L_{t+s}, M_{t+s+1}/P_{t+s})$$

subject to the sequence of budget constraints

$$\begin{aligned} w_{t+s}L_{t+s} &+ \int_0^1 D_t(j)\,dj + S_t d_{t+s} + (Q_{t+s}-1)\frac{B_{t+s}}{P_{t+s}} + tr_{t+s} - C_{t+s}^p - T_t \\ &= \frac{B_{t+s+1}-B_{t+s}}{P_{t+s}} + v_{t+s}(S_{t+s+1}-S_{t+s}) + \frac{M_{t+s+1}-M_{t+s}}{P_{t+s}}. \end{aligned} \tag{4.65}$$

The first-order conditions for the optimal choice of private consumption C_t^p, the number of stocks S_{t+1}, bond holdings B_{t+1} and money M_{t+1} are:

$$\Lambda_t = \phi(C_t - \chi\bar{C}_{t-1})^{-\sigma}(\Xi_t)^{\frac{1}{1-1/\rho_C}-1}\left(C_t^p\right)^{-\frac{1}{\rho_C}}, \tag{4.66a}$$

$$v_t\Lambda_t = \beta\mathbb{E}_t\Lambda_{t+1}(v_{t+1}+d_{t+1}), \tag{4.66b}$$

$$\Lambda_t = \beta \mathbb{E}_t \Lambda_{t+1} \frac{Q_{t+1}}{\pi_{t+1}}, \tag{4.66c}$$

$$\Lambda_t = \mathbb{E}_t \left[\beta \frac{\Lambda_{t+1}}{\pi_{t+1}} + \gamma_0^M \left(\frac{M_{t+1}}{P_t} \right)^{-\gamma_1^M} \right], \tag{4.66d}$$

where Ξ is defined as above, $\Xi_t = \phi \left(C_t^p\right)^{1-1/\rho_c} + (1-\phi)G_t^{1-1/\rho_c}$. The marginal utility from consumption is represented by (4.66a). The three optimality conditions (4.66b), (4.66c), and (4.66d) describe the optimal savings behavior with respect to the three assets equity, government bonds, and real money. The expected return of these three assets must be equal to each other and amount to $\mathbb{E}_t \{\Lambda_t/(\beta\Lambda_{t+1})\}$. The return from equity is equal to the dividend yield plus the increase in stock prices, $(v_{t+1}+d_{t+1}-v_t)/v_t$. Government bonds yield the nominal return $Q_{t+1} - 1$, while money has a nominal return of zero, but also provides marginal utility $\gamma_0^M \left(\frac{M_{t+1}}{P_t}\right)^{-\gamma_1^M}$ in period t.

4.5.2.7 Monetary Policy

The central bank sets the nominal interest rate Q_{t+1} according to the Taylor rule

$$Q_{t+1} = Q_t^{\theta^R} \left(\frac{\pi}{\beta}\right)^{1-\theta^R} \left(\frac{\pi_t}{\pi}\right)^{\theta^\pi} \left(\frac{Y_t}{Y}\right)^{\theta^Y} e^{\epsilon_t^Q}, \quad \theta^R \in [0,1), \quad \epsilon_t^Q \sim N(0, \sigma^Q). \tag{4.67}$$

The elasticity of Q_{t+1} with respect to the deviation of the inflation factor π_t from its steady-state value π will be chosen such that the equilibrium is determinate. Usually, this requires $\theta^\pi > 1$.[48]

Seignorage is transferred lump-sum to the households:

$$P_t Tr_t = M_{t+1} - M_t. \tag{4.68}$$

Since we normalized the number of households to one, individual transfers are equal to aggregate transfers, $tr_t(h) = Tr_t$.

[48] An inflation reaction coefficient in excess of unity prevents self-fulfilling expectations with respect to the path of inflation. See, for example, Bullard and Mitra (2002). The intuition for this behavior is quite simple. Assume that aggregate demand and prices increase and that the other reaction coefficients are equal to zero, $\theta^R = \theta^Y = 0$. If the reaction coefficient θ^π were less than one, nominal interest rates would rise less than prices so that the real interest rate would decline. As a consequence, aggregate demand would increase further and inflation went up even more. The monetary policy clearly would become unstable. While we exclude this kind of behavior in our model, we, however, refrain from imposing a lower zero bound on the net nominal interest rate $Q_t - 1$, as it rarely becomes binding in our simulations.

4.5.2.8 Government

The government consumes final goods in the amount G_t, which are financed by taxes (recall that we set the equilibrium supply of government bonds equal to zero). We assume that the government spending rule G_t follows an AR(1) process

$$\ln G_t = (1-\rho^G)\ln G + \rho^G G_{t-1} + \rho^Y (\ln Y_{t-1} - \ln Y) + \epsilon_t^G, \quad \epsilon_t^G \sim N(0,\sigma^G), \tag{4.69}$$

where G denotes steady-state government consumption. Notice that we have introduced an additional additive term, $\rho^Y (\ln Y_{t-1} - \ln Y)$. Accordingly, the government reacts to the state of the economy as described by the percentage deviation of output Y_{t-1} from its steady-state value Y. Since the implementation of fiscal policy takes time, we assume that the government can only react with a lag of one period. In addition, government consumption is subject to a shock that cannot be controlled by the government. For example, government expenditures in Germany increased after 1989 due to reunification of West and East Germany. In the following, we will first analyze the case with $\rho^Y = 0$ before we compute the value of ρ^Y that maximizes welfare.

In addition, the government cannot control the parameter ρ^G that governs the adjustment of government expenditures after a shock. As one possible reason, consider the decision-making process about budgets in a government. To cut government expenditures, various bureaucratic agencies (ministries) that seek to increase their individual budgets have to agree to cut them instead. These government agencies also do not have any incentive to reduce the budget to their actual needs but usually maximize their well-being using the amount of their budgets. In addition, both politicians and citizens grow accustomed to spending levels. These so-called *ratchet effects* help to explain the significant inertia in budget cuts.

In equilibrium, the government budget is balanced:

$$P_t G_t = P_t T_t. \tag{4.70}$$

4.5.2.9 General Equilibrium

In equilibrium, factor and product markets clear, and bonds are in zero supply, $B_t = 0$.

In addition, consolidating the household and the government budgets results in

$$g_t \tilde{Y}_t + \int_0^1 \left(\frac{P_t(j)}{P_t} - g_t \right) Y_{jt}\, dj = C_t^p + I_t + G_t, \tag{4.71}$$

which reduces to the usual goods market equilibrium condition

$$Y_t = C_t^p + I_t + G_t. \tag{4.72}$$

4.5.2.10 Equilibrium Dynamics

The equilibrium conditions of the model consist of the firm's optimality conditions, the production function, the capital accumulation equation, the economy's resource constraint implied by the household's budget constraint, the wage-setting equations, the household's optimality conditions, and the Taylor rule (4.67). We disregard the solution for the stock of real balances (the first-order condition of the household with respect to real money M_{t+1}/P_t), as this condition is only used to determine the real money supply that is necessary to support the interest rate policy Q_t of the central bank. Therefore, we can remove it from the equilibrium conditions such that the full model is described by the following 24 equations in the 24 variables Y_t, $\tilde{Y}_t$, C_t^p, C_t, Ξ_t, I_t, L_t, K_t, w_t, $\tilde{w}_t$, g_t, Q_t, π_t, q_t, p_{At}, s_t^y, s_t^w, Λ_t, Γ_{1t}, Γ_{2t}, Δ_{1t}, Δ_{2t}, d_t and v_t:

$$\Lambda_t = \phi(C_t - \chi C_{t-1})^{-\sigma} (\Xi_t)^{\frac{1}{1-1/\rho_C}-1} \left(C_t^p\right)^{-\frac{1}{\rho_C}} \tag{4.73a}$$

$$C_t = \left[\phi \left(C_t^p\right)^{1-1/\rho_C} + (1-\phi)G_t^{1-1/\rho_C}\right]^{\frac{1}{1-1/\rho_C}}, \tag{4.73b}$$

$$\Xi_t = \phi \left(C_t^p\right)^{1-1/\rho_C} + (1-\phi)G_t^{1-1/\rho_C}, \tag{4.73c}$$

$$w_t = (1-\alpha) g_t \frac{\tilde{Y}_t}{L_t}, \tag{4.73d}$$

$$\tilde{Y}_t = Z_t L_t^{1-\alpha} K_t^{\alpha}, \tag{4.73e}$$

$$Y_t = C_t^p + I_t + G_t, \tag{4.73f}$$

$$s_t^y Y_t = \tilde{Y}_t. \tag{4.73g}$$

$$q_t = \frac{1}{\Phi'(I_t/K_t)}, \tag{4.73h}$$

$$d_t = g_t \tilde{Y}_t - w_t L_t - I_t, \tag{4.73i}$$

$$p_{At} = \frac{\epsilon_y}{\epsilon_y - 1} \frac{\Gamma_{1t}}{\Gamma_{2t}}, \tag{4.73j}$$

$$\tilde{w}_t = \frac{\epsilon_w}{\epsilon_w - 1} \frac{\Delta_{1t}}{\Delta_{2t}}, \tag{4.73k}$$

$$w_t^{1-\epsilon_w} = (1-\varphi_w)\tilde{w}_t^{1-\epsilon_w} + \varphi_w \left(\frac{\pi_{t-1}}{\pi_t} w_{t-1}\right)^{1-\epsilon_w}, \tag{4.73l}$$

$$K_{t+1} = (1-\delta)K_t + \Phi(I_t/K_t)K_t, \tag{4.73m}$$

$$\Lambda_t v_t = \beta \mathbb{E}_t \Lambda_{t+1}(v_{t+1} + d_{t+1}), \tag{4.73n}$$

$$\Lambda_t = \beta \mathbb{E}_t \frac{\Lambda_{t+1} Q_{t+1}}{\pi_{t+1}}, \tag{4.73o}$$

$$q_t = \beta \mathbb{E}_t \frac{\Lambda_{t+1}}{\Lambda_t} \left\{ \alpha g_{t+1} Z_{t+1} L_{t+1}^{1-\alpha} K_{t+1}^{\alpha-1} - \frac{I_{t+1}}{K_{t+1}} \right.$$

$$\left. + q_{t+1} \left[\Phi \left(\frac{I_{t+1}}{K_{t+1}} \right) + 1 - \delta \right] \right\}, \tag{4.73p}$$

$$1 = (1-\varphi_y) p_{At}^{1-\epsilon_y} + \varphi_y (\pi_{t-1}/\pi_t)^{1-\epsilon_y}. \tag{4.73q}$$

$$s_t^y = (1-\varphi_y) p_{At}^{-\epsilon_y} + \varphi_y (\pi_{t-1}/\pi_t)^{-\epsilon_y} s_{t-1}^y, \tag{4.73r}$$

$$\Gamma_{1t} = g_t \Lambda_t Y_t + (\beta \varphi_y) \mathbb{E}_t \left(\frac{\pi_t}{\pi_{t+1}} \right)^{-\epsilon_y} \Gamma_{1t+1}, \tag{4.73s}$$

$$\Gamma_{2t} = \Lambda_t Y_t + (\beta \varphi_y) \mathbb{E}_t \left(\frac{\pi_t}{\pi_{t+1}} \right)^{1-\epsilon_y} \Gamma_{2t+1}, \tag{4.73t}$$

$$(s_t^w)^{1+\frac{1}{\nu_1}} = (1-\varphi_w) \left(\frac{\tilde{w}_t}{w_t} \right)^{-\epsilon_w (1+\frac{1}{\nu_1})} + \varphi_w \left(\frac{\pi_{t-1} w_{t-1}}{\pi_t w_t} \right)^{-\epsilon_w (1+\frac{1}{\nu_1})} (s_{t-1}^w)^{1+\frac{1}{\nu_1}}, \tag{4.73u}$$

$$\Delta_{1t} = \nu_0 \left(\frac{\tilde{w}_t}{w_t} \right)^{-\epsilon_w (1+\frac{1}{\nu_1})} L_t^{1+\frac{1}{\nu_1}} + (\beta \varphi_w) \mathbb{E}_t \left(\frac{\pi_t \tilde{w}_t}{\pi_{t+1} \tilde{w}_{t+1}} \right)^{-\epsilon_w (1+\frac{1}{\nu_1})} \Delta_{1t+1}, \tag{4.73v}$$

$$\Delta_{2t} = \Lambda_t \left(\frac{\tilde{w}_t}{w_t} \right)^{-\epsilon_w} L_t + (\beta \varphi_w) \mathbb{E}_t \left(\frac{\tilde{w}_t}{\tilde{w}_{t+1}} \right)^{-\epsilon_w} \left(\frac{\pi_t}{\pi_{t+1}} \right)^{1-\epsilon_w} \Delta_{2t+1}, \tag{4.73w}$$

$$Q_{t+1} = (Q_t)^{\theta^R} \left(\frac{\pi}{\beta} \right)^{1-\theta^R} \left(\frac{\pi_t}{\pi} \right)^{\theta^\pi} e^{\epsilon_t^R}. \tag{4.73x}$$

4.5.2.11 Stationary Equilibrium

As usual, the stationary equilibrium is defined by setting the shocks equal to their unconditional means and by assuming that $x_{t+1} = x_t = x$ for all variables x in the model. In this case, $P_{At} = P_t$, and therefore, $p_A = 1$. Similarly, the dispersion variables of prices and wages are equal to one, $s^y = s^w = 1$, because all wholesale producers charge the same price and all household members receive the same wage. Therefore, $\tilde{Y} = Y$ in steady state. Moreover, $Q_{t+1} = Q_t$, implying

$$Q = \frac{\pi}{\beta}.$$

Accordingly, the steady-state nominal interest rate set exogenously by the central bank determines the steady-state inflation rate.

Equations (4.73j), (4.73s), and (4.73t) (using $\pi_t = \pi_{t+1} = \pi$ and $p_A = 1$) simplify to

$$g = \frac{\epsilon_y - 1}{\epsilon_y}, \tag{4.74a}$$

meaning that the mark-up

$$\frac{1}{g} = \frac{1}{1 - \frac{1}{\epsilon_y}}$$

increases with more inelastic goods demand or, equally, a lower price elasticity of demand ϵ_y.

To derive the steady-state capital stock with the help of the assumption that labor supply is equal to $L = 0.3$ (which will later help to determine the utility parameter ν_0), we transform equation (4.73p):

$$\frac{K}{L} = \left(\frac{\alpha g}{1/\beta - (1-\delta)}\right)^{\frac{1}{1-\alpha}} \tag{4.74b}$$

such that for a given L, the stationary stock of capital equals

$$K = \left(\frac{\alpha g}{1/\beta - (1-\delta)}\right)^{\frac{1}{1-\alpha}} L. \tag{4.74c}$$

Consequently, output Y is determined by (4.73e). With the help of the steady-state value for Y, we will calibrate G such that the government consumption share in GDP is equal to 20%, $G/Y = 0.20$.

Given the properties of the adjustment cost function Φ, equation (4.54) implies

$$I = \delta K, \tag{4.74d}$$

and we obtain the stationary value of private consumption from the resource constraint, $C_t^p = Y_t - I_t - G_t$. Given the solution for C^p, we can compute the solution for C with the help of (4.73c) and for Λ from (4.73a) (using $\bar{C}_{t-1} = C$).

The stationary real wage w follows from equation (4.73d). Equation (4.73l) implies that $\tilde{w} = w$. Γ_1 and Γ_2 are implied by (4.73s) and (4.73t). Ψ_1 and Ψ_2 are implied by (4.73u) and (4.73v), and thus, we are able to determine the parameter ν_0 with the help of (4.73k):

$$\nu_0 = \frac{\epsilon_w - 1}{\epsilon_w} \Lambda w L^{-\frac{1}{\nu_1}}. \tag{4.74e}$$

Dividends d follow from equation (4.73h). The stationary share price derives from (4.73n):

$$v = \frac{\beta}{1-\beta}d. \tag{4.74f}$$

4.5.2.12 Calibration

To calibrate the parameters of our model, we choose standard values from the literature. Our parameters are summarized in Table 4.5. The values for α, β, δ, σ, ρ^Z, σ^Z, ρ^G, and σ^G are taken from Table 4.3.

The new preference parameters are chosen as follows: The habit parameter χ is set equal to 0.65, as in Christiano, Eichenbaum, and Evans (2005). The Frisch

Table 4.5 Calibration of the New Keynesian model

Parameter	Value	Description
β	0.99	Subjective discount factor
$1/\sigma$	1/2	Intertemporal elasticity of substitution
χ	0.65	Habit parameter
ϕ	1.0	Relative weight of private and public consumption in effective consumption
ρ_c	0.5	Elasticity of Substitution between private and public consumption
γ_1^M	0.1776	Elasticity of marginal utility of real money
γ_0^M	0.331	Weight of utility from real money in total utility
L	0.3	Steady-state labor supply
ν_1	0.20	Frisch labor supply elasticity
α	0.36	Share of capital in value added
δ	0.025	Rate of capital depreciation
$1/\zeta$	1/3.0	Elasticity of investment with respect to q
φ_y	0.5	Share of wholesale producers who cannot adjust prices
φ_w	0.5	Share of workers who cannot adjust wages
ϵ_y	6.0	Price elasticity of demand for intermediate goods
ϵ_w	6.0	Wage elasticity of labor demand
ρ^Z	0.95	Autocorrelation of TFP shock
σ^Z	0.0072	Standard deviation of innovations of TFP shock
G/Y	0.20	Share of government spending in steady-state production
ρ^G	0.90	Autocorrelation parameter in a government spending shock
σ^G	0.01	Standard deviation of innovations in a government spending shock
ρ^Y	0	Reaction coefficient on output in gov. consumption rule
θ^π	1.50	Coefficient of inflation in Taylor rule
θ^R	0.90	Autocorrelation parameter in Taylor rule
θ^Y	0.25	Coefficient of output in Taylor rule
σ^Q	0.0252	Standard deviation of innovations to Taylor rule
$\pi-1$	0.5%	Quarterly inflation rate

labor supply elasticity is set equal to $\nu_1 = 0.2$. As we argued in the previous section, this value for ν_1 is in the middle range of the values estimated by micro-econometric studies. In our benchmark, we set $\phi = 1.0$, meaning that government consumption does not affect utility. As a consequence, the elasticity of public and private consumption does not affect the results, and we set it equal to $\rho_c = 0.5$. In our sensitivity analysis, we also consider the value $\phi = 1.3$.

The utility parameters from real money balances, γ_0^M and γ_1^M, are only needed if we intend to compute household utility. For the computation of the results in this section, money demand is residual (for a given nominal interest rate Q_t), and thus, we can neglect it. Since, however, the reader might be interested in using the model in this section for further studies, for example in a welfare analysis, we describe the calibration of these parameters. To determine θ_i^M, $i = 1, 2$, we use two identifying restrictions. First, the annual velocity of money is set to 6.0 in steady state in accordance with the empirical evidence presented in Heer, Maußner, and McNelis (2011). Second, we impose the condition that the semi-interest elasticity of money is equal to 5.95%, as computed in Dotsey and Ireland (1996). To calibrate these two parameters, we have to solve a non-linear equation routine $f(\gamma_1^M) = 0$ in the variable γ_1^M. In particular, we start with an initial value for γ_1^M and compute the value of γ_0^M from the first-order condition of the household with respect to next-period money holdings, (4.66d), using $PY/M = 1.5$ (the quarterly velocity of money in steady state). Next, we increase steady-state inflation by 10 percentage points, use the Taylor rule for monetary policy in steady state to compute the corresponding nominal interest rate Q, and solve the model to obtain the new velocity of money $P'Y'/M'$. Following Dotsey and Ireland (1996), we compute the percentage change in the income velocity after an absolute increase in the interest rate by 0.10 as a proxy for the semi-interest rate elasticity of money demand and consider its deviation from the empirical value as our non-linear equation:

$$f(\gamma_M^1) = \frac{\ln\left(\frac{P'Y'}{M'}\right) - \ln\left(\frac{PY}{M}\right)}{0.10} - 5.95.$$

If the value of $f(\gamma_1^M)$ is equal to 0, we are done. Otherwise, we have to provide a new guess for γ_1^M until $f(\gamma_1^M)$ converges to zero. We start with two initial values $\left(\gamma_1^M\right)^1 = 0.10$ and $\left(\gamma_1^M\right)^2 = 0.20$ and use the *secant method*[49] with a smoothing parameter $\kappa = 0.1$ to find the solution $\gamma_1^M = 0.1776$[50]:

$$\left(\gamma_1^M\right)^{i+2} = \left(\gamma_1^M\right)^{i+1} - \kappa \frac{\left(\gamma_1^M\right)^{i+1} - \left(\gamma_1^M\right)^i}{f(\left(\gamma_1^M\right)^{i+1}) - f(\left(\gamma_1^M\right)^i)} f(\left(\gamma_1^M\right)^i).$$

[49]See Fig. 2.15 and Eq. (2.59) in Appendix 2.1.

[50]The sequence does not converge for the standard secant method with $\kappa = 1$.

We assume that both the average goods prices and wage contracts last 2 quarters on average, implying $\varphi_y = \varphi_w = 0.5$. The price elasticity of demand for intermediate goods, $\epsilon_y = 6$, is set in accordance with Christiano, Eichenbaum, and Evans (2005). Our value of ϵ_y implies a mark-up of 20% on marginal costs. We also set the wage elasticity of labor demand ϵ_w equal to 6.0. The elasticity of investment with respect to Tobin's q is assumed to amount to 1/3, implying $\zeta = 3.0$. The value of this elasticity is in accordance with Jermann (1998), who chooses a somewhat smaller elasticity of 0.23.

The remaining parameters that we need to calibrate are the parameters of the Taylor rule (4.67): $\theta^\pi = 1.50$, $\theta^R = 0.90$, $\theta^Y = 0.25$, and $\sigma^Q = 0.0252$. θ^π and θ^Y are set as in Taylor (1993). The interest rate rule implies high inertia with $\theta^R = 0.90$, as in Walsh (2005). Finally, we use the steady-state inflation factor $\pi = 1.005$ to imply an annual inflation rate of approximately 2.0%.

The parameters a_1, a_2, and ν_0 are computed with the help of the steady-state conditions as described above. The calibration and solution of the model is computed with the help of the GAUSS program *Ch4_newkeynesian.g* that can be downloaded from my homepage.

4.5.3 Impulse Responses

Figures 4.19 and 4.20 describe the impulse responses of output Y_t and the technology level Z_t/government spending shock (top-left panel), the demand components effective consumption C_t, private consumption C_t^p, investment I_t, and government consumption G_t (top-right panel), the real wages w_t, labor supply L_t, and the nominal interest rate Q_t (bottom-left panel), Tobin's q q_t, the inflation factor π_t, the capital stock K_t, and the mark-up, $1/g_t$ (bottom-right panel) that result from a one-standard-deviation shock to the technology level and government consumption, respectively. The case of a shock to the nominal interest rate is illustrated and discussed in Appendix 4.5 because the focus of this book is on fiscal rather than monetary policy. In the illustrated case, we set $\phi = 1$, meaning that government consumption does not enter utility. In addition, we first analyze the case in which output does not affect government consumption, $\rho^Y = 0$, in the policy rule (4.69).

4.5.3.1 Technology Shock

To obtain a better understanding of the different mechanisms at work, let us start by considering the effects of a technology shock as illustrated in Fig. 4.19. Following an unexpected increase in the technology level, the productivity of labor and capital increase, and thus, demand for both factors also increases. Households demand higher wages, and thus, the real wage and income increase. Since households intertemporally smooth utility, private consumption is increased over many periods, and the households also have to increase savings. In addition, capital adjustment costs also drive up Tobin's q.

On impact in period $t = 2$, hours L decrease by 0.4%. This behavior is in accordance with results from empirical studies such as Galí (1999) and Basu,

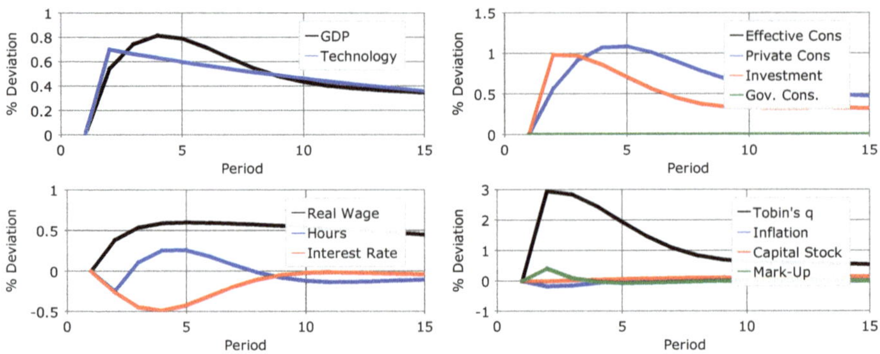

Fig. 4.19 Impulse responses to a technology shock in the New Keynesian model

Fernald, and Kimball (2006) and is typical in New Keynesian models with sticky prices and adjustment costs of capital. The increase in the real wage is not sufficient to offset the labor tax implied by the rise in the adjustment costs of capital. In addition, the real wage is sluggish, and thus, the (incentive) effect of higher wages only manifests itself after 2–3 quarters, when most of the adjustment has taken place. Recall that during the first period after the shock, only half of the households will be able to secure an increase in wages in the presence of wage staggering. The remaining households only benefit from the technology increase over the course of the subsequent periods, such that labor L gradually increases in period $t = 3, \ldots$.

To provide a more intuitive argument why a technology shock initially lowers labor supply in the New Keynesian model, consider the case of completely rigid prices. Assume, in particular, that prices remain constant in period $t = 2$ when the shock occurs. If firms cannot lower their prices, demand for final goods remains also constant ceteris paribus such that production does not increase. Since productivity Z_t, however, has increased, the firms have to decrease the use of the input factors labor and/or capital. Since the adjustment of capital is costly, firms prefer to reduce the use of labor. In the medium run, prices and wages are flexible so that the New Keynesian model behaves like the RBC model. Labor, capital, production, and consumption all increase. Since the households optimally smooth their consumption intertemporally, they already increase their consumption in period $t = 2$ when the shock occurs. As a consequence, aggregate demand increases and this effect increases the demand for labor. Similarly, investment demand also increases as the firms build up their capital stock over time. In addition, we assume that prices are not completely rigid so that (some) firms can lower their prices to increase their sales. Accordingly, aggregate demand and, hence, production increase. As is evident from the inspection of the upper left panel of Fig. 4.19, the effect on aggregate demand is lower than the increase in productivity in period 2. Therefore, the firm optimally reduces its labor demand. The increase in aggregate goods demand is higher if prices are more flexible. For the case of fully flexible prices (as in the standard RBC model), the increase in goods demand and, therefore, production is higher than the increase in productivity so that labor input has to increase as well.[51]

In addition, marginal costs and, hence, inflation decline. Since the technology level Z_t increases, firms in the intermediate goods sector can produce at lower marginal costs. As the prices in the wholesale sector only adjust gradually, the mark-up of wholesale prices P_t over the prices P_{yt} of intermediate goods increases or, equally, g_t falls. The response of the mark-up also helps to explain the hump-shaped response of the real wage because the real wage increases as P_{yt} approaches P_t over time (recall that real wages depend on the inverse of the mark-up according to $w_t = g_t(1-\alpha)Z_t(K_t/L_t)^{\alpha}$). In the long run, producers in the wholesale sector adjust their prices downward, and we observe negative inflation (the blue line in the bottom-right panel of Fig. 4.19).

The central bank adjusts the nominal interest rate Q_t downward. The central bank follows its Taylor rule (4.73x). While the higher output increases the nominal interest rate, the lower inflation rate reduces it. The net effect is negative, and thus, the nominal interest rate declines (see the red line in the bottom-left panel of Fig. 4.19). The decline in the nominal interest rate is even larger than the decline in the inflation rate, and thus, real interest rates also decrease. Therefore, the monetary policy reinforces the effect of the technology shock on private consumption and leisure, as the households intertemporally reallocate their consumption and leisure and increase both variables in periods with low interest rates according to their Euler condition (4.66c).

4.5.3.2 Government Consumption Shock

Figure 4.20 illustrates the effects of a government consumption shock, $\epsilon_2^G = 0.01$. Given the autoregressive process of government consumption G_t, G_t (the green line in the top-right panel) gradually declines over time after the impact in period $t = 2$, from 1% to 0. Notice that the positive impulse responses of output and the negative impulse responses of investment are in accordance with empirical observations. The government multiplier amounts to 0.6 on impact, meaning that, again, we can closely replicate empirical evidence on the multiplier in the present case of sticky prices.[52] However, private consumption and wages decline after an increase in government consumption.

---

[51] The reader is invited to experiment with the values of the parameters $\{\varphi_y, \varphi_w, \epsilon_y, \epsilon_w, \zeta, \chi\}$ in the GAUSS program *Ch4_newkeynesian.g* in order to study the sensitivity of the labor impulse responses.

[52] Farhi and Werning (2016) demonstrate in a standard New Keynesian model that the multiplier increases and exceeds unity in case of a liquidity trap (in which interest rates hit zero). Extending their model to the open economy, these authors find that the fiscal multiplier is smaller and below one for a country in a currency union. Erceg and Linde (2012) find that the fiscal multiplier is below one in an economy with fixed-exchange rates and, in accordance with the Mundell-Flemming model, above the one with flexible exchange rates. They show that their latter result is sensitive with respect to the slope of the Phillips curve and the presence of a persistent liquidity trap.

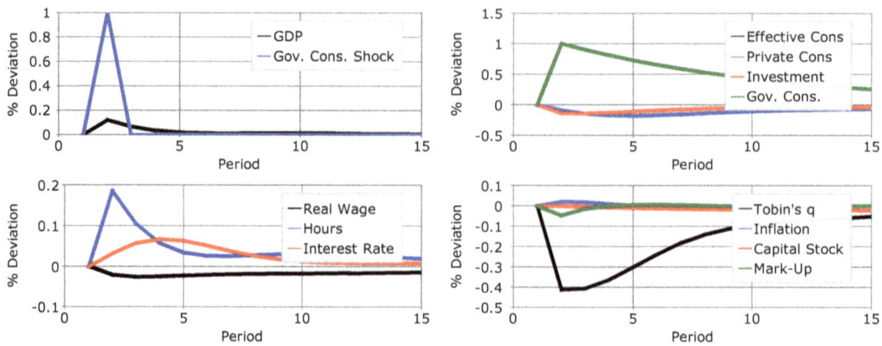

Fig. 4.20 Impulse responses to a government consumption shock in the New Keynesian model

The economic intuition for the impulse responses of the economic variables is as follows: Higher government consumption increases taxes and reduces wealth. As a consequence, labor L rises.[53] Consequently, the marginal product of labor and, hence, wages decline. Because of their reduced disposable income, households both consume and save less, meaning that investment declines, and hence, the capital stock and the marginal product of labor decrease. This effect reinforces the decline in wages.

As the marginal product of labor falls, marginal costs of production increase, resulting in higher prices. Accordingly, the increase in total demand drives up inflation. Monetary policy reinforces the effect of higher government consumption on private consumption and labor supply in our model. Following its Taylor rule, the central bank increases its interest rate Q_t on nominal government debt. For the illustrated case, $\theta^Y = 0.25$ and $\theta^\pi = 1.5$, the increase in the nominal interest rate Q_t is larger than the increase in the inflation rate, meaning that real interest rates also increase, resulting in lower private consumption and higher labor supply by the households, ceteris paribus.[54]

To generate a positive response to higher government consumption in private consumption (in accordance with empirical evidence), we set $\phi = 1.3$. Evidently, our utility function implies that for $\phi > 1$, higher public consumption decreases

[53]Of course, the response of hours depend on our assumption that taxes are lump-sum rather than proportional to wage income.

[54]Linnemann and Schabert (2003) show analytically how the central bank's rule affects the consequences of higher government consumption for labor, output, and prices. In response to higher (government) demand, labor demand increases, while the wealth effect drives up labor supply. The strength of the demand effect depends on the response of the real interest rate, which is governed by the monetary policy rule. When the rise in the real interest rate is dampened by an interest rate rule (as in our case), output and inflation can increase. They also show that, if the central bank follows a simple money-growth rule, fiscal expansions could be both deflationary and contractionary.

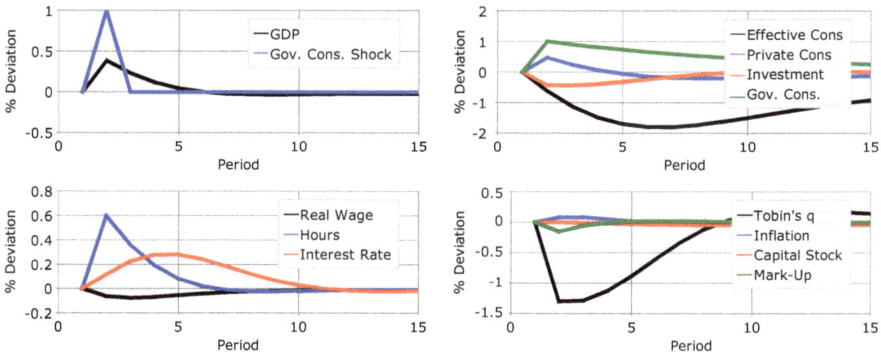

Fig. 4.21 Impulse responses to a government consumption shock in the New Keynesian model with very sticky wages and $\phi = 1.3$

household utility.[55] The results are displayed in Fig. 4.21. For this choice of ϕ, private consumption increases by 0.45% following a 1% shock to government consumption. Furthermore, most of the qualitative impulse response functions are in accordance with empirical observations, i.e., output, private consumption,[56] inflation, and the nominal interest rate all increase on impact, while investment declines. Only the response of wages is at odds with the evidence presented at the beginning of the chapter.[57]

4.5.4 Second Moments

Table 4.6 displays the second moments that result from the simulation of the model for $\phi = 1.3$ (with the corresponding impulse response functions displayed in Fig. 4.21). The relative volatility of investment and output is accordance with empirical observations, while private consumption and employment (hours) are excessively volatile.

[55]Ni (1995) provides empirical evidence that the estimates of the coefficient of public consumption in utility, $(1-\phi)/\phi$, are of small magnitude, with their signs depending on the measure of interest rates. If he uses net-of tax real taxes in his GMM estimation, he finds a negative coefficient, which corresponds to $\phi > 1$ above.

[56]The positive response of private consumption is even more pronounced if adjustment costs are smaller, e.g., with a capital adjustment cost parameter $\zeta = 0$. In this case, firms reduce their capital stock more rapidly and investment declines more strongly, meaning that more resources are freed up for private consumption.

[57]Heer and Scharrer (2018) present a model that is in accordance with all the empirical impulse responses of output, labor, demand components, and factor prices. For this reason, they introduce both rule-of-thumb consumers and a variable price of capital in terms of the consumption goods into an otherwise standard New Keynesian model.

Table 4.6 Second moments for the New Keynesian model with $\phi = 1.3$

Variable	s_x	s_x/s_Y	r_{xY}	r_{xG}	$r_{xx_{-1}}$
Production Y	1.18	1.00	1.00	0.27	0.71
Effective consumption C	4.38	3.72	0.72	−0.33	0.88
Private consumption C^p	1.36	1.15	0.92	0.26	0.80
Investment I	2.57	2.18	0.69	−0.22	0.56
Real wage w	0.77	0.66	0.62	−0.13	0.83
Hours L	1.35	1.14	0.66	0.38	0.53
Inflation π	0.68	0.51	0.38	0.14	0.77
Public consumption G	1.22	1.04	0.27	1.00	0.64
Mark-Up $1/g$	0.59	0.50	−0.37	−0.16	0.35

Notes: s_x: = Standard deviation of HP filtered simulated time series x, s_x/s_Y: = Standard deviation of the variable x relative to the standard deviation of output Y, r_{xY}: = Cross-correlation of the variable x with output Y, r_{xG}: = Cross-correlation of the variable x with government consumption G, $r_{xx_{-1}}$: autocorrelation of the variable x

With regard to the correlations of the economic variables, all demand components, investment, private consumption, and government consumption are procyclical, as observed empirically. The pro-cyclicity of government demand in our model can be explained with the help of the impulse response functions in Fig. 4.21. After a positive shock to government consumption, both G_t and Y_t increase above their steady-state levels for many periods and are characterized by positive comovement. Notice that the other model correlations are also in accordance with the model impulse responses, e.g., hours and inflation are positively correlated with government consumption, while investment and real wages are negatively correlated with G_t. The only variable whose correlation is not in accordance with empirical observations is the mark-up. Empirically, the mark-up is procyclical, while it is countercyclical in the model.[58]

4.5.5 Stabilization Policies

In this section, we conduct a simple policy experiment: How should the government respond to a cyclical increase in GDP? Should it consume in a procyclical or countercyclical way? To do so, we reconsider the government spending rule (4.69) for different policy parameters ρ^Y and study how they affect output volatility. For your convenience, let us restate (4.69):

$$\ln G_t = (1 - \rho^G) \ln G + \rho^G G_{t-1} + \rho^Y (\ln Y_{t-1} - \ln Y) + \epsilon_t^G.$$

[58]Nekarda and Ramey (2013) present evidence that the price-cost markup is procyclical or at best acyclical, which causes problems for standard New Keynesian models.

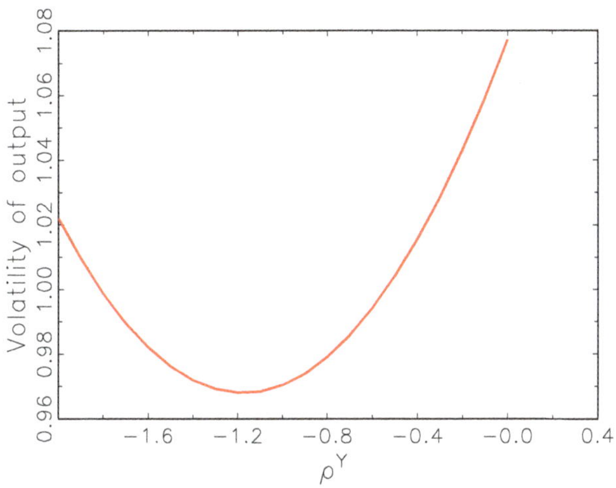

Fig. 4.22 Effect of policy rule parameter ρ^Y on output volatility

We computed the solution for the parameters $\rho^Y = 0.1, -0.1, \ldots, -2.0$ with the help of the GAUSS program *Ch4_new_keynes_stabil.g*. The program is available on my web page.

In Fig. 4.22, we display how the volatility of output changes for different values of ρ^Y.[59] The policy parameter that minimizes output fluctuations is equal to $\rho^Y = -1.2$. As a consequence, output volatility decreases by 10%, from 1.08% to 0.97%, as measured by the standard deviation of log output.[60]

The effects of a government spending rule $\rho^Y = -1.2$ (such that the government reduces public consumption for higher output Y_t) are illustrated in Figures 4.23 and 4.24, which present the impulse responses for shocks ϵ^G and ϵ^Z to government consumption and technology, respectively. We consider the case with $\phi = 1.3$ in which private consumption increases with higher government spending. Comparing Fig. 4.23 with Fig. 4.21, we find that, with $\rho^Y = -1.2$, the government reduces its consumption more rapidly after a positive shock ϵ^G. A government consumption shock increases all demand components (public and private consumption) except investment, meaning that output increases. With a negative reaction coefficient ρ^Y,

[59]Since we simulate time series of output with the help of a random number generator for the three shocks ϵ^Z, ϵ^G, and ϵ^Q, the results do not lie exactly on the curves displayed in Fig. 4.22. To smooth the curve, we fitted a polynomial of order two to the data points using a simply OLS regression. The estimation is contained in the GAUSS program *Ch4_new_keynes_stabil.g*.

[60]At this point, we refrain from deriving the optimal fiscal policy because it would take us too far into the field of numerical methods. Using perturbation methods of higher order, Schmitt-Grohé and Uribe (2007) derive optimal monetary and fiscal policy rules. For the fiscal policy rule, they consider a tax rule that sets total taxes as a function of government liabilities and the fiscal deficit. They find that whether the fiscal policy rule is active or passive does not significantly affect welfare.

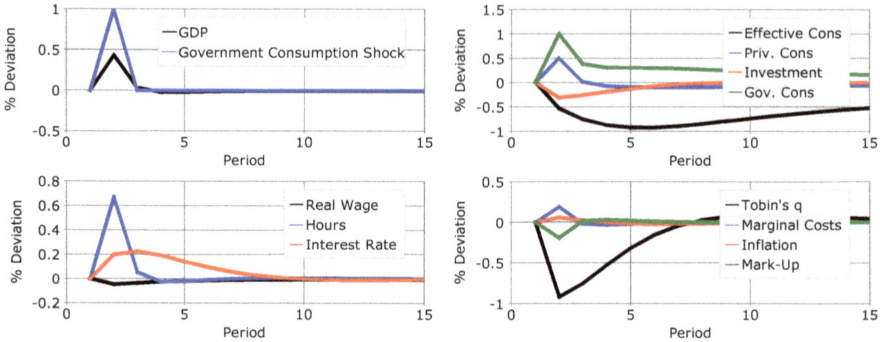

Fig. 4.23 Impulse responses to a government consumption shock with $\rho^Y = -1.2$

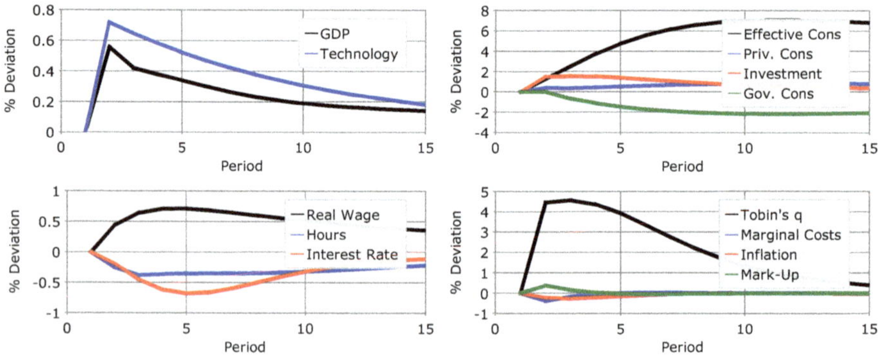

Fig. 4.24 Impulse responses to a technology shock with $\rho^Y = -1.2$

therefore, the rule results in lower government consumption in period $t = 3, \ldots$ after the shock in period $t = 2$.

In the case of a technology shock, output increases, and hence, the government reduces consumption. Comparing Fig. 4.24 with Fig. 4.19, we find that the response of output is reduced after period $t = 3, \ldots$ in the case with $\rho^Y = -1.2$. Although the coefficient is negative and government consumption falls in response to a technology shock, the correlation of output and government remains positive and equals 0.42.

How does the volatility-minimizing spending rule depend on monetary policy? To determine this, we conduct two experiments: (1) We set the shock to the interest rate rule equal to 0, $\epsilon_t^Q \equiv 0$ for all t. In this case, the parameter that minimizes output fluctuations is equal to -1.4. (2) We set the reaction coefficient of the Taylor rule equal to zero, $\theta^Y = 0$, such that monetary policy does not respond to the output gap. In this case, the volatility-minimizing coefficient also amounts to -1.4. Apparently, the output-variance-minimizing coefficient ρ^Y of the fiscal policy rule is rather insensitive to monetary policy and we find that fiscal policy should be countercyclical.[61]

In conclusion, however, we advise the reader to be careful about drawing firm policy conclusions from this policy experiment for at least five reasons. (1) We have not considered the welfare implications of such a rule.[62] (2) We have unrealistically assumed that additional government expenditures are financed by non-distortionary lump-sum taxes.[63] (3) We have taken monetary policy as exogenous, while it seems natural to search for the optimal stabilization policy using both monetary and fiscal policy as in Schmitt-Grohé and Uribe (2007). (4) Unlike the interest rates set by the central bank, fiscal policy parameters cannot be changed overnight. (5) We neglect political economy considerations. Politicians may wish to increase spending during a recession but not wish to cut public expenditures during a boom.

Appendix 4.1: Reverse Shooting

To compute the transition dynamics in the Numerical Example of Sect. 4.3, we first have to set the length of the transition period. When is the transition complete? We will assume that the dynamics are complete when the state variables are reasonably close to the new steady-state values. For practical reasons, we will stop searching for a better length of the transition if the divergence between the value of the state variable K_t in the last period of the transition from the steady-state value is less than 0.01% or if the value of the divergence is small, e.g., less than 10^{-5} in absolute value. The latter number is used when the state variable is very small and 0.01% of the state variable would be close to the accuracy of the solution algorithm (of the non-linear equation solver).

As we will discover in the remainder of the book, the transition in the neoclassical growth model occurs relatively fast compared to that in the other benchmark model that we consider in this book, the overlapping generations (OLG) model. Often, a transition length of fewer than 50 periods is sufficient, while we need to consider more than 100 periods in the OLG model in later chapters. We will choose 40 periods for the present case.

In essence, we have to solve for the dynamics of the three endogenous variables K_t, C_t^p, and L_t for $t = 1, \ldots, 40$ given $K_0 = \bar{K}$, $L_0 = \bar{L}$ and $C_0^p = \bar{C}^p$ for the

[61]As another sensitivity analysis, we considered the case with $\phi = 1.0$, such that public consumption does not affect household utility. In this case, the output-volatility-minimizing fiscal policy is specified with $\rho^Y = -1.3$.

[62]Our microfounded model has the benefit that we can quantitatively compare the welfare of different stabilization policies. In the present model, the equilibrium is not Pareto-efficient because various welfare distortions are present. First, firms in the wholesale sector operate as monopolistic competitors. Second, there is both price and wage dispersion.

[63]We will introduce income taxes and debt in the upcoming Chaps. 5 and 7, respectively.

initial values and $L_{41} = \tilde{L}$, $K_{41} = \tilde{K}$, and $C_{41}^p = \tilde{C}^p$ for the final values in case 1 and $L_{41} = \bar{L}$, $K_{41} = \bar{K}$, and $C_{41}^p = \bar{C}^p$ in case 2. More formally, we have to solve a two-point boundary value problem.

Let us first consider case 1 with a permanent change in government consumption $G_t = \tilde{G}$ for $t = 1, \ldots$. For the solution, we will make use of the first-order conditions, which we restate for your convenience:

$$w_t = \frac{\kappa}{\phi}\left(\frac{C_t}{1-L_t}\right)^{\frac{1}{\rho}} (\Xi_t)^{1-\frac{1}{1-1/\rho_C}} \left(C_t^p\right)^{\frac{1}{\rho_C}}, \tag{4.75a}$$

$$\beta(1+r_{t+1}-\delta) = \left(\frac{X_{t+1}}{X_t}\right)^{1-\frac{1-\sigma}{1-1/\rho_C}} \left(\frac{C_{t+1}}{C_t}\right)^{\frac{1}{\rho}} \left(\frac{\Xi_{t+1}}{\Xi_t}\right)^{1-\frac{1}{1-1/\rho_C}} \left(\frac{C_{t+1}^p}{C_t^p}\right)^{\frac{1}{\rho_C}}, \tag{4.75b}$$

$$K_{t+1} = (1-\delta)K_t + K_t^\alpha L_t^{1-\alpha} - G_t - C_t^p, \tag{4.75c}$$

where effective consumption C_t and the real interest rate r_t are given by (4.21) and (4.9b), and X_t is defined as follows:

$$X_t \equiv C_t^{1-\frac{1}{\rho}} + \kappa(1-L_t)^{1-\frac{1}{\rho}}. \tag{4.76}$$

In the first period of the transition at $t = 1$, the government unexpectedly increases public consumption G_t. As a consequence, the household can only adjust its behavior in period $t = 1$, not before. The beginning-of-period capital stock K_1, therefore, is predetermined by the decision of the household in period $t = 0$. The household can only choose consumption C_1^p, labor L_1, and the next-period capital stock K_2. As a consequence, we have to solve for the 119 unknowns $\{K_t\}_{t=2}^{40}$, $\{L_t\}_{t=1}^{40}$, and $\{C_t^p\}_{t=1}^{40}$. We require that K_{40} diverges by less than 0.01% from $\tilde{K} = 1.2946$.

The algorithm that we will apply is called *reverse shooting*. We will provide a guess of K_{40} that is close to the new steady-state value $K_{41} = \tilde{K}$. We can compute the values L_{40} and C_{40}^p with the help of the first-order conditions (4.75a) and (4.75c) for period $t = 40$ and using the steady-state values $K_{41} = \tilde{K}$, $L_{41} = \tilde{L}$, and $C_{41}^p = \tilde{C}^p$ for the values of the endogenous variables in period $t+1 = 41$. The problem is to solve a system of two non-linear equations in two unknowns, L_{40} and C_{40}^p. As the initial guess that we have to provide to the non-linear equation solver routine used in the Gauss program *Ch4_subs_private_pub_dyn.g*, we simply take the new steady-state values $\tilde{L}$ and $\tilde{C}^p$, respectively.

With the help of the values $\{K_{40}, L_{40}, C_{40}^p\}$, we can compute the values $\{K_{39}, L_{39}, C_{39}^p\}$ using the three non-linear equations (4.75a), (4.75b), and (4.75c). More generally, we can compute $\{K_t, L_t, C_t^p\}$ with the help of $\{K_{t+1}, L_{t+1}, C_{t+1}^p\}$ in the same way for $t = 39, 38, \ldots, 1$ providing $\{K_{t+1}, L_{t+1}, C_{t+1}^p\}$ as an initial value to the non-linear equation solver. In the final iteration, we have $\{K_1, L_1, C_1^p\}$. If K_1 is close to its value in the initial steady state $\bar{K}$ we are done. Otherwise, we have to adjust our guess of K_{40} and retry.

How do we choose an initial guess K_{40}? This is not a trivial task. First, we know that K_1 is smaller than K_{41} from our steady-state computation in the Gauss program *Ch4_subs_private_pub.g*. Since we know from the standard continuous-time Ramsey model that the transition path is saddle-point stable and that the speed of convergence declines during the transition, we would expect K_{40} to lie very close to K_{41}.[64] For this reason, we perform a grid search of the optimal start value K_{40} in the interval $[\bar{K} + 0.9(\tilde{K} - \bar{K}), \tilde{K}]$. We choose an equispaced grid of $nk = 1,000$ points. For the lower values in this grid, the capital stock falls below zero in fewer than 40 periods during the recursive iteration over $\{K_t, L_t, C_t^p\}$, and we subsequently have to exclude these values. As it turns out, the optimal point (that produces a value of K_1 closest to $\bar{K}$) is very close to $\tilde{K}$, and again, we use a much tighter grid, $[\bar{K} + 0.999(\tilde{K} - \bar{K}), \tilde{K}]$, and find $K_{40} = 1.2945418$ as our initial guess. Given this value, we simply find the solution to the non-linear equation $f(K_1) = K_1 - \bar{K} = 0$.

How does our algorithm perform? We compute a value of K_1 that diverges by less than 10^{-8} from the old steady-state value $\bar{K} = 1.2882145$. Therefore, we have very accurately computed the transition dynamics. Let us conclude this section with some qualifying remarks:

1. If we had chosen a higher number for the transition periods than 40, we might have failed to compute the transition dynamics. Why? Let us consider the final values of K_{39} and K_{40}. We have computed the values of $K_{39} = 1.2945418$ and $K_{40} = 1.2945502$ for an accuracy of 10^{-8}. The values are very close to one another, and the transition is basically complete after 30–35 periods. If we had chosen an even higher number, say 100, we would not have been able to provide an initial guess for K_{100} that is different from $K_{101} = \tilde{K}$ given machine accuracy. If we, however, use $K_{100} = K_{101}$, our algorithm will just compute the steady-state values for L_{100} and C_{100}^p in the first step and for all other values $\{K_t, L_t, C_t^p\}$ for $t = 99, \ldots, 1$. If we, however, consider a smaller value, e.g., $K_{100} = 1.2946334$ instead of $\tilde{K} = 1.2946335$, we will obtain a negative value for K_t in our recursive iteration for $t > 1$ and be unable to find a solution.
2. Is it possible to solve for the dynamics if we select K_2 as a starting value and iterate forward (so-called *forward shooting*)? Yes. The reason is as follows: Given K_1 and K_2, we can solve for C_1^p and L_1 using (4.75a) and (4.75c). In the next step, we seek to solve for $\{K_3, C_2^p, L_2\}$. Therefore, we use (4.75a) and (4.75c) for period $t = 2$ and (4.75b) for period $t = 1$. The solution $\{K_3, C_2^p, L_2\}$ is used in the next step. We iterate forward until we have found

[64] See, for example, Chapter 2 in Barro and Sala-i-Martin (2003) for a derivation of the transition dynamics in the continuous-time neoclassical growth model and, in particular, Section 2.6.6 for the speed of convergence.

the solution for $\{K_{41}, C^p_{40}, L_{40}\}$ and compare K_{41} to the new steady-state value $\tilde{K}$. If we are close, we are done. Otherwise, we have to specify a new guess for $\tilde{K}$.[65]

In Chap. 6, you will encounter a problem (an OLG model with pay-as-you-go pensions) where reverse shooting is possible, while forward shooting is not. In practice, reverse shooting is often used even if forward shooting is possible. The reason is because it is often easier to provide an initial guess for the capital stock in the last period rather than the first period of the transition because the speed of convergence declines and you have to search in a smaller neighborhood around the final steady state than in the case for a guess of the capital stock in the initial period. If you use forward shooting, you have to search in a larger neighborhood of the initial steady state. This is particularly cumbersome if your state variable is not a single variable but is multi-dimensional.

3. In the case of a temporary government shock, we know that the new steady-state values are equal to the old steady-state values. In this case, however, the capital stock K_t approaches K_{41} from above. Therefore, we have to provide a guess for K_{40} that is larger than $K_{41} = \bar{K}$. The rest of the computational procedure is completely equivalent to the case of a permanent increase in government consumption.

Appendix 4.2: Frisch Labor Supply Elasticity for Cobb-Douglas Utility

The *Frisch labor supply elasticity* or *intertemporal labor supply elasticity* $\eta_{L,w}$ is defined as the percentage change in the labor supply in response to a 1% increase in the wage given a constant marginal utility of consumption u_C:

$$\eta_{L,w} \equiv \frac{dL}{dw}|_{u_C=\text{const}} \frac{w}{L}. \tag{4.77}$$

Let utility $u(C, L)$ be a function of consumption C and labor supply L. Ignoring taxes, contributions, and pensions, the first-order condition of the household with respect to its labor supply is given by:

$$-u_L(C, L) = w u_C(C, L). \tag{4.78}$$

According to (4.78), the disutility from working another time unit is equal to the utility from the additional consumption that can be afforded by the additional wage income from working an additional time unit. The total differential of (4.78) for a

[65]You will be asked to compute the solution in Problem 4.1.

constant $u_C(C, L)$ is given by

$$-u_{LL}dL - u_{CL}dC = u_C dw. \tag{4.79}$$

Furthermore, $du_C = 0$ implies

$$u_{CC}dC + u_{CL}dL = 0,$$

or

$$dC = -\frac{u_{CL}}{u_{CC}}dL.$$

Inserting the last equation into (4.79), we obtain

$$\eta_{L,w} \equiv \frac{dL}{dw}|_{u_C=\text{const}}\frac{w}{L} = \frac{u_C}{\frac{u_{CL}^2}{u_{CC}} - u_{LL}}\frac{w}{L}, \tag{4.80}$$

and for the Cobb-Douglas utility function with $u(C, L) = \frac{\left(C^\iota(1-L)^{1-\iota}\right)^{1-\sigma}-1}{1-\sigma}$, the Frisch labor supply elasticity is

$$\eta_{L,w} = \frac{1-\iota(1-\sigma)}{\sigma}\frac{1-L}{L}.$$

Appendix 4.3: Microfoundations of Calvo Price Setting

Let us consider a wholesale firm j with the relative price $P_{t+s}(j)/P_{t+s}$.[66] In period t, the firm received the signal to choose its optimal relative price $p_{At} = P_{At}/P_t$ and, since then, has not received a signal to do so again up to period $t+s$. Between period t and $t+s$, the price of good j, $P_t(j)$, increases with the lagged inflation rate $\pi_t, \pi_{t+1}, \ldots, \pi_{t+s-1}$ in each period $t+1, t+2, \ldots, t+s$, while the aggregate

[66] Appendices 4.3 and 4.4 were afforded in large parts by Alfred Maußner and are based upon the exposition in Heer, Maußner, and Ruf (2017). A more detailed description of the derivation of the microfoundations of Calvo price staggering can be found in Maußner (2000). I would like to thank Alfred for his thoughtful comments and support that have greatly helped to improve the presentation of the material in this chapter. All remaining errors are mine.

price level P_t increases with the inflation rates $\pi_{t+1}, \pi_{t+2}, \ldots, \pi_{t+s}$:

$$\frac{P_{t+s}(j)}{P_{t+s}} = \frac{\pi_{t+s-1}\cdots\pi_t}{\pi_{t+s}\cdots\pi_{t+1}} p_{At} = \frac{\pi_t}{\pi_{t+s}} p_{At}.$$

Accordingly, the firm will choose p_{At} in period t to maximize discounted dividends (after inserting the demand function (4.43) into dividends (4.46)):

$$\mathbb{E}_t \sum_{s=0}^{\infty} (\beta\varphi_y)^s \frac{\Lambda_{t+s}}{\Lambda_t} \left[\left(\frac{\pi_t}{\pi_{t+s}} p_{At} \right)^{1-\epsilon_y} Y_{t+s} - g_{t+s} \left(\frac{\pi_t}{\pi_{t+s}} p_{At} \right)^{-\epsilon_y} Y_{t+s} \right],$$

where $(\varphi_y)^s$ denotes the probability that the firm cannot adjust its price for s consecutive periods.

Differentiating this equation with respect to p_{At} results in the first-order condition:

$$0 = \mathbb{E}_t \sum_{s=0}^{\infty} (\beta\varphi_y)^s \frac{\Lambda_{t+s}}{\Lambda_t} \left[(1-\epsilon_y) \left(\frac{\pi_t}{\pi_{t+s}} \right)^{1-\epsilon_y} Y_{t+s} p_{At}^{-\epsilon_y} + \epsilon_y g_{t+s} \left(\frac{\pi_t}{\pi_{t+s}} \right)^{-\epsilon_y} Y_{t+s} p_{At}^{-\epsilon_y - 1} \right],$$

where we used the theorem that the derivative operator can be interchanged with the expectational operator and the sum operator.

The first-order equation can be re-written as

$$p_{At} = \frac{\epsilon_y}{\epsilon_y - 1} \frac{\Gamma_{1t}}{\Gamma_{2t}}, \tag{4.81a}$$

$$\Gamma_{1t} = \mathbb{E}_t \sum_{s=0}^{\infty} (\beta\varphi_y)^s \left(\frac{\pi_t}{\pi_{t+s}} \right)^{-\epsilon_y} g_{t+s} \Lambda_{t+s} Y_{t+s} = g_t \Lambda_t Y_t + (\beta\varphi_y) \mathbb{E}_t \left(\frac{\pi_t}{\pi_{t+1}} \right)^{-\epsilon_y} \Gamma_{1t+1}, \tag{4.81b}$$

$$\Gamma_{2t} = \mathbb{E}_t \sum_{s=0}^{\infty} (\beta\varphi_y)^s \left(\frac{\pi_t}{\pi_{t+s}} \right)^{1-\epsilon_y} \Lambda_{t+s} Y_{t+s} = \Lambda_t Y_t + (\beta\varphi_y) \mathbb{E}_t \left(\frac{\pi_t}{\pi_{t+1}} \right)^{1-\epsilon_y} \Gamma_{2t+1}. \tag{4.81c}$$

Γ_{1t} and Γ_{2t} are simply auxiliary variables whose behaviors are described by (stochastic) first-order difference equations. Therefore, they are easily amenable to the solution with the linearization methods described in Appendix 2.3.

The price index (4.45) implies

$$P_t^{1-\epsilon_y} = (1-\varphi_y) P_{At}^{1-\epsilon_y} + \varphi_y P_{Nt}^{1-\epsilon_y} = (1-\varphi_y) P_{At}^{1-\epsilon_y} + \varphi_y (\pi_{t-1} P_{t-1})^{1-\epsilon_y}.$$

The second equality follows from the updating rule (4.47) and the fact that the non-optimizers are a random sample of optimizers and non-optimizers. Dividing by P_t on both sides yields:

$$1 = (1-\varphi_y) p_{At}^{1-\epsilon_y} + \varphi_y (\pi_{t-1}/\pi_t)^{1-\epsilon_y}. \tag{4.81d}$$

Finally, consider the definition of $\tilde{Y}_t$ given in (4.50):

$$\tilde{Y}_t = \int_0^1 Y_t(j)\, dj = \int_0^1 \left(\frac{P_t(j)}{P_t}\right)^{-\epsilon_y} Y_t\, dj = \left(\frac{\tilde{P}_t}{P_t}\right)^{-\epsilon_y} Y_t,$$

with the definitions

$$\tilde{P}_t^{-\epsilon_t} \equiv \int_0^1 P_t(j)^{-\epsilon_y}\, dj,$$

and

$$s_t^y \equiv \left(\frac{\tilde{P}_t}{P_t}\right)^{-\epsilon_y}.$$

Therefore,

$$\tilde{Y}_t = s_t^y Y_t. \tag{4.81e}$$

Using the same reasoning for $\tilde{P}_t$ as for the price index P_t above results in the following first-order difference equation for the dispersion of individual prices $P_t(j)$ in the wholesale sector:

$$s_t^y = (1-\varphi_y) p_{At}^{-\epsilon_y} + \varphi_y (\pi_{t-1}/\pi_t)^{-\epsilon_y} s_{t-1}^y. \tag{4.81f}$$

Appendix 4.4: Microfoundations of Wage Setting

Consider the real wage $W_t(h)/P_t$ of a household member h who has set his wage optimally in period t to $\tilde{w}_t = W_{At}/P_t$ and who has not been able to do so again until period s. Between period t and $t+s$, the nominal wage of the household member h increases with the lagged inflation rate $\pi_t, \pi_{t+1}, \ldots, \pi_{t+s-1}$ in each period $t+1$, $t+2, \ldots, t+s$, while the aggregate price level P_t increases with the inflation rates $\pi_{t+1}, \pi_{t+2}, \ldots, \pi_{t+s}$:

$$\frac{W_{Nt+s}}{P_{t+s}} = \frac{\prod_{i=1}^{s} \pi_{t+i-1} W_{At}}{\prod_{i=1}^{s} \pi_{t+i} P_t} = \frac{\pi_t}{\pi_{t+s}} \tilde{w}_t.$$

The demand for his type of labor service follows from (4.58)

$$L_{t+s}(h) = \left(\frac{(\pi_t/\pi_{t+s})\tilde{w}_t}{w_{t+s}}\right)^{-\epsilon_w} L_{t+s},$$

where $w_{t+s} = W_{t+s}/P_{t+s}$ denotes the real wage prevailing in period $t+s$. Accordingly, the Lagrangian for the optimal real wage is represented by:

$$\begin{aligned}
\mathscr{L} =& \mathbb{E}_t \sum_{s=0}^{\infty} (\beta\varphi_w)^s \left\{ \frac{(C_{t+s}(h) - \chi \bar{C}_{t+s})^{1-\sigma} - 1}{1-\sigma} \right. \\
& - \frac{\nu_0}{1+\frac{1}{\nu_1}} \left[\left(\frac{(\pi_t/\pi_{t+s})\tilde{w}_t}{w_{t+s}}\right)^{-\epsilon_w} L_{t+s} \right]^{1+\frac{1}{\nu_1}} \\
& + \gamma_0^M \frac{\left(\frac{M_{t+1}(h)}{P_t}\right)^{1-\gamma_1^M} - 1}{1-\gamma_1^M} \\
& \left. + \Lambda_{ht+s} \left[\frac{\pi_t}{\pi_{t+s}} \tilde{w}_t \left(\frac{(\pi_t/\pi_{t+s})\tilde{w}_t}{w_{t+s}}\right)^{-\epsilon_w} L_{t+s} + RMT \right] \right\},
\end{aligned}$$

where $(\varphi_w)^s$ denotes the probability that the household cannot adjust its wage optimally for s periods and RMT_t is a placeholder for the remaining terms of the household budget constraint (4.61).

The first-order condition with respect to $\tilde{w}_t$ is represented by

$$\begin{aligned}
0 = \mathbb{E}_t \sum_{s=0}^{\infty} (\beta\varphi_w)^s & \left\{ \epsilon_w \nu_0 \tilde{w}_t^{-\epsilon_w(1+\frac{1}{\nu_1})-1} \left(\frac{(\pi_t/\pi_{t+s})}{w_{t+s}}\right)^{-\epsilon_w(1+\frac{1}{\nu_1})} L_{t+s}^{1+\frac{1}{\nu_1}} \right. \\
& \left. + (1-\epsilon_w)\Lambda_{ht+s}\tilde{w}_t^{-\epsilon_w} w_{t+s}^{\epsilon_w} \left(\frac{\pi_t}{\pi_{t+s}}\right)^{1-\epsilon_w} L_{t+s} \right\}.
\end{aligned}$$

Using $\Lambda_{ht+s} = \Lambda_{t+s}$, this equation can be arranged to read as

$$\tilde{w}_t = \frac{\epsilon_w}{\epsilon_w - 1} \frac{\Delta_{1t}}{\Delta_{2t}}, \tag{4.82a}$$

where

$$\begin{aligned}
\Delta_{1t} &= \nu_0 \mathbb{E}_t \sum_{s=0}^{\infty} (\beta\varphi_w)^s \left(\frac{\pi_t \tilde{w}_t}{\pi_{t+s} w_{t+s}}\right)^{-\epsilon_w(1+\frac{1}{\nu_1})} L_{t+s}^{1+\frac{1}{\nu_1}}, \\
&= \nu_0 \left(\frac{\tilde{w}_t}{w_t}\right)^{-\epsilon_w(1+\frac{1}{\nu_1})} L_t^{1+\frac{1}{\nu_1}} + (\beta\varphi_w)\mathbb{E}_t \left(\frac{\pi_t \tilde{w}_t}{\pi_{t+1}\tilde{w}_{t+1}}\right)^{-\epsilon_w(1+\frac{1}{\nu_1})} \Delta_{1t+1},
\end{aligned} \tag{4.82b}$$

$$\Delta_{2t} = \mathbb{E}_t \sum_{s=0}^{\infty} (\beta\varphi_w)^s \Lambda_{t+s} \left(\frac{\tilde{w}_t}{w_{t+s}}\right)^{-\epsilon_w} \left(\frac{\pi_t}{\pi_{t+s}}\right)^{1-\epsilon_w} L_{t+s},$$

$$= \Lambda_t \left(\frac{\tilde{w}_t}{w_t}\right)^{-\epsilon_w} L_t + (\beta\varphi_w)\mathbb{E}_t \left(\frac{\tilde{w}_t}{\tilde{w}_{t+1}}\right)^{-\epsilon_w} \left(\frac{\pi_t}{\pi_{t+1}}\right)^{1-\epsilon_w} \Delta_{2t+1}. \quad (4.82c)$$

Again, Δ_{1t} and Δ_{2t} are simply auxiliary variables whose behaviors are described by (stochastic) first-order difference equations. Therefore, they are easily amenable to the solution with the linearization methods described in Appendix 2.3.

The wage index (4.59) implies

$$W_t^{1-\epsilon_w} = (1-\varphi_w) W_{At}^{1-\epsilon_w} + \varphi_w (\pi_{t-1} W_{t-1})^{1-\epsilon_w}$$

and thus, the real wage equals

$$w_t^{1-\epsilon_w} = (1-\varphi_w)\tilde{w}_t^{1-\epsilon_w} + \varphi_w \left(\frac{\pi_{t-1}}{\pi_t} w_{t-1}\right)^{1-\epsilon_w}. \quad (4.82d)$$

Finally, consider the index

$$\tilde{L}_t^{1+\frac{1}{\nu_1}} = \int_0^1 L_t(h)^{1+\frac{1}{\nu_1}} dh,$$

in the families of current-period utility functions. Using labor demand function (4.58), this index can be re-written as

$$\tilde{L}_t^{1+\frac{1}{\nu_1}} = L_t^{1+\frac{1}{\nu_1}} \int_0^1 \left(\frac{W_t(h)}{W_t}\right)^{-\epsilon_w(1+\frac{1}{\nu_1})} dh.$$

Therefore,

$$\tilde{L}_t = s_t^w L_t. \quad (4.82e)$$

implies the definition of the wage dispersion measure (4.63).

Next, we need to derive the dynamics of the wage dispersion measure s_t^w. For this reason, consider

$$\bar{W}_t^{-\epsilon_w(1+\frac{1}{\nu_1})} = \int_0^1 W_t(h)^{-\epsilon_w(1+\frac{1}{\nu_1})} dh$$

$$= (1-\varphi_w)(W_{At})^{-\epsilon_w(1+\frac{1}{\nu_1})} + \varphi_w(\pi_{t-1} W_{Nt-1})^{-\epsilon_w(1+\frac{1}{\nu_1})}.$$

Accordingly, wage dispersion s_t^w is described by the following equation:

$$(s_t^w)^{1+\frac{1}{\nu_1}} = \left(\frac{\tilde{W}_t}{W_t}\right)^{-\epsilon_w(1+\frac{1}{\nu_1})} = \left(\frac{\tilde{W}_t/P_t}{W_t/P_t}\right)^{-\epsilon_w(1+\frac{1}{\nu_1})} = \left(\frac{\tilde{w}_t}{w_t}\right)^{-\epsilon_w(1+\frac{1}{\nu_1})}.$$

Using the same line of argument employed to derive (4.81f) yields the dynamic equation for the measure of wage dispersion s_t^w:

$$(s_t^w)^{1+\frac{1}{\nu_1}} = (1-\varphi_w)\left(\frac{\tilde{w}_t}{w_t}\right)^{-\epsilon_w(1+\frac{1}{\nu_1})} + \varphi_w\left(\frac{\pi_{t-1}w_{t-1}}{\pi_t w_t}\right)^{-\epsilon_w(1+\frac{1}{\nu_1})}(s_{t-1}^w)^{1+\frac{1}{\nu_1}}. \tag{4.82f}$$

Appendix 4.5: Monetary Policy Analysis in the New Keynesian Model

Figure 4.25 illustrates the impulse response functions to an interest rate shock of one percentage point in the benchmark New Keynesian model with $\phi = 1.0$. In accordance with empirical observations, restrictive monetary policy decreases all private demand, investment and private consumption; moreover, output and labor decline. As expected, inflation also decreases under a restrictive monetary policy.

The reasons are as follows: Following an increase in the nominal interest rate Q_t, prices only adjust slowly so that also the real interest rate in the economy increases. As a consequence, firms reduce their investment, and Tobin's q decreases. In addition, households postpone consumption to later periods in accordance with

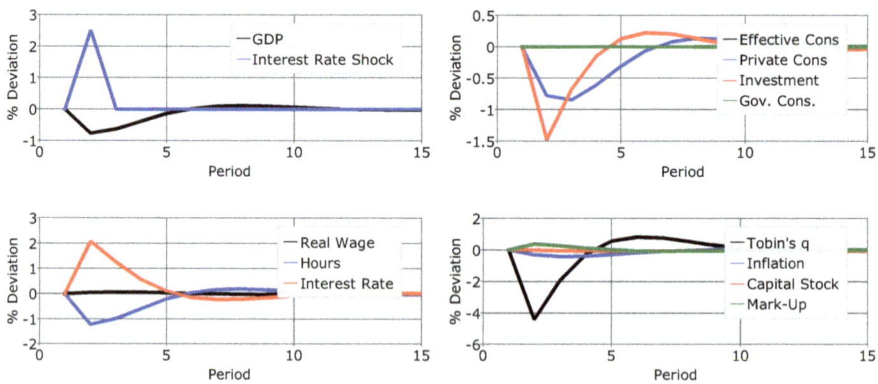

Fig. 4.25 Impulse responses to an interest rate shock

their Euler condition (4.66c). Therefore, demand decreases, and prices fall. Since wholesale producers are slow to adjust their prices, the mark-up increases.

Appendix 4.6: Data Sources

The time series on government expenditures that we use in this chapter are attached as separate Excel files to my Matlab/Gauss programs.

- **Government Expenditures** General government total expenditure as percent of GDP (National currency). The data for Fig. 4.1 are retrieved from the IMF's World Economic Outlook database (Accessed on 15 February 2018).
  http://www.imf.org/external/pubs/ft/weo/2017/02/weodata/index.aspx.
  The German data for the years 1980–1990 are retrieved from the Deutsche Bundesbank (Accessed on 15 February 2018).
  http://www.bundesbank.de/Navigation/DE/Statistiken/Zeitreihen_Datenbanken/zeitreihen_datenbank.html.
  The US data for the years 1980–2000 and the data for Fig. 4.2 are retrieved from 'National Accounts at a Glance': 6. General Government (Accessed on 15 February 2018).
  https://stats.oecd.org/Index.aspx?DataSetCode=NAAG#.
- **Pensions** Government, Public expenditure, Pension expenditure from OECD Factbook 2015–2016, which can be obtained via the Internet at the web site (Accessed on 15 February 2018)
  http://www.oecd-ilibrary.org/economics/oecd-factbook-2015-2016/public-and-private-expenditure-on-pensions-as-a-percentage-of-gdp-2011_factbook-2015-graph171-en.
- **Health Expenditures** are retrieved from the OECD database 'OECD iLibrary' (Accessed on 15 February 2018).
  http://stats.oecd.org/BrandedView.aspx?oecd_bv_id=health-data-en&doi=data-00349-en.
- **Education Expenditures** are retrieved from the OECD publication 'Education at a Glance: 2014', Table B2.3 – Expenditure on educational institutions as a percentage of GDP, by source of fund and level of education (2011) (Accessed on 15 February 2018).
  http://www.oecd-ilibrary.org/education/education-at-a-glance-2014/expenditure-on-educational-institutions-as-a-percentage-of-gdp-by-source-of-fund-and-level-of-education-2011_eag-2014-table124-en.
- **Defense Expenditures** are retrieved from the OECD publication 'National Accounts at a Glance: 2015', General government, Table 24.1. General government expenditure by function, 'Defence' and 'Public order and safety', Percentage of GDP, 2012 (Accessed on 15 February 2018).

http://www.oecd-ilibrary.org/economics/national-accounts-at-a-glance-2015/total-general-government-expenditure-by-function_na_glance-2015-table35-en.

- **Government Consumption** The series on government consumption is constructed with the help of the data from FRED provided by the Federal Reserve Bank of St. Louis (Accessed on 28 October 2015). In particular, I used the series GCEC1 Real Government Consumption Expenditures & Gross Investment, Billions of Chained 2009 Dollars, Quarterly, Seasonally Adjusted Annual Rate, and subtracted real government investments. I computed real government investment from the series A782RC1Q027SBEA Gross Government Investment, divided by the implicit price deflator (2009=100) of GDP (GDPDEF).

Problems

4.1. Solve the transition dynamics for the Numerical Problem in Sect. 4.3 by *forward shooting* as described in Appendix 4.1.

4.2. Derive the Frisch labor supply elasticity (4.39) of the Cobb-Douglas utility function (4.28).

4.3. In applied work, researchers often select model parameters that are not empirically observable or only estimated with a high degree of uncertainty by optimizing the behavior of the RBC model. Using the model in Sect. 4.4, use a grid search over $\phi \in [0, 1.5]$ and $\rho^C \in [0.3, 1.3]$ to find the minimum distance (i.e., the minimum of the squared deviations) of the theoretical second moments (as implied by the model) from the empirical second moments (as presented in Table 4.1).

4.4. Use the preferences

$$u(C^p, L) = \frac{(C^p)^{1-\sigma}(1-L)^{1+\vartheta}}{1-\sigma}, \quad \sigma > 1, \vartheta > 0,$$

and recompute the RBC model with stochastic government. Set $\sigma = 2.0$, and calibrate ϑ such that steady-state labor supply is equal to $L = 0.3$. Does private consumption increase after an increase in government consumption?

4.5. Derive (4.72) using the individual budget constraint, the government budget, and the firms' first-order equations. Apply Euler's theorem according to which the aggregate output is equal to the sum of all factor payments for a constant-returns-to-scale technology under perfect competition.

4.6. Derive the equilibrium dynamics (4.73) for the New Keynesian model in Sect. 4.5. Assume, however, that firms in the wholesale sector who cannot optimally choose their price adjust their nominal price P_{NT} according to the average inflation rate:

$$P_{Nt} = \pi P_{Nt-1}.$$

References

Abel, A. B. (1990). Asset pricing under habit formation and catching-up-with-the-Joneses. *American Economic Review, 80*, 38–42.

Aiyagari, R. S., Christiano, L. J., & Eichenbaum, M. (1994). The output, employment, and interest effect of government consumption. *Journal of Monetary Economics, 30*, 73–86.

Altonij, J. G. (1986). Intertemporal substitution in labor supply: Evidence from micro data. *Journal of Political Economy, 94*, S176–S215.

Ambler, S., & Paquet, A. (1996). Fiscal spending shocks, endogenous government spending, and real business cycles. *Journal of Economic Dynamics and Control, 20*, 237–256.

Aschauer, D. A. (1985). Fiscal policy and aggregate demand. *American Economic Review, 75*, 117–121.

Auerbach, A. J., & Gorodnichenko, Y. (2011). *Fiscal multipliers in recession and expansion.* NBER working paper, 17447.

Barro, R. J. (1981). Output effects of government purchases. *The Journal of Political Economy, 89*(6), 1086–1121.

Barro, R. J., & Sala-i-Martin, X. (2003). *Economic growth* (2nd ed.). Cambridge, MA: MIT Press.

Basu, S., Fernald, J. G., & Kimball, M. S. (2006). Are technology improvements contractionary?. *American Economic Review, 96*, 1418–1448.

Baxter, M., & King, R. G. (1993). Fiscal policy in general equilibrium. *American Economic Review, 91*, 167–185.

Blanchard, O. J., & Perotti, R. (2002). An empirical characterization of the dynamic effects of changes in government spending and taxes on output. *Quarterly Journal of Economics, 177*, 1329–1368.

Brandner, P., & Neusser, K. (1992). Business cycles in open economies: Stylized facts for Austria and Germany. *Review of World Economics, 128*(1), 67–87.

Brinca, P., Holter, H. A., Krusell, P., & Malafry, L. (2016). Fiscal multipliers in the 21st century. *Journal of Monetary Economics, 77*, 53–69.

Bullard, J., & Mitra, K. (2002). Learning about monetary policy rules. *Journal of Monetary Economics, 49*(6), 1105–1129.

Calvo, G. A. (1983). Staggered prices in a utility-maximizing framework. *Journal of Monetary Economics, 12*, 383–398.

Christiano, L. J., & Eichenbaum, M. (1992). Current Real-Business-Cycle theories and aggregate labor market fluctuations. *American Economic Review, 82*(3), 430–450.

Christiano, L. J., Eichenbaum, M., & Evans, C. L. (1997). Sticky prices and limited participation models of money: A comparison. *European Economic Review, 41*, 1201–1249.

Christiano, L. J., Eichenbaum, M., & Evans, C. L. (2005). Nominal rigidities and the dynamic effects of a shock to monetary policy. *Journal of political Economy, 113*(1), 1–45.

Clarida, R., Gali, J., & Gertler, M. (2000). Monetary policy rules and macroeconomic stability: Evidence and some theory. *Quarterly Journal of Economics, 115*(1), 147–180.

Constandinides, G. M. (1990). Habit formation: A resolution of the equity premium. *Journal of Political Economy, 98*, 519–543.

Cooley, T. F., & Prescott, E. C. (1995). Economic growth and business cycles. In T. F. Cooley (Ed.), *Frontiers of business cycle research* (pp. 1–64). Princeton: Princeton University Press.

Davis, S. J., & Kahn, J. A. (2008). Interpreting the Great Moderation: Changes in the volatility of economic activity at the macro and micro levels. *Journal of Economic Perspectives, 22*(4), 155–180.
Devereux, M. B., Head, A. C., & Lapham, B. J. (1996). Monopolistic competition, increasing returns, and the effects of government spending. *Journal of Money, Credit, and Banking, 28*(2), 233–254.
Dixit, A., & Stiglitz, J. (1977). Monoplistic competition and optimum product diversity. *American Economic Review, 67*(3), 297–308.
Domeij, D., & Floden, M. (2006). The labor supply elasticity and borrowing constraints: Why estimates are biased. *Review of Economic Dynamics, 9*, 242–262.
Dotsey, M., & Ireland, P. (1996). The welfare cost of inflation in general equilibrium. *Journal of Monetary Economics, 37*, 29–47.
Dynan, K. E., Elmendorf, D. W., & Sichel, D. (2006). Can financial innovation help to explain the reduced volatility of economic activity? *Journal of Monetary Economics, 53*(1), 123–150.
Erceg, C. J., Henderson, D. W., & Levin, A. T. (2000). Optimal monetary policy with staggered wage and price contracts. *Journal of Monetary Economics, 46*(2), 281–313.
Erceg, C. J., & Linde, J. (2012). Fiscal consolidation in an open economy. *American Economic Review, 102*(3), 186–191.
Farhi, E., & Werning, I. (2016). Fiscal multipliers: Liquidity traps and currency unions. In J. B. Taylor & H. Uhlig (Eds.), *Handbook of the macroeconomics* (pp. 2418–2492). Amsterdam: Elsevier.
Fatás, A., & Mihov, I. (2001). *The effects of fiscal policy on consumption and employment: Theory and evidence*. CEPR Discussion Papers 2760, C.E.P.R. Discussion Papers.
Fehr, H., Kallweit, M., & Kindermann, F. (2013). Should pensions be progressive? *European Economic Review, 63*, 94–116.
Féve, P., Matherod, J., & J.-G. Sahuc (2012). *A pitfall with DSGE-based, estimated government spending multipliers*. IDIE Working Paper, 708.
Fisher, J. D. M., & Peters, R. (2010). Using stock returns to identify government spending shocks. *Economic Journal, 120*(544), 414–436.
Galí, J. (1999). Technology, employment, and the business cycle: Do technology shocks explain aggregate fluctuations? *American Economic Review, 89*, 249–271.
Galí, J., & Lopez-Salido, J. D. (2007). Understanding the effects of government spending on consumption. *Journal of the European Economic Association, 5*(1), 227–270.
Hanushek, E. A., & Woessmann, L. (2011). The economics of international differences in educational achievement. In E. A. Hanushek, S. Machin, & L. Woessmann (Eds.), *Handbook of the Economics of Education* (Vol. 3, pp. 89–200). Amsterdam: Elsevier.
Heer, B., & Maußner, A. (2009). *Dynamic general equilibrium modeling: Computational methods and applications* (2nd ed.). Heidelberg: Springer.
Heer, B., Maußner, A., & McNelis, P. (2011). The money-age distribution: Empirical facts and the limits of three monetary models. *Journal of Macroeconomics, 33*, 390–405.
Heer, B., Maußner, A., & Ruf, H. (2017). Q-Targeting in New Keynesian models. *Journal of Business Cycle Research, 13*, 189–224.
Heer, B., Rohrbacher, S., & Scharrer, C. (2017). Aging, the Great Moderation and business-cycle volatility in a life-cycle model. *Macroeconomic Dynamics, 21*, 362–383.
Heer, B., & Scharrer, C. (2018). The age-specific burdens of short-run fluctuations in government spending. *Journal of Economic Dynamics and Control, 90*, 45–75.
Hindriks, J., & Myles, G. D. (2006). *Intermediate public economics*. Cambridge, MA: MIT Press.
Jaimovich, N., & Siu, H. E. (2009). The young, the old, and the restless: Demographics and business cycle volatility. *American Economic Review, 99*, 804–826.
Jermann, U. J. (1998). Asset pricing in production economies. *Journal of Monetary Economics, 41*, 257–275.
Killingsworth, M. R. (1983). *Labor supply*. Cambridge, MA: Cambridge University Press.
Kwan, Y. K. (2006). The direct substitutability between government and private consumption in East Asia. NBER working paper, 12431.

Linnemann, L., & Schabert, A. (2003). Fiscal policy in the New Neoclassical synthesis. *Journal of Money, Credit and Banking, 35*(6), 911–929.

Lucas, R. E. (1976). Econometric policy evaluation: A critique. *Carnegie-Rochester Series on Public Policy, 1*, 19–46.

Lugauer, S. (2012). Estimating the effect of the age distribution on cyclical output volatility across the United States. *The Review of Economics and Statistics, 94*(4), 896–902.

Lugauer, S., & Redmond, M. (2012). The age distribution and business cycle volatility: International evidence. *Economics Letters, 117*(3), 694–696.

MaCurdy, T. E. (1981). An empirical model of labor supply in a life-cycle setting. *Journal of Political Economy, 89*, 1059–1085.

Maußner, A. (2000). *The analytics of New Keynesian Phillips curves*. Discussion Paper Series Institute of Economics, University of Augsburg, 313.

Monacelli, T., & Perotti, R. (2008). *Fiscal policy, wealth effects, and markups*. NBER working paper, 14584.

Monacelli, T., Perotti, R., & Trigari, A. (2010). Unemployment fiscal multipliers. *Journal of Monetary Economics, 57*, 531–553.

Murphy, D., & Walsh, K. J. (2016). *Spending shocks and interest rates*. Working Paper.

Nekarda, C., & Ramey, V. A. (2013). The cyclical behavior of the price-cost markup. Working paper, mimeo.

Ni, S. (1995). An empirical analysis on the substitutability between private consumption and government purchases. *Journal of Monetary Economics, 36*, 593–605.

Ramey, V. A. (2011). Identifying government spending shocks: It's all in the timing. *The Quarterly Journal of Economics, 126*, 1–50.

Ramey, V. A., & Shapiro, M. D. (1998). Costly capital reallocation and the effects of government spending. *Carnegie-Rochester Conference Series on Public Policy, 48*(1), 145–194.

Ravn, M. O., Schmitt-Grohé, S., & Uribe, M. (2012). Consumption, government spending, and the real exchange rate. *Journal of Monetary Economics, 59*, 215–234.

Rotemberg, J. J., & Woodford, M. (1992). Oligopolistic pricing and the effects of aggregate demand on economic activity. *Journal of Political Economy, 100*, 1153–1207.

Schmitt-Grohé, S., & Uribe, M. (2007). Optimal simple and implementable monetary and fiscal rules. *Journal of Monetary Economics, 54*, 1702–1725.

Stock, J. H., & Watson, M. W. (2003). Has the business cycle changed and why? In M. Gertler & K. Rogoff (Eds.), *National Bureau of Economic Research macroeconomics annual 2002* (pp. 159–230). Cambridge, MA: MIT Press.

Taylor, J. B. (1993). Discretion versus policy rules in practice. *Carnegie-Rochester Conference Series on Public Policy, 33*, 195–214.

Trabandt, M., & Uhlig, H. (2011). The Laffer curve revisited. *Journal of Monetary Economics, 58*, 305–327.

Uhlig, H. (2007). Explaining asset prices with external habits and wage rigidities in a DSGE model. *American Economic Review: Papers and Proceedings, 97*, 239–243.

von Weizsäcker, C. C. (1971). Notes on endogenous change of tastes. *Journal of Economic Theory, 3*, 345–372.

Walsh, C. E. (2005). Labor market search, sticky prices, and interest rate policy. *Review of Economic Dynamics, 8*, 829–849.

Walsh, C. E. (2010). *Monetary theory and policy*. Cambridge, MA: MIT Press.

Income Taxation 5

5.1 Introduction

After a brief survey of the empirical findings on income taxation in the US and German economies in Sect. 5.2,[1] you learn about the substantial welfare costs that are associated with the taxation of labor income. In Sect. 5.3, these costs are computed in both partial and general equilibrium. As one result, the deadweight loss of labor income taxation in Germany is found to be twice as high as the one in the US. In Sect. 5.4, the seminal result from optimal taxation that capital income should not be taxed in the long run is derived and discussed critically. Section 5.5 estimates the US Laffer curve and shows that the US government, in contrast to many European governments, can still raise its revenues from labor and capital income taxation by approximately 10% of GDP. In Sect. 5.6, the quantitative effects of higher taxes on economic growth are derived in a Dynamic General Equilibrium (DGE) model and are shown to be substantially higher than those typically found in growth regressions. Finally, we demonstrate that stochastic taxes improve the time series properties of the real business cycle (RBC) model with respect to the volatility of aggregate demand components and the dynamics of labor and wages in Sect. 5.7.

5.2 Empirical Regularities

Tax revenues vary considerably across OECD countries. As presented in Fig. 5.1, within the OECD, the United States has among the lowest shares of revenue in

[1]The two countries were chosen because they are (1) relatively large in size and (2) characterized by substantial differences in their income tax schedules. In addition, these two countries feature prominently in the quantitative analysis of Prescott (2004) that we reference in the following.

B. Heer, *Public Economics*, Springer Texts in Business and Economics,
https://doi.org/10.1007/978-3-030-00989-2_5

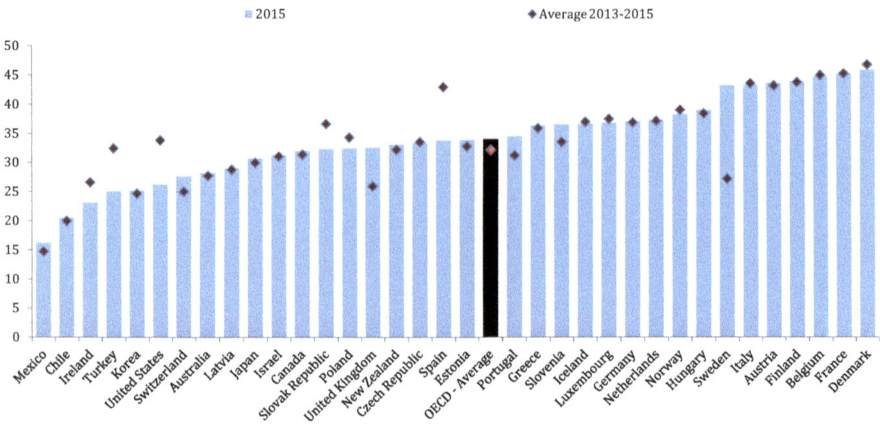

Fig. 5.1 Tax revenue relative to GDP in OECD countries in 2015

GDP, amounting to 26.2% in 2015.[2] The Scandinavian countries of Denmark and Finland together with France have among the highest taxes in the OECD, amounting to 45.9%, 43.9%, and 45.2% of GDP, respectively.

Notice that although the GIIPS countries (Greece, Italy, Ireland, Portugal, and Spain) are characterized by government shares of 50% or above (see Chap. 4),[3] they have considerably lower revenue shares in the range of 27–36%. Only Italy has a revenue share of 44%. As we will point out in more detail in Chap. 7, which looks at public debt, many of these countries including Spain and Greece were characterized by budget deficits of 10% or even above in the aftermath of the financial crisis 2007–2008.

Figure 5.2 plots tax revenue of selected OECD countries over time. Total tax revenue are upward sloping for the Eurozone countries France, Italy, and Spain (and, to a smaller extent, Germany) as well as Japan, while there is no discernable trend in tax revenue (as % of GDP) in the UK and US. In most countries, tax revenue fell during and after the financial crisis with the exception of Italy.[4] Most of the OECD countries are characterized by a progressive income tax schedule. Consequently, if average income falls during a recession, tax revenue decrease stronger than GDP.

The composition of the US tax revenue (as % of GDP) in 2015 is presented in Table 5.1. The largest component is the tax on personal income which constitutes

[2]The data are retrieved from the OECD as described in Appendix 5.2. Tax revenue is defined as the revenues collected from taxes on income and profits, social security contributions, taxes levied on goods and services, payroll taxes, taxes on the ownership and transfer of property, and other taxes.

[3]Spain's share of government expenditure share in GDP is generally just short of 50% and only amounted to 44% during 2013–2015.

[4]Italy raised its VAT rate by 1 percentage point in both 2011 and 2013.

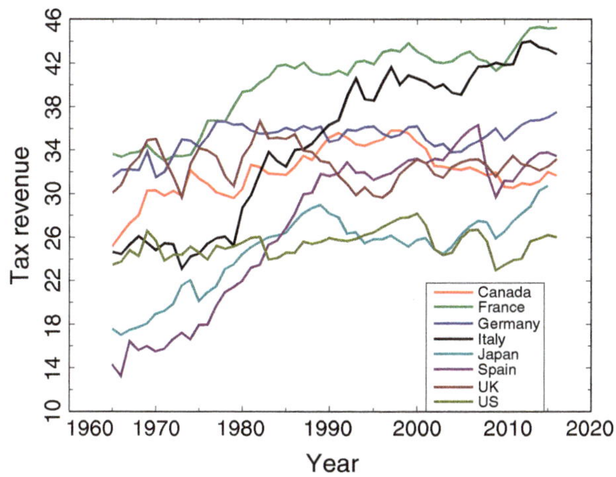

Fig. 5.2 Tax revenue (% of GDP), 1965–2016

Table 5.1 US tax composition in 2015 (% of GDP)

Tax	Revenue
Total taxes	26.22
Tax on personal income	10.63
Tax on corporate profits	2.23
Social security contributions	6.20
Tax on payroll	0.00
Tax on property	2.70
Tax on goods and services	4.46

Notes: Data are retrieved from the OECD. See Appendix 5.2 for a description of the data source

40.5% of total tax revenue. The second largest component is represented by social security contributions, which we will consider separately in the next chapter.

Taxes on goods and services are levied on product sales and include value-added taxes. In the US, sales taxes are not imposed uniformly. In 2014, for example, five US states (Alaska, Delaware, Montana, New Hampshire and Oregon) did not impose any sales taxes, while California had the highest sales tax at 7.5%. In many European countries, value-added taxes contribute a much larger share in total revenues than in the US economy. Italy, for example, imposes a value-added tax rate of 22% at present and total taxes on goods and services amount to 11.7% of GDP or 26.9% of total tax revenue during 2013–2015.

The US property tax includes the inheritance tax and is complemented by the gift tax, and these accrue if an estate is transferred. Due to substantial exemptions, only the top 0.2% of estates in the US are taxed. In 2015, for example, only estates exceeding $5.43 million were subject to the estate tax. The amount excluded from

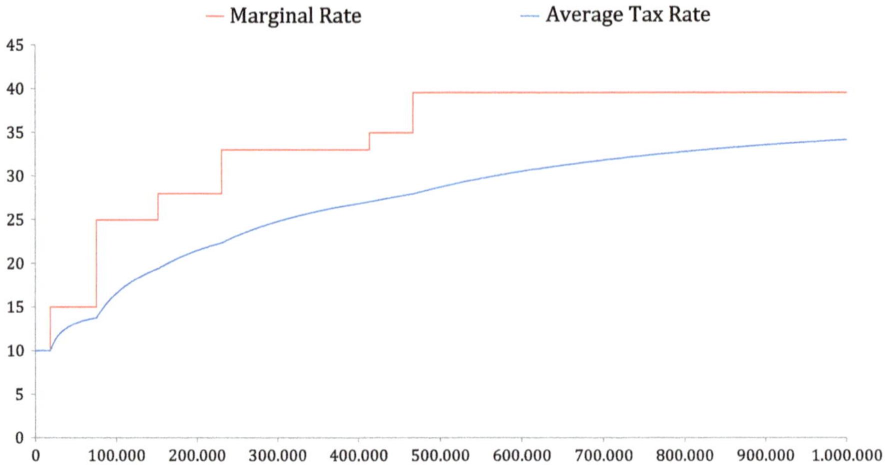

Fig. 5.3 Marginal and average US income tax rates, 2016

taxation was raised gradually, from $675,000 in 2001 to its present value, which explains the declining share of the estate tax in total tax revenues in recent years.

The US income tax schedule (for married households) is presented in Fig. 5.3. While the marginal tax rate τ' describes the additional tax resulting from a $1 increase in taxable income, the effective tax rate τ reports the average tax rate (total income tax divided by total taxable income). Taxable income is defined as the total income less allowable deductions, where income is broadly defined and includes wage income, rental income, and interest and dividend income, among other categories. Most business expenses are deductible, while individuals may also deduct a personal allowance and certain personal expenses. The definition of taxable income differs considerably across countries. For example, in the US, home mortgage interest can be deducted from total income (if the dwelling is not rented out and does not generate rental income reported in tax filings), while this is not possible in Germany.

The US income tax system is progressive. As a consequence, individuals with higher incomes pay both a higher effective and a higher marginal income tax rate. In 2016, the lowest tax bracket was $0–18,550, with an income tax rate equal to 10.0%, while the highest income bracket with a tax rate equal to 39.6% starts at an income equal to $466,950 (for a married couple filing jointly). Even a couple with a high income lying in the range $151,900–231,450 only pays a marginal income tax rate equal to 28.0%. Notice that in the US, only 4.2% of households earned an income in excess of $200,000 in 2010 according to US census data, meaning that most taxpayers face a marginal tax rate below 28.0%.[5]

[5]Take care to distinguish between individual and household income (or wealth). For example, the OECD uses the following conversion system when comparing households with different sizes: a

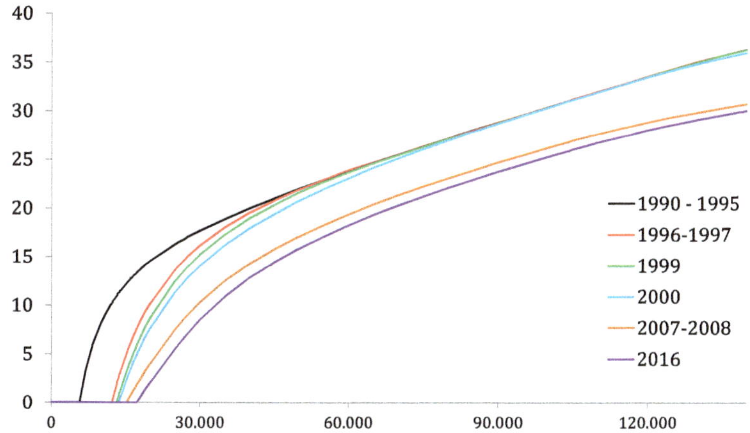

Fig. 5.4 Average income tax rate, Germany (married household)

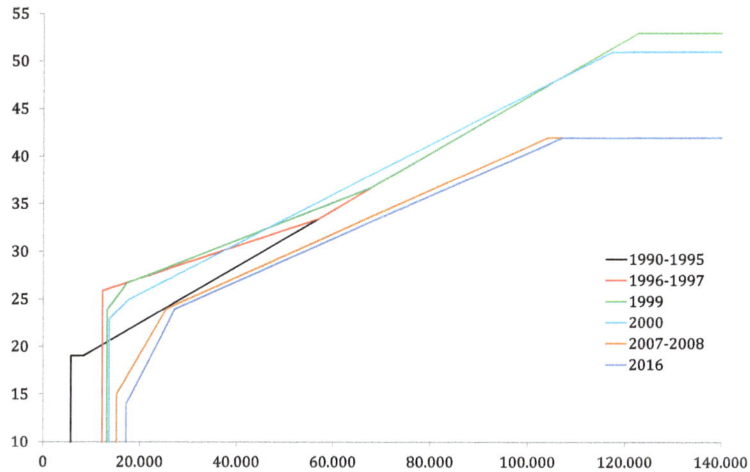

Fig. 5.5 Marginal income tax rate, Germany (married household)

For comparison, we also present the average and marginal tax rates for married households[6] in Germany in Figs. 5.4 and 5.5. The German income tax system is clearly much more progressive than the US income tax system. While the average and marginal income tax rates amount to 17% and 25% for an income of $100,000

household consisting of one individual is weighted by measure 1.0, while a household with two individuals and no children is weighted by the measure 1.6. Therefore, if the former has an income equal to $100,000 and the latter has a total income of $160,000, both households are reported to have a household income of $100,000.

[6]In 2015, 59% of both US and German households were married couples.

in the US, the corresponding income tax rates for an income of €90,000 (assuming an exchange rate of 1.11$/€) amount to approximately 24% and 37%. For a joint income below the taxable income threshold of €17,306, a couple does not pay any taxes in 2016. The marginal income tax rate increases much faster in Germany than in the US, and for a joint income equal to €107,332, a German couple has to pay a marginal tax rate equal to 44.3% in 2016.[7] The highest income tax rate in Germany (as of 2016) is imposed on incomes exceeding €521,066 and amounts to 47.5%.

In this chapter, we will analyze the welfare and distortionary effects of income taxation. For this reason, we also need to account for other costs that increase the tax wedge, including social security contributions.[8] In Germany, pension contributions amount to 35.65%, which are levied on wage income up to a threshold income that depends on the residence of the taxpayer (East versus West Germany). For pension contributions, the threshold amounts to approximately €60,000 (in the East) and €71,000 (in the West), while the threshold for health insurance contributions is uniform and amounted to €48,600 in 2014.[9] According to the OECD Economic Outlook 2014, the tax wedge on the average wage income in the German and US economies amounted to 49.3% and 31.6% (for a household composed of a single individual with no children), respectively.[10]

To characterize the progressivity of a tax system, we often use the yield elasticity, which is defined as follows:

$$\eta_{T,Y} = \frac{dT(Y)}{dY}\frac{Y}{T(Y)} = \frac{\tau'}{\bar{\tau}}, \tag{5.1}$$

where we speak of a regressive, proportional, or regressive tax system if:

$$\eta_{T,Y} = \begin{cases} < 1 & : \text{ regressive tax code} \\ = 1 & : \text{ proportional} \\ > 1 & : \text{ progressive.} \end{cases}$$

[7]The tax rate is composed of the ordinary income tax rate equal to 42.0% and a surcharge of 5.5% on the taxes, which is called the "Solidaritätszuschlag". This surcharge was first imposed in 1992 to finance the additional government expenditures resulting from German reunification in 1989. As of this writing, this surcharge remains in effect.

[8]The tax wedge is defined as the deviation from the equilibrium price or quantity as a result of the taxation of a good (or production factor). In the present case, we look at the factor 'labor' and its price in the form of the wage.

[9]The contribution rates for pensions and health amounted to 18.7% and 16.85% in 2014, including both the employee's and employer's shares. Chapter 6 will focus on the effects of a pension system and optimal social security reform.

[10]Prescott (2004) applies income tax rates of 59% and 40% for the German and US economies during the period 1993–1996. In particular, he also includes consumption taxes τ^c in his computations. For this reason, consider the budget constraint $(1+\tau^c)c = (1-\tau)wl$, where the household consumes its total net income from working l hours and receiving net wage $(1-\tau)w$. Accordingly, the tax wedge amounts to $1-(1-\tau)/(1+\tau^c)$. Since the value added tax in Germany is equal to 19%, while it is 7.5% or less in the US depending on the state, the difference in the tax wedge between these two countries is even larger after accounting for consumption taxes.

The yield elasticity provides useful information for tax authorities on how an increase in GDP translates into additional tax revenues.[11]

Thus far, we have assumed that all sources of income, e.g., labor income, interest income, or rents, are treated equally by the national tax codes. In most OECD countries, capital gains, for example, are included in the definition of taxable income. In the case of the US, capital gains are also taxable, but capital losses can only be deducted from taxable income up to a certain threshold.

Labor and capital income, however, are subject to different amounts of allowances and exemptions and have to be paid by households and the corporate sector (which differ in their tax treatment). In Germany, for example, in 2015, the first €801 of annual interest and dividend income was tax-exempt for each individual (meaning that a married couple can have up to €1,602 in tax-exempt interest and dividend income). Consequently, capital and labor income are not burdened equally by the US or German tax system.

Figure 5.6 presents the effective capital and labor income tax rates for the US during the period 1948–2008 as computed by Gomme, Ravikumar, and Rupert (2011).[12] Clearly, capital income was taxed more heavily than labor income in the last century; however, in recent years, the tax rates on capital and labor income have converged in the US. The average capital and labor income tax rates amount to 41% and 23%, and we will use these values in the models in this chapter.

Using the data on US labor and capital income tax rates from Fig. 5.6, we have computed business cycle statistics for the US economy as presented in Table 5.2.[13] Both the cyclical components of the capital and the labor income tax rates τ^L and τ^K are positively correlated with output. However, in the more stable subperiod after the Korean War, 1956–2008, the tax rates are less procyclical, and correlations with output decline. In addition, both tax rates are positively correlated with labor, while labor income taxes are positively correlated with and capital income taxes are uncorrelated with government consumption.

[11]An alternative measure to characterize the progressivity of the tax system is presented by the residual elasticity, where the residual is defined as the net income after taxes $Y^n = Y - T(Y)$:

$$\eta_{Y^n,Y} = \frac{dY^n}{dY}\frac{Y}{Y^n} = \frac{1-\tau'}{1-\bar{\tau}}.$$

This measure provides important information to the participants in the wage bargaining process, i.e., employees, unions, and employers.

[12]The figures and the business cycle statistics in Table 5.2 are computed with the help of the GAUSS program *Ch5_data.g*.

[13]For this reason, we have also taken the logarithm of the two income tax rates and applied the HP filter with weight $\lambda = 1600$.

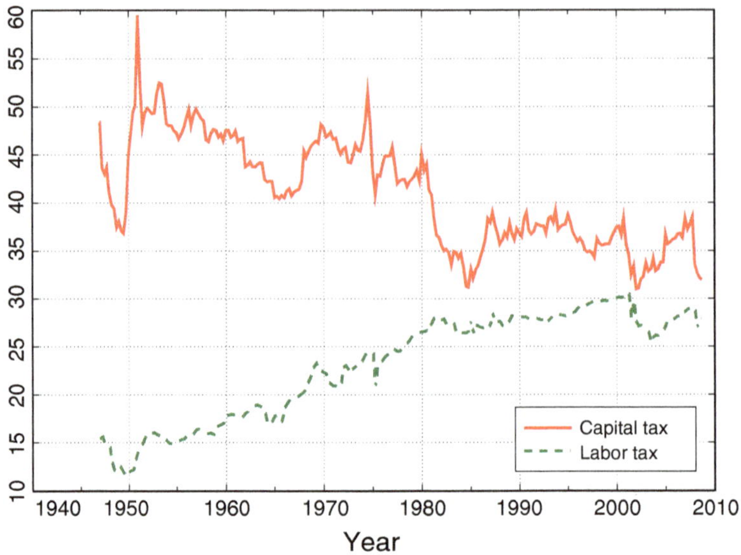

Fig. 5.6 US income tax rates, 1948–2008 (in percentage points)

Table 5.2 US business cycle statistics

Variable	s_x	$r_{x,Y}$	$r_{x,G}$	r_{x,τ^K}	r_{x,τ^L}
1947–2008					
Output Y	1.66	1.000	0.186	0.346	0.390
Public consumption G	3.30	0.186	1.000	0.083	0.541
Private consumption C^p	1.28	0.760	−0.215	0.256	0.055
Hours L	1.83	0.878	0.079	0.368	0.417
Capital tax τ^K	4.75	0.346	0.083	1.000	0.497
Labor tax τ^L	3.98	0.390	0.055	0.497	1.000
1956–2008					
Output Y	1.53	1.000	−0.005	0.222	0.237
Public consumption G	1.36	−0.005	1.000	−0.009	0.152
Private consumption C^p	1.23	0.873	−0.169	0.209	0.127
Hours L	1.76	0.874	−0.169	0.326	0.366
Capital tax τ^K	3.90	0.221	−0.009	1.000	0.443
Labor tax τ^L	2.97	0.237	0.152	0.443	1.000

Notes: s_x:= Standard deviation of the time series x in percentages, where $x \in \{Y, G, C^p, L, \tau^K, \tau^L\}$. Empirical time series were HP filtered with weight 1600. r_{xY}:= Cross-correlation of the variable with output, r_{xG}:=Cross-correlation of the variable with government consumption, $r_{x\tau^K}$:= Cross-correlation of the variable with the capital income tax, $r_{x\tau^L}$:= Cross-correlation of the variable with the labor income tax

5.3 Labor Income Tax

The labor income tax distorts household labor supply decisions. Depending on the strength of the substitution relative to the income effect, the labor supply may increase or decrease. Although we also observe regions of the wage with a backward-bending labor supply curve, the substitution effect associated with a tax increase usually dominates. As a consequence, output and welfare also decline with higher labor income taxes, ceteris paribus.[14] In later sections, we will see that the change in the labor supply is also associated with a change in the economic growth rate in some types of growth models.

We will first derive the partial equilibrium effects of a change in the labor supply, which is the subject of standard textbooks on public economics. In this analysis, the wage rate is held constant. Since individuals change their labor supply, total labor supply will adjust, and the equilibrium wage (before taxes) does not remain constant. In addition, consumption demand declines and, hence, labor demand is also reduced. This general equilibrium effect is integrated in the analysis in the second part of this section.

5.3.1 Partial Equilibrium

Figure 5.7 presents the effects of a labor income tax on labor supply and demand. The individual labor supply l^s is a function of the net wage after taxes $(1-\tau^L)w$, while labor demand l^d is a function of the gross wage w. Taxes are imposed at a proportional tax rate τ^L, and the labor supply and demand curves are graphed as functions of the gross wage w. In the initial equilibrium at point e, taxes are equal to zero, $\tau^L = 0$, and the equilibrium point is (l_0, w_0). If the state imposes a proportional labor income tax τ^L, the labor supply curve shifts upward by a factor of $1/(1-\tau^L)$ to $l^{s\prime}$ (the argument on the ordinate is the gross wage w), and the new equilibrium point is g at (l_1, w_1). At this point, the tax revenues are equal to $\tau^L w_1 l_1$, which is equal to the area of the yellow rectangle $agdc$. While the consumer rent declines by the area $ageb$, the producer surplus falls by the area $bedc$. The difference between the gain in taxes and the losses in surpluses for the consumer and producer is denoted as the *excess burden* and equals the triangle gde.

Notice further that it does not matter who pays the tax. The *economic incidence*, i.e., the agent who bears the cost, does not depend on the *legal incidence*, i.e., the agent who is legally obliged to pay the tax. If we identify the economic incidences of the producer and the worker by the increase in the producer wage, $w_1 - w_0$, and the decline in the worker's wage, $w_0 - (1-\tau^L)w_1$, respectively, we recognize that

[14]Welfare does not need to fall if another distortion is reduced simultaneously, e.g., if an increase in the labor income tax results in a decline in another distorting tax or if the tax revenues are used for welfare-improving government spending.

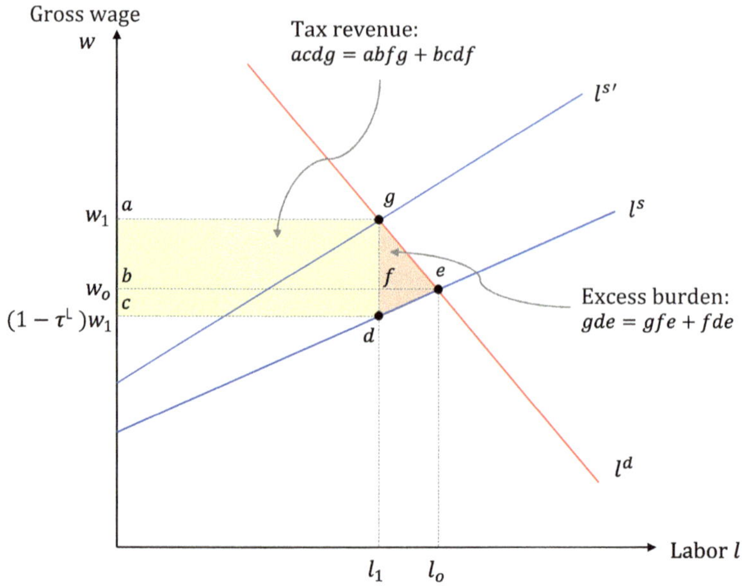

Fig. 5.7 Equilibrium in the labor market and the effect of a labor income tax

it is independent of the legal incidence in Walrasian labor markets.[15] The economic incidence instead depends on the elasticities of labor demand and labor supply with respect to the wage. If labor supply (labor demand) becomes more elastic, the economic loss borne by the worker decreases (increases). In the case of a perfectly elastic labor supply (labor demand), the gross wage increases by the same amount as the tax increase (does not change).[16]

The welfare loss or, equivalently, excess burden can be computed with the help of the equivalent compensation. To see this, consider Fig. 5.8, where we derive the optimal labor supply. Assume that the household has an exogenous income of I such that total income is equal to $Y = (1 - \tau^L)w^0 l + I$ for the wage rate w^0.[17] We assume that the household lives for just one period and, therefore, consumes all

[15]The legal incidence may affect the economic incidence, for example, in a labor market with a minimum wage. If the minimum wage is defined as the wage that is paid by the employer to the worker, the new equilibrium point depends on who actually pays the taxes.

[16]In the case of a perfectly elastic labor supply, the labor supply curve l^s is horizontal and a labor income tax rate τ^L implies a horizontal shift of this curve to $l^{s\prime}$. Evidently, the complete economic incidence falls on the producer.

[17]To be consistent with our previous notation, we keep denoting individual labor supply by l and aggregate labor supply by L. In the Ramsey model with a representative agent, individual and aggregate labor supply coincided. In the following, we will also introduce compensated labor supply which we will denote by h.

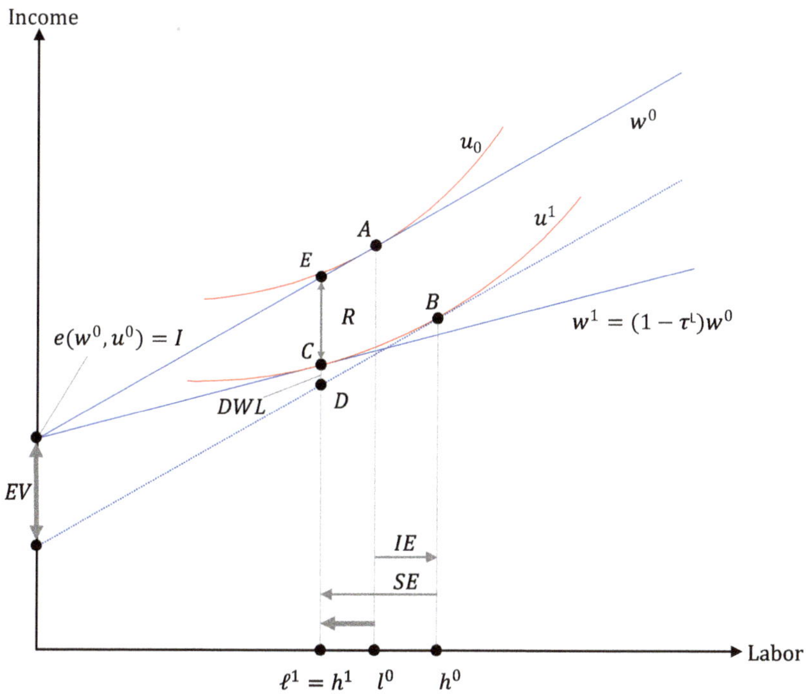

Fig. 5.8 Welfare effects of a labor income tax

its income. The budget constraint that is presented by the straight line originating at point $(0, I)$ rotates clockwise if a proportional tax τ^L is imposed.

The optimal labor supply is found at the point where the indifference curve is tangent to the budget constraint at point A. Notice that the indifference curve is an increasing function in the (l, Y)-space because labor l is a bad and not a good. If the household increases labor l, it needs to be compensated with higher income (alternatively, consumption) to hold utility constant. An imposition of a tax τ^L results in the new equilibrium at point C. We distinguish the substitution and income effects. Because of the reduction in income, the household increases its labor supply to point B due to the income effect. Because of the substitution effect (due to lower net wages), the household reduces its labor supply from point B to point C. Whether the net effect is positive or negative depends on the preferences of the household, and both cases are possible depending on the relative strength of the income and substitution effects.

The welfare loss of the household can be measured by equivalent compensation EV, which is the difference in the expenditure functions $e(w, u)$ for the initial wage w^0 and the utility levels u^0 and u^1 associated with points A and B, respectively[18]:

$$EV = e(w^0, u^0) - e(w^0, u^1) \tag{5.2}$$

To compute the equivalent variation, note the following[19]:

$$\begin{aligned} EV &= e(w^0, u^0) - e(w^0, u^1) \\ &= I - e(w^0, u^1) \\ &= e(w^1, u^1) - e(w^0, u^1) \\ &= \int_{w^0}^{w^1} \frac{\partial e(w, u^1)}{\partial w} dw \\ &= \int_{w^0}^{w^1} -h(w, u^1) dw \\ &= \int_{w^1}^{w^0} h(w, u^1) dw. \end{aligned}$$

Accordingly, the equivalent variation is equal to the area under the compensated (Hicksian) labor supply curve h with utility level u^1.

To compute the excess burden (also called the deadweight loss) DWL, we need to subtract tax revenues, $R = \tau^L w^0 l$, from EV. The deadweight loss is depicted by the distance between points C and D in Fig. 5.8.[20]

The compensated labor supply h (for constant utility level u^1) and the Marshallian labor supply l are presented in Fig. 5.9.[21] Assume that the old wage prior to the imposition of the tax is equal to w^0, while the new wage is given by $w^1 = (1 - \tau^L)w^0$. For the derivation of the labor supply curves in Fig. 5.9, we consider the case that the Marshallian and compensated labor supply coincide at the

---

[18]Remember from microeconomics that the expenditure function specifies the minimum amount of money that is needed to achieve a given level of utility $\bar{u}$.

[19]In the derivation, we use the following property of the expenditure function: $\frac{\partial e(w,u)}{\partial w} = -h(w, u)$. This result is derived from applying the envelope theorem to the Lagrangian associated with the minimization of expenditures for given level of utility $\bar{u}$:

$$\mathcal{L} = Y - wh + \mu \left[u(Y, h) - \bar{u} \right].$$

Here, h denotes the compensated (Hicksian) labor supply.

[20]For the derivation of the DWL, we follow the exposition in Keuschnigg (2005), pp. 62–64.

[21]Recall that the Marshallian labor supply curve is derived from maximizing utility subject to the budget constraint. Notice that $l^0 < h^0$ due to the income effect that is considered in the case of l but not in the case of the compensated labor supply h.

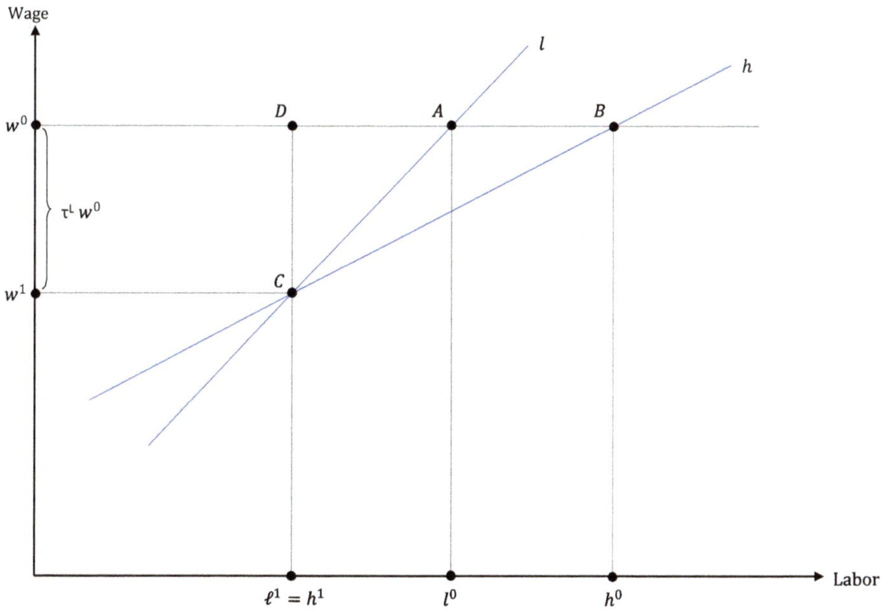

Fig. 5.9 Compensated (Hicksian) and Marshallian labor supply, l and L

wage w^1, $h^1 = l^1$ with implied utility u^1. At the new equilibrium point (l^1, w^1), the compensated labor supply elasticity is given by

$$\eta_{h,w} = \frac{\Delta h / h^1}{\Delta w / w^1},$$

where $\Delta w = w^1 - w^0 = -\tau^L w^0$ denotes the change in the net wage rate. The deadweight loss DWL is equal to the surface of the triangle BDC and follows from:

$$DWL = EV - R = \frac{1}{2}\Delta w \cdot \Delta h = \frac{1}{2}\frac{\tau^L}{1-\tau^L} \cdot \eta_{h,w} \cdot R \tag{5.3}$$

where we used

$$\Delta h = \eta_{h,w}\frac{\Delta w}{w^1}h^1 = \eta_{h,w}\frac{-\tau^L w^0}{w^1}h^1 = -\eta_{h,w}\frac{\tau^L}{1-\tau^L}h^1,$$

and $R = \tau^L w^0 l_1 = -\Delta w \cdot l_1 = -\Delta w \cdot h_1$. Notice that, importantly, the deadweight loss increases non-linearly with the tax rate τ^L. Therefore, a tax increase from 30% to 40% results in higher welfare losses than a tax increase from 10% to 20%.

Table 5.3 Deadweight losses relative to revenues: Partial equilibrium

τ^L (%)	DWL/R (%)
10	1.7
25	5.0
40	10.0
50	15.0
59	21.6

To obtain a quantitative estimate of the welfare losses from labor income taxes, we use the empirical value of the compensated labor supply elasticity equal to $\eta_{h,w} \approx 0.30$.[22] The deadweight loss that results from tax revenues R is presented in Table 5.3 for various tax rates τ^L.

Therefore, if we use the estimates of Prescott (2004) for the tax wedges in the US and Germany, 40% and 59%, the deadweight losses amount to 10.0% and 21.6%, respectively, and distortions in the German economy due to labor income taxes (and social security contributions) are more than twice as large as those in the US economy.

5.3.2 General Equilibrium

In the previous section, we considered the welfare losses in a partial equilibrium model. In particular, we held the position of the labor demand curve constant. In the following, we embed the welfare analysis in a general equilibrium model in which wages (and savings) are endogenous and compute the change in the welfare results. In particular, we choose the Ramsey model that we introduced in Chap. 2. In addition, we specify a (parameterized) utility function to allow us to express welfare effects in consumption equivalent changes.

Assume that the households maximize intertemporal utility[23]

$$U = \sum_{t=0}^{\infty} \beta^t u(C_t, L_t), \tag{5.4}$$

where β, again, denotes the discount factor, with $\beta < 1$. The household inelastically supplies L_t units of labor, and its total endowment is normalized to one, and thus, $1 - L_t$ denotes leisure.

[22] Chetty, Guren, Manoli, and Weber (2011) provide a summary review of empirical studies on the labor supply elasticity, including studies on both the *compensated* and *Frisch* labor supply elasticities.

[23] Since we study the behavior of a representative household, we identify the individual labor supply with the aggregate labor supply and denote both variables by L_t in the following.

Instantaneous utility is specified as follows:

$$u(C, 1-L) = \frac{\left(C^{\iota}(1-L)^{1-\iota}\right)^{1-\sigma}}{1-\sigma}, \tag{5.5}$$

where $1/\sigma$ denotes the intertemporal elasticity of substitution, and ι and $1-\iota$ are the relative weights of consumption and leisure in utility.

The household owns the capital stock K_t in period t, which evolves according to

$$K_{t+1} = (1-\delta)K_t + I_t. \tag{5.6}$$

Capital K_t depreciates at rate δ. The household lends the capital stock to the firms, which pay real interest rate r_t. The household faces wage rate w_t and labor income taxes τ_t^L, meaning that its net labor income is equal to $(1-\tau_t^L)w_t L_t$. Its net income is spent on private consumption C_t and savings S_t, which are equal to the increase in capital holdings, $S_t = K_{t+1} - (1-\delta)K_t$. Consequently, the household budget constraint is

$$(1-\tau_t^L)w_t L_t + r_t K_t = C_t + K_{t+1} - (1-\delta)K_t. \tag{5.7}$$

The first-order conditions follow from the derivation of the Lagrangian

$$\mathscr{L} = \sum_{t=0}^{\infty} \beta^t \left[\frac{\left(C_t^{\iota}(1-L_t)^{1-\iota}\right)^{1-\sigma} - 1}{1-\sigma} + \lambda_t \left((1-\tau_t^L)w_t L_t + (1+r_t-\delta)K_t - C_t - K_{t+1} \right) \right] \tag{5.8}$$

with respect to C_t, L_t, and K_{t+1}, taking government consumption G_t and taxes τ_t^L as exogenous. In particular, the first-order conditions are given by:

$$\lambda_t = \iota C_t^{\iota(1-\sigma)-1}(1-L_t)^{(1-\iota)(1-\sigma)}, \tag{5.9a}$$

$$\lambda_t(1-\tau_t^L)w_t = (1-\iota)C_t^{\iota(1-\sigma)}(1-L_t)^{(1-\iota)(1-\sigma)-1}, \tag{5.9b}$$

$$\lambda_t = \beta\lambda_{t+1}(1+r_{t+1}-\delta). \tag{5.9c}$$

Goods and factor markets are characterized by perfect competition. We, again, assume that production is described by a Cobb-Douglas technology:

$$Y_t = K_t^{\alpha} L_t^{1-\alpha}. \tag{5.10}$$

Firms rent capital from the households. Therefore, wages and the real interest rate are given by:

$$w_t = (1-\alpha)K_t^{\alpha} L_t^{-\alpha}, \tag{5.11a}$$

$$r_t = \alpha K_t^{\alpha-1} L_t^{1-\alpha}. \tag{5.11b}$$

Finally, the government budget is assumed to balance, meaning that

$$G_t = \tau_t^L w_t L_t. \tag{5.12}$$

In equilibrium, the resource constraint of the economy is presented by[24]

$$Y_t = C_t + G_t + K_{t+1} - (1-\delta)K_t. \tag{5.13}$$

In steady state, all economic variables are constant, and thus, the following six conditions in the six variables K, L, C, w, r, and τ^L hold for given exogenous government expenditures G:

$$\frac{1}{\beta} = 1 + r - \delta, \tag{5.14a}$$

$$(1-\tau^L)w = \frac{1-\iota}{\iota}\frac{C}{1-L}, \tag{5.14b}$$

$$w = (1-\alpha)K^\alpha L^{-\alpha}, \tag{5.14c}$$

$$r = \alpha K^{\alpha-1} L^{1-\alpha}, \tag{5.14d}$$

$$K^\alpha L^{1-\alpha} = C + G + \delta K, \tag{5.14e}$$

$$G = \tau^L w L. \tag{5.14f}$$

To compute quantitative welfare effects, we need to calibrate the model. Let us assume that periods are equal to 1 year, meaning that the discount rate is approximately equal to $\beta = 0.96$, implying a real interest rate (net of depreciation) $r - \delta$ of approximately 4% annually. Assuming an annual depreciation rate of 10%, $\delta = 0.10$, we can compute r with the help of (5.14a) and the capital-labor ratio K/L with the help of (5.14d). Assuming a steady-state labor supply equal to $L = 0.30$, we can compute the capital stock from $K/L \times 0.3$. Production is computed as $Y = K^\alpha L^{1-\alpha}$. We set the steady-state labor tax rate equal to the US average rate during the period 1956–2008, implying $\tau^L = 0.23$. Consequently, government expenditures, $G = \tau^L w L$, are equal to 14.72% of GDP, $G = 0.1472 \cdot Y$, and consumption can be computed from the resource constraint, $C = Y - \delta K - G = 0.303$. Finally, we can solve (5.14b) for $\iota = 0.3423$. The steady state and the following results for the welfare losses from labor income taxation are computed with the help of the GAUSS program *Ch5_welfare_taul.g*.

As our first exercise in the neoclassical growth model, we compute the partial equilibrium effect of a one-percentage-point tax increase to $\tilde{\tau}^L = 0.24$, where the wage rate w is held constant (assuming a perfectly elastic labor demand), and the loss in income only decreases consumption, not savings (which is exactly

[24] You can derive (5.13) by substituting (5.11) and (5.12) into (5.7), noticing that $Y_t = w_t L_t + r_t K_t$.

the assumption used in the partial equilibrium analysis in the previous section). Accordingly, the new optimal labor supply is provided by:

$$L' = 1 - \frac{1-\iota}{\iota}\frac{C'}{(1-\tilde{\tau}^L)w}$$
$$C' = (1-\tilde{\tau}^L)wL' + (r-\delta)K,$$

If we solve this equation, we find that the household changes its optimal consumption and labor supply from $(L, C) = (0.3, 0.303)$ to $(L', C') = (0.2994, 0.2998)$. We can also compute the *consumption equivalent* change Δ from

$$u\left((1+\Delta)C, L)\right) = u(C', L')$$

implying

$$\Delta = \left(\frac{u(C, 1-L)}{u(C', 1-L')}\right)^{-\frac{1}{\iota(1-\sigma)}} - 1 = -1.07\%.$$

Notice that the welfare change is quantitatively significant. A one-percentage-point increase in the labor income tax rate decreases welfare by approximately 1.1% of consumption. Of course, we need to consider that, by assumption, the additional government expenditures generated by the additional revenues are simply waste and do not have any effect on either utility of the household or productivity. Therefore, to give a fair evaluation of the welfare costs, we instead compute the deadweight loss as the difference between the consumption loss, $1.07 \times C = 0.003244$, and the additional tax revenues or, equivalently, government expenditures, $\Delta G = 0.003238$. Accordingly, the deadweight loss amounts to $DWL = 0.000005$ or, relative to the additional revenues, $\frac{DWL}{\Delta G} = 0.17\%$.

Next, we consider general equilibrium effects under the assumption that both savings and factor prices w and r are endogenous. Accordingly, we have to solve the steady-state conditions (5.14) for the new value of τ^L. As our new steady-state values, we compute $L'' = 0.2995$ and $C'' = 0.2998$. The welfare loss is almost the same and amounts to $\Delta = -1.10\%$ and is presented in Table 5.4.[25]

[25]Take care when you compare the general equilibrium effects in Table 5.4 with those resulting from the partial equilibrium analysis reported in Table 5.3. For the partial equilibrium effect, (5.3) provides an estimation of the average welfare costs from the imposition of a tax, while, in the general equilibrium model, we computed the marginal welfare costs of a one-percentage-point increase in the tax rate. One can show that the marginal deadweight loss in the partial equilibrium model is equal to

$$\frac{dDWL}{dR} = \frac{\frac{\tau^L}{1-\tau^L}\eta_{h,w}}{1 - \frac{\tau^L}{1-\tau^L}\eta_{h,w}}.$$

Table 5.4 General equilibrium welfare effects of a 1% labor income tax increase

τ^L (%)	Δ steady state (%)	DWL/R (%)	Δ incl. transition (%)
10	−0.9	3.4	−0.9
23	−1.1	6.8	−1.1
40	−1.3	16.4	−1.3
50	−1.6	29.1	−1.5
59	−1.9	54.7	−1.8

To apply our general equilibrium computation to the characteristics of the US and German economies, we use the tax wedges proposed by Prescott (2004) and set $\tau^L \in \{0.40, 0.59\}$ in the two cases. For the US economy (characterized by a tax wedge of 40%), the deadweight loss relative to the tax revenues rises to $\frac{dDWL}{dG} = 16.4\%$. For the case of Germany (with a tax wedge equal to 59%), the distortions are exorbitantly high, and the losses from increases in taxes from 59% to 60% are equal to 54.7% of the additional tax revenues.

In our final step, we also consider the transition dynamics after a permanent change in the tax rate, which we neglected when considering the partial equilibrium effect (where we assumed that K would remain constant) and the general equilibrium steady-state analysis. To do so, we assume that the labor income tax rate is changed in period $t = 0$ and that the change is unanticipated, meaning that the economy is in the old steady state prior to period $t = 0$. In period 0, the capital stock is predetermined, $K_0 = K$.

The transition paths of the endogenous variables K, L, Y, C, and G are illustrated in Fig. 5.10.[26] As presented in the top-left panel, the capital stock gradually declines to its new steady-state value.[27] The other variables adjust much faster to the new steady-state values. Labor, as illustrated in the top-right panel, actually undershoots its long-run equilibrium value because the household is able to pay for its consumption from reduced savings. Since the marginal utility of leisure increases with consumption, leisure $1 - L$ increases to a larger extent than in the new steady state.

Although it is difficult to see in Fig. 5.10, government consumption (= tax revenue) only adjusts gradually after the initial jump in period $t = 1$. Therefore, when we compare the steady-state welfare effects with the welfare effects during

For example in the German economy with $\tau^L = 0.59$, the marginal deadweight loss in partial equilibrium, therefore, is equal 56.2% and is close to the general equilibrium effect reported in Table 5.4.

[26]The transition is computed using the method of reverse shooting described in Appendix 4.1. The method is implemented in the Gauss program *Ch5_welfare_taul.g*.

[27]In our computational algorithm, we set the number of transition periods equal to 40, which appears to be sufficient time for the capital stock to converge.

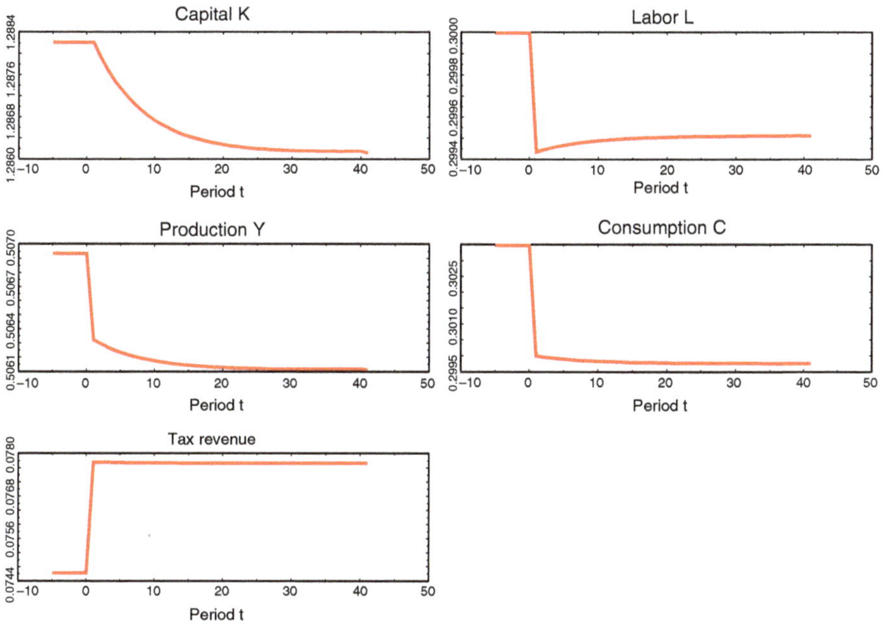

Fig. 5.10 Transition dynamics after a permanent increase of τ^L

the transition in all periods $t = 0, 1 \ldots$, we compare two situations with different government expenditure paths.

To compute the change in welfare over the entire transition period, we compute lifetime utility from (5.4) for the paths of consumption and labor illustrated in Fig. 5.10 for $\tau_t^L = 0.23$, $t = 0, 1, \ldots$ with constant labor L and consumption C and compare it to the lifetime utility (for tax rates $\tau_0^L = 0.23$ and $\tau_t^L = 0.24$ for $t = 1, \ldots$). The former is simply calculated as

$$\frac{\left(C^{\iota}(1-L)^{1-\iota}\right)^{1-\sigma}}{(1-\sigma)(1-\beta)},$$

while we need to use computational methods to calculate lifetime utility for $\tau^L = 0.24$. Again, we can compute the consumption equivalent change Δ:

$$\sum_{t=0}^{\infty} \beta^t u((1+\Delta)C_t, L_t) = \sum_{t=0}^{\infty} \beta^t u(C_t', L_t'),$$

where L and C are the steady-state values of labor and consumption for $\tau^L = 0.23$, and L'_t and C'_t denote the equilibrium values of labor and consumption for $\tau_t^L = 0.24$ for $t = 1, 2, \ldots$. The resulting consumption equivalent change, $\Delta = -1.07\%$, is again very close to our estimate from the steady-state evaluation ($\Delta = -1.10\%$). Notice that the welfare loss is smaller if we also account for the transition because consumption and leisure are higher during the initial phase of the transition than in the new steady state.

At this point, we offer a word of caution. The three estimates of the consumption equivalent change from the partial equilibrium, the steady state in general equilibrium, and the complete transition in general equilibrium are very close to one another and in range between -1.07% and -1.10%. This need not be the case in general (for other models). In fact, we will discuss a finding from the literature according to which the consideration of the transition not only significantly changes the quantitative results but may also overturn the steady-state welfare effect and change the sign.

The present analysis is also close in spirit to Prescott (2004), in which the Nobel laureate Edward C. Prescott attempts to answer the following question: *"Why do Americans work so much more than Europeans?"*. In the early 1970s, Europeans and Americans were working nearly the same amount. For example, Germans worked 5% more hours than the Americans during the period 1970–1974, according to Table 1 in Prescott (2004). During the 1990s, however, Americans worked approximately 50% more than Europeans. For example, Americans worked 56% more than Italians and 33% more than Germans. Prescott finds that different marginal tax rates on labor income and consumption are sufficient to fully explain this finding.[28] Therefore, he studies a neoclassical growth model just like that above and uses the same preferences for households in the US and Europe; however, the tax rates are set as those prevailing in the individual countries.[29] As one important conclusion from this study, we find that the tax policy implies significant distortions in the real economy, and the focus on supply-side economic policy in the US during the 1980s helped to increase employment.[30] In Problem 5.2, you are asked to recompute some of the results from Prescott's study.

---

[28]In his analysis, Prescott emphasized that it is important to consider the marginal rather than the average tax rates for consumption, labor, capital, and investment.

[29]If, instead, the explanation for the observed puzzle were that Europeans were lazier than Americans, the parameter ι in the above utility function should be different for the households in the individual countries.

[30]A similar result is presented by Chakraborty, Holter, and Stepanchuk (2015), who analyze the effects of both income taxes and the divorce rate in an OLG model. In their cross-country comparison of the US with 17 EU countries, they find that the lower income tax rates and higher divorce rates in the US explain approximately 45% of the higher labor supply in the US.

5.4 Capital Income Tax

In this section, we will first analyze the effects of a capital income tax τ^K on equilibrium values of consumption, savings, employment, and output. Next, we will present the *Chamley-Judd* result that it is optimal not to tax capital income in the long run and study the effects of depreciation-deductibility.

5.4.1 Distortionary Effects of Capital Taxes

To derive the general equilibrium effects of capital taxation, we extend the model in the previous section by introducing a capital income tax τ^K that is levied upon interest income, $\tau^K r_t K_t$. We will distinguish two cases: (1) The tax law allows for the deductibility of the depreciation costs of capital. Therefore, the household can subtract the amount $\tau^K \delta K_t$ from its capital income taxes, and the budget constraint (5.7) changes to

$$(1-\tau_t^L)w_t L_t + (1-\tau^K)r_t K_t + \tau^K \delta K_t = C_t + K_{t+1} - (1-\delta)K_t,$$

(2) Alternatively, we consider the case in which depreciation is not tax-deductible. Accordingly, we use the following two budget constraints:

$$C_t + K_{t+1} - K_t = \begin{cases} (1-\tau_t^L)w_t L_t + (1-\tau^K)(r_t-\delta)K_t, & \text{case 1,} \\ (1-\tau_t^L)w_t L_t + (1-\tau^K)r_t K_t - \delta K_t, & \text{case 2.} \end{cases} \tag{5.15}$$

Consequently, the first-order condition in the form of the Euler equation also needs to be adjusted

$$\lambda_t = \begin{cases} \lambda_{t+1}\beta\left[1+(1-\tau^K)\,(r_{t+1}-\delta)\right], & \text{case 1,} \\ \lambda_{t+1}\beta\left[1+(1-\tau^K)r_{t+1}-\delta\right], & \text{case 2.} \end{cases} \tag{5.16}$$

In addition, government expenditures also include capital income taxes, and thus, the balanced government budget is given by

$$G_t = \begin{cases} \tau_t^L w_t L_t + \tau^K(r_t-\delta)K_t, & \text{case 1,} \\ \tau_t^L w_t L_t + \tau^K r_t K_t, & \text{case 2.} \end{cases} \tag{5.17}$$

All other equilibrium conditions of the model with capital taxation are identical to those of the model in the previous section, and the steady state is characterized by the following equations in the six variables K, L, C, w, r, and τ^K for given government expenditures G and labor income taxes τ^L:

$$\frac{1}{\beta} = \begin{cases} 1+(1-\tau^k)\,(r-\delta), & \text{case 1,} \\ 1+(1-\tau^k)r-\delta, & \text{case 2,} \end{cases} \tag{5.18a}$$

$$(1-\tau^L)w = \frac{1-\iota}{\iota}\frac{C}{1-L}, \tag{5.18b}$$

$$w = (1-\alpha)K^\alpha L^{-\alpha}, \tag{5.18c}$$

$$r = \alpha K^{\alpha-1}L^{1-\alpha}, \tag{5.18d}$$

$$K^\alpha L^{1-\alpha} = C + G + \delta K, \tag{5.18e}$$

$$G = \begin{cases} \tau^L wL + \tau^K (r-\delta)K, & \text{case 1,} \\ \tau^L wL + \tau^K rK, & \text{case 2.} \end{cases} \tag{5.18f}$$

To derive numerical results, we use the same calibration as in Sect. 5.3.2. In addition, we need to calibrate the capital income tax rate τ^K, which we set equal to 41% (see the empirical evidence presented in Sect. 5.2).[31] The steady state and the following results for the welfare losses from capital income taxation are computed with the help of the GAUSS program *Ch5_welfare_tauk.g*. The two different tax scenarios 1 and 2 have a significant effect on tax revenues and, hence, equilibrium government expenditures. If capital depreciation is tax-deductible, government consumption G amounts to 20.8% of GDP, which is close to the values observed empirically in the US economy (see Chap. 4). In case 2, the government share G rises to 29.5% of GDP.

In the following, we consider the equilibrium effects of a change in the capital income tax τ^K. To balance the government budget (5.17), the labor income tax τ^L adjusts while government expenditures G remain constant in each of the two cases. Since we assume in our model that government consumption G is a pure waste and does not increase either utility or productivity, it would not make sense to compare the welfare of tax policies with different levels of G.

Figure 5.11 presents the effects of the capital income tax rate τ^K on the steady-state values of the model. If the capital tax rate is reduced from the present level of 41.0%, the labor income tax rate τ^L has to increase to balance the budget. For example, if capital taxes are abolished, $\tau^K = 0$, the labor income tax rate τ^L increases from 23.0% to 30.3%. Since the net return on capital (after taxes) increases, households increase their savings (the substitution effect dominates the income effect), and the capital stock K increases by 29%, from 0.963 to 1.245. Labor supply L, however, falls by 4% from 0.300 to 0.289. There are two opposing effects of lower capital taxes on the labor supply. On the one hand, higher capital increases the wage rate w because the marginal product of labor increases. On the other hand, higher labor income taxes τ^L reduce the net wage. Since the latter effect dominates, labor supply decreases (again, the substitution effect dominates the income effect of lower net wages). The increase in capital is more pronounced than the decrease in labor, and as a result, production rises by 16.0%, from 0.633

[31] Again, we calibrate the utility parameter $\iota = 0.3355$ ($\iota = 0.3256$) such that the steady-state labor supply is equal to 30% in case 1 (case 2), $L = 0.30$.

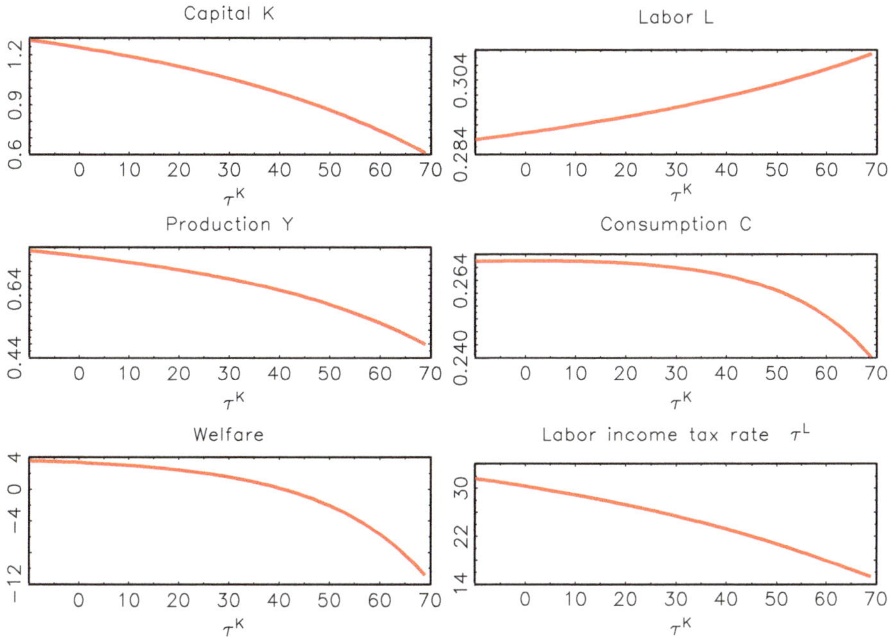

Fig. 5.11 Steady-state effects of capital income tax rate τ^K in case 1 (depreciation tax-deductible)

to 0.734. Since part of the higher production has to be invested to keep the capital stock constant in steady state, the increase in consumption C is smaller and only amounts to 1.9% as C rises from 0.265 to 0.270.

How do capital income taxes affect welfare as measured by the steady-state utility of the household? Lower capital income taxes reduce labor and increase consumption, meaning that utility rises unanimously in case 1. In fact, it is optimal to subsidize capital income. The welfare increase that is associated with the abolition of capital income taxes τ^K amounts to 3.4% of total consumption as measured by the consumption equivalent change.

Figure 5.12 illustrates the effects of capital income taxes in case 2 where deprecation is not tax-deductible. In this case, again, capital increases and labor decreases for lower values of τ^K. Total production increases over the total range considered, $-10\% \leq \tau^K \leq 70\%$. Notice, however, that consumption is now a concave function that peaks at approximately $\tau^K = 0\%$. For lower values of the capital income tax rate, the increase in production is not sufficient to offset the higher depreciation of capital, and thus, consumption falls. Consequently, there are two opposing effects of very low capital income taxes on utility. While labor supply falls, and hence, leisure and utility increase, consumption and, therefore, utility decrease. The optimal capital income tax rate τ^K as presented in the lower-left panel of Fig. 5.12 is approximately zero in case 2. In fact, we will show in the next section that the optimal capital tax rate is zero in case 2. The welfare effects of

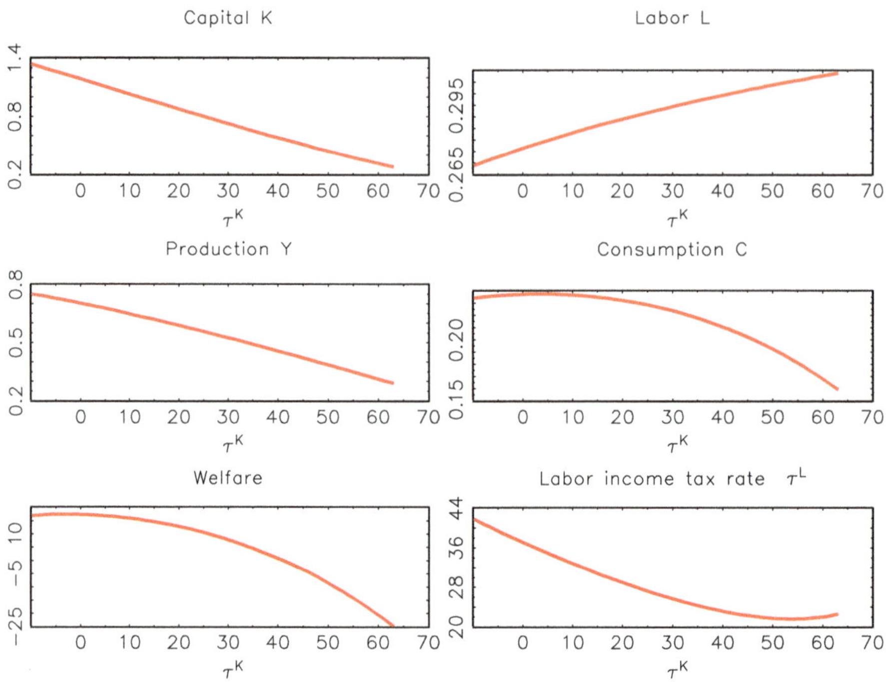

Fig. 5.12 Steady-state effects of capital income tax rate τ^K in case 2 (depreciation not tax-deductible)

the optimal capital tax are much higher than in case 1 and amount to approximately 17.3%.

5.4.2 Optimal Capital Taxation: Chamley-Judd Result

In the quantitative analysis in the previous section, we found that capital income taxes should be equal to zero if depreciation of capital is not tax-deductible. In this section, we present the famous *Chamley-Judd result* that *long-run capital income taxes (and, similarly, wealth taxes) should indeed be equal to zero* in this case and sketch the proof of this theoretical result.

Chamley (1986) and Judd (1985) independently analyzed the optimal policy of a social planner who maximizes the welfare of an infinitely lived household via the choice of a time path for taxes. The government cannot resort to non-distortionary lump-sum taxes to finance an exogenously given path of government expenditures but instead has to use proportional labor and capital income taxes. They find that whenever the economy converges to a balanced growth path, capital income taxes τ^K must converge to zero.

To show their optimal taxation result, let us revisit the Ramsey model from the previous section. However, we do not need to assume specific functional forms for utility and production as above but only assume that utility from consumption and leisure $u(C, L)$ is concave and production $F(K, L)$ is concave and characterized by constant returns to scale. Therefore, the Chamley-Judd result holds for very general forms of preferences and technology.

We assume that deprecation is not tax-deductible, meaning that the household budget constraint (5.15) for case 2 holds. Households maximize utility $u(C_t, L_t)$, and thus, the maximization of the Lagrangian

$$\mathscr{L} = \sum_{t=0}^{\infty} \beta^t \Big[u(C_t, L_t) + \lambda_t \Big((1-\tau_t^L) w_t L_t \\ + \Big(1 + (1-\tau_t^K) r_t - \delta\Big) K_t - C_t - K_{t+1} \Big) \Big]$$

implies the first-order conditions

$$\lambda_t (1-\tau_t^L) w_t = -\frac{\partial u}{\partial L_t} = -u_{L_t}, \tag{5.19a}$$

$$\lambda_t = \frac{\partial u}{\partial C_t} = u_{C_t}, \tag{5.19b}$$

$$\lambda_t = \beta \lambda_{t+1} \Big(1 + (1-\tau_{t+1}^K) r_{t+1} - \delta\Big). \tag{5.19c}$$

We can eliminate λ_t from the above equations to obtain

$$0 = u_{L_t} + u_{C_t} (1-\tau_t^L) w_t, \tag{5.20a}$$

$$0 = \beta u_{C_{t+1}} \Big(1 + (1-\tau_t^K) r_{t+1} - \delta\Big) - u_{C_t}. \tag{5.20b}$$

To derive the optimal tax, $\left\{\tau_t^L, \tau_t^K\right\}_{t=0}^{\infty}$, for given exogenous government expenditures, $\{G_t\}_{t=0}^{\infty}$, the government maximizes the intertemporal utility of the household (5.4), subject to the following constraints:

1. The government budget (5.17) for case 2 is balanced. As a consequence, the aggregate resource constraint holds:

$$C_t + G_t + K_{t+1} = F(K_t, L_t) + (1-\delta) K_t. \tag{5.21}$$

To derive (5.21), we assumed that production is characterized by constant returns to scale and factor and goods markets are competitive such that the Euler's theorem holds:

$$F(K_t, L_t) = w_t L_t + r_t K_t.$$

2. The household maximizes intertemporal utility, and thus, its consumption, savings, and labor supply are described by (5.20) for given exogenous tax rates, $\{\tau_t^L, \tau_t^K\}_{t=0}^{\infty}$.
3. Firms maximize profits such that the factor prices are equal to their marginal products, $w_t = \frac{\partial F}{\partial L_t}$, $r_t = \frac{\partial F}{\partial K_t}$.

Consequently, the government maximizes the following Lagrangian:

$$\begin{aligned} \max_{\tau_t^K, \tau_t^L, C_t, L_t, K_{t+1}} \quad & \sum_{t=0}^{\infty} \beta^t \Big\{ u(C_t, L_t) + \\ & \psi_t \left[\tau_t^K F_{K_t} K_t + \tau_t^L F_{L_t} L_t - G_t \right] \\ & + \theta_t \left[F(K_t, L_t) + (1-\delta) K_t - C_t - G_t - K_{t+1} \right] \\ & + \mu_{1t} \left[u_{L_t} + u_{C_t} (1 - \tau_t^L) w_t \right] \\ & + \mu_{2t} \left[\beta u_{C_{t+1}} \left(1 + (1 - \tau_{t+1}^K) r_{t+1} - \delta \right) - u_{C_t} \right] \Big\}. \end{aligned} \tag{5.22}$$

The solution to this problem, $\{\tau_t^L, \tau_t^K\}_{t=0}^{\infty}$, is called the *Ramsey policy* and is derived in greater detail in Appendix 5.1. In particular, it is easy to show that the following condition holds in steady state:

$$[\psi + \theta] \tau^K F_K = 0, \tag{5.23}$$

which is true only for $\tau^K = 0$.

Notice the following:

- The result of zero capital income taxation holds only in the long run. During the transition, it may be optimal to set $\tau_t^K > 0$. In particular, it is optimal in the first period to tax capital income because the capital stock is in fixed supply when the tax policy is announced. To understand this result, notice that only the capital income tax in period $t + 1$ (but not in period t) affects the intertemporal optimization behavior of the household in period t, as indicated directly by (5.19c). More formally, assume that the government has to satisfy the following intertemporal budget constraint:

$$\sum_{t=0}^{\infty} \left[\frac{\tau_t^L w_t L_t + \tau_t^K r_t K_t}{\prod_{s=0}^{t} (1 + r_s)} \right] = \sum_{t=0}^{\infty} \left[\frac{G_t}{\prod_{s=0}^{t} (1 + r_s)} \right], \tag{5.24}$$

where we set $r_0 \equiv 0$. In addition, K_0 is given. One can show[32] that the optimal solution is given by a tax policy whereby the government should raise as much taxes from capital income in the initial period $t = 0$, $\tau_0^K r_0 K_0$, as possible.

However, this policy suffers from the curse of time inconsistency.[33] If the households knew that it would be optimal for the government to prohibitively tax capital in period $t = 1$, they would not have accumulated any in period 0. Moreover, the government has a problem committing to a policy of zero capital income taxation in the long run because at the beginning of each period t, it has an incentive to deviate from its policy and prohibitively tax capital income. If the public learns from the government's past behavior, it would predict that the government will again tax capital in period $t+1$; therefore, the public will not invest. As a consequence, the government cannot use a prohibitive tax in any period t.

- It is easy to show[34] that it is also optimal to set a wealth tax τ^V equal to zero. With a wealth tax, the budget constraint is represented by

$$C_t + K_{t+1} = (1-\tau_t^L)w_t L_t + (1 + r_t - \delta)\,K_t - \tau_t^V K_t. \tag{5.25}$$

  Again, one can show that in steady state, the effects of a capital income tax and a wealth tax are identical if the rates satisfy

$$\tau^K r = \tau^V.$$

- In the US economy, capital income taxes averaged 41% during the period 1948–2008, although they have declined in recent decades. The Chamley-Judd result suggests that it is optimal to substantially reduce capital income taxation. In our computation exercise, we find that considerable welfare effects equal to multiple percentage points of total consumption accrue in the long run. Thus, what might keep politicians from decreasing capital income taxes? Besides political arguments, are there any economic arguments in favor of capital income taxes?

  To derive their results, Chamley (1986) and Judd (1985) analyze a representative agent model. If, instead, households are heterogeneous, capital income taxes may help to redistribute from wealth-rich households, which are characterized by a low marginal utility of consumption, to wealth-poor households, which have a much higher marginal utility of consumption. Consequently, average utility may increase.

---

[32]For a formal proof, see Chapter 2 in Kocherlakota (2010).

[33]More formally, a time-consistent policy is a policy in a multi-period problem that is optimal in the present period and remains optimal in future time periods. The main reference for the presentation of the time-inconsistency problem is provided by Kydland and Prescott (1977). Fischer (1980) presents the problem of time-inconsistent fiscal policy in a two-period model. A good textbook illustration of the Fischer model and its implications for optimal tax policy is presented in Chapter 6.2 of Wickens (2011).

[34]You will be asked to show these results in Problem 5.3.

In addition, Chamley (1986) and Judd (1985) assume that households are infinitely lived. If one considers a finite lifetime and overlapping generations, the older households may leave bequests to their children in the form of both human and physical capital. To ensure equality of opportunity and allow a household with poor parents (but, perhaps, with high learning abilities and intelligence) to invest in education, it may be welfare-improving to tax inheritances and redistribute among the young generation by providing better public education.

Moreover, many studies introduce heterogeneous households into the representative agent Ramsey model to study the welfare effects of income taxation, e.g., the distributional effects of capital income taxes or the optimal degree of progressivity in the income tax schedule. In these models, households are heterogeneous in their productivity and assets. In this vein, Domeij and Heathcote (2004) show that household heterogeneity and market incompleteness imply different welfare effects of tax changes. In fact, it is optimal not to reduce the capital income tax rate in this case. A capital income tax redistributes income from the wealth-rich to the wealth-poor, where the latter are usually characterized by lower income and consumption.[35] Therefore, as pointed out above, wealth-poor households have a higher marginal utility of income than do wealth-rich households, and thus, such a tax increase increases the mean utility in the economy.

For the US economy, Conesa and Krueger (2006) find that the optimal progressivity of income taxation (of both labor and capital income) is rather flat and well approximated by a constant income tax of 17.2%. Here, the disincentive effect of a more progressive income tax on the labor supply of the most productive workers is so detrimental to aggregate output and income that the redistributive effect does not compensate for it. In Heer and Trede (2003), we compare two revenue-neutral income tax reform proposals, (i) a flat-rate income tax and (ii) a consumption tax, in a general equilibrium model with an elastic labor supply and progressive income taxation. The model is calibrated to the German economy in 1996 such that the endogenous labor income distribution as computed from our model is equal to the empirical labor income distribution in Germany. Both tax reform proposals result in a moderate increase in aggregate employment and a strong increase in aggregate savings. Importantly, the two reform proposals imply significant steady-state welfare gains that are equivalent to increases in total consumption of 3.6% and 8.2%, respectively.

[35]Notice that income and wealth are not perfectly correlated. Budría Rodriguez, Díaz-Giménez, and Quadrini (2002), for example, find that the correlation between labor income and wealth only amounts to 0.27 in the US economy.

5.5 The Laffer Curve

During the recent financial crisis, many governments increased their debt (relative to GDP) to unprecedented levels since World War II. The larger debt service that is associated with this development has raised the question of whether government debt and expenditures are sustainable. To answer this question, one has to consider the amount of revenues that the government is able to generate from taxes.

One concept that has prominently been used in US economic policy since the 1970s is the *Laffer curve*.[36] This curve illustrates the amount of tax revenues that is associated with different levels of the income tax rate. It is hump-shaped, as illustrated in Fig. 5.13. Let the tax revenues be equal to $R = \tau Y$, where τ and Y denote the income tax rate and taxable income, respectively. If the tax is zero, $\tau = \tau_0 = 0\%$, revenues are equal to zero, $R = 0$. If the tax rate τ increases, therefore, revenues R also increase. Given increasing revenues, however, the tax base Y will shrink because households have less incentive to generate income. If income is taxed at $\tau = \tau_{100} = 100\%$, income Y and, hence, revenues R drop to zero.

For the politician, it is important to identify the location of the tax rate associated with the present income tax system on the Laffer curve. If the tax rate is higher than

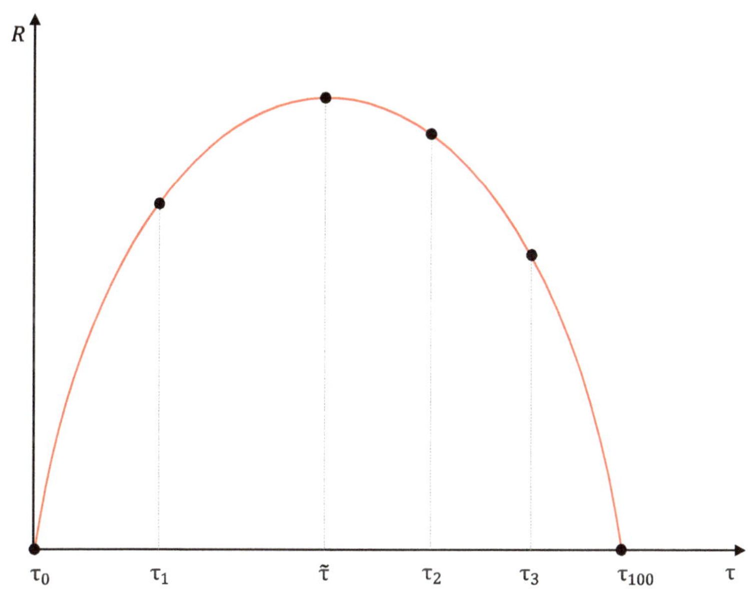

Fig. 5.13 Laffer curve

[36] Arthur Betz Laffer was a member of Reagan's Economic Policy Advisory Board (1981–1989) and a 2016 campaign advisor of Donald Trump.

$\tilde{\tau}$, for example at $\tau = \tau_3$, the government should cut tax rates. We will find below that the US income tax is at a point τ_1 to the left of the Laffer curve peak $\tilde{\tau}$. As a consequence, the US government can still raise tax revenues R. In addition, we will separately analyze tax revenues for labor and capital income taxes because these two forms of income are taxed differently in the US (see Fig. 5.6). In our analysis in this section, we find that the two tax rates are approximately half of those associated with the peak and that tax revenues from labor and capital income can be raised by approximately 50–70%.

To derive these results, we use a simplified version of the model proposed by Trabandt and Uhlig (2011).[37] To compare our results to theirs, we also adopt their calibration with respect to the functional forms of utility and production and their parameterization for the US economy. In addition to the ingredients of the model in the previous section, here, the government issues debt, and the economy is growing.

5.5.1 The Model

5.5.1.1 Households

Households maximize intertemporal utility

$$U_0 = \sum_{t=0}^{\infty} \beta^t \left[u(C_t, L_t) + \varphi(G_t) \right], \tag{5.26}$$

where instantaneous utility is a function of consumption C and labor L:

$$u(C, L) = \frac{1}{1-\sigma} \left(C^{1-\sigma} \left[1 - \nu_0 (1-\sigma) L^{1+1/\nu_1} \right]^{\sigma} - 1 \right). \tag{5.27}$$

The functional form of utility is very convenient. (1) Utility from government consumption $\varphi(G_t)$ is additive, meaning that there is no direct effect of G_t on the marginal utility from consumption C or leisure $1 - L$. (2) This form of utility is in accordance with the fact that labor L is constant in the long run on the balanced growth path. $1/\sigma$ denotes the constant intertemporal elasticity of substitution, while ν_1 is equal to the Frisch labor supply elasticity.[38]

The household holds two forms of assets, government bonds B_t and capital K_t. The real return on bond r_t^B is set by the government, while the return on capital is denoted by r_t. The household invests I_t in the accumulation of capital:

$$K_{t+1} = (1-\delta) K_t + I_t. \tag{5.28}$$

[37] In particular, we neglect income from abroad.

[38] See Appendix 4.2 for the definition of the Frisch labor supply elasticity.

The household budget constraint is presented by:

$$(1+\tau^C)C_t+I_t+B_{t+1} = (1-\tau_t^L)w_t L_t+(1-\tau_t^K)(r_t-\delta)K_t+\delta K_t+(1+r_t^B)B_t+Tr_t, \tag{5.29}$$

where τ^C and Tr_t denote the constant consumption tax rate and government transfers to the households in period t. Depreciation is tax-deductible.

The first-order conditions are represented by

$$\lambda_t(1+\tau^C) = C_t^{-\sigma}\left[1-\nu_0(1-\sigma)L_t^{1+1/\nu_1}\right]^{\sigma}, \tag{5.30a}$$

$$\lambda_t(1-\tau_t^L)w_t = \nu_0\sigma\left(1+\frac{1}{\nu_1}\right)C_t^{1-\sigma}\left[1-\nu_0(1-\sigma)L_t^{1+1/\nu_1}\right]^{\sigma-1}L_t^{1/\nu_1}, \tag{5.30b}$$

$$\lambda_t = \beta\lambda_{t+1}\left[1+(1-\tau_{t+1}^K)(r_{t+1}-\delta)\right], \tag{5.30c}$$

$$\lambda_t = \beta\lambda_{t+1}(1+r_{t+1}^B). \tag{5.30d}$$

From (5.30c) and (5.30d), it follows that the two assets B_t and K_t must yield the same return after taxes,

$$r_{t+1}^B = (1-\tau_{t+1}^K)(r_{t+1}-\delta).$$

5.5.1.2 Production

Production is Cobb-Douglas in the two production factors capital K_t and labor L_t:

$$Y_t = A_t K_t^{\alpha} L_t^{1-\alpha}. \tag{5.31}$$

Total factor productivity A_t grows at the exogenous rate γ_A[39]:

$$A_t = A_0(1+\gamma_A)^t. \tag{5.32}$$

[39]Notice that, different from the production function (3.37), we did not introduce A_t as labor productivity, but as total factor productivity. These two specifications are equivalent for the Cobb-Douglas production function if the growth of labor productivity γ is related to γ_A according to

$$1+\gamma = (1+\gamma_A)^{\frac{1}{1-\alpha}}.$$

The above equation follows from

$$Y_t = A_t K_t^{\alpha} L_t^{1-\alpha} = K_t^{\alpha}\left(A_t^{\frac{1}{1-\alpha}} L_t\right)^{1-\alpha}.$$

In Sect. 3.4 we also showed that, in steady state, output, capital and consumption all grow at the rate γ so that $(1+\gamma_A)^{\frac{1}{1-\alpha}}$ denotes the stationary growth factor.

Define $\psi = (1+\gamma_A)^{1/(1-\alpha)}$, where ψ denotes the stationary growth factor. Accordingly, stationary output is given by:

$$\tilde{Y}_t \equiv \frac{Y_t}{\psi^t} = \frac{A_0(1+\gamma_A)^t K_t^{\alpha} L_t^{1-\alpha}}{\psi^t} = A_0 \tilde{K}_t^{\alpha} L_t^{1-\alpha}, \tag{5.33}$$

where $\tilde{K}_t \equiv K_t/\psi^t$. Without loss of generality, we set the total factor productivity in period 0 to unity, $A_0 = 1$.

In factor market equilibrium with competitive goods and factor markets, profit-maximizing firms set the wage and the interest rate equal to their marginal products:

$$\tilde{w}_t = (1-\alpha)\frac{\tilde{Y}_t}{L_t}, \tag{5.34a}$$

$$r_t = \alpha \frac{\tilde{Y}_t}{\tilde{K}_t}. \tag{5.34b}$$

Again, we define the stationary variable $\tilde{w}_t = w_t/\psi^t$.

5.5.1.3 Government

The government finances government expenditures with taxes and debt according to[40]:

$$G_t + Tr_t + r_t^B B_t = T_t + B_{t+1} - B_t, \tag{5.35}$$

with taxes T_t represented by

$$T_t = \tau^C C_t + \tau^L w_t L_t + \tau^K (r_t - \delta) K_t. \tag{5.36}$$

5.5.1.4 Equilibrium Conditions

In equilibrium, the resource constraint of the economy holds:

$$Y_t = C_t + G_t + I_t. \tag{5.37}$$

[40]We will analyze government debt in greater detail in Chap. 7.

The first-order conditions, budget constraints and equilibrium conditions can be expressed in stationary variables. The steady state is described by the following seven equations in the seven endogenous variables r^B, $\tilde{K}$, $\tilde{C}$, $\tilde{Y}$, L, $\tilde{w}$, and $\tilde{I}$[41]:

$$\frac{1+\tau^C}{1-\tau^L}\frac{1+1/\nu_1}{1-\alpha}\frac{\tilde{C}}{\tilde{Y}} = \frac{1}{\sigma\nu_0 L_t^{1+1/\nu_1}} + 1 - \frac{1}{\sigma}, \tag{5.38a}$$

$$1+r^B = \frac{\psi^\sigma}{\beta}, \tag{5.38b}$$

$$\frac{\tilde{Y}}{L} = \left(\frac{\tilde{K}}{\tilde{Y}}\right)^{\frac{\alpha}{1-\alpha}}, \tag{5.38c}$$

$$\tilde{w}_t = (1-\alpha)\frac{\tilde{Y}_t}{L_t}, \tag{5.38d}$$

$$\frac{\tilde{Y}}{\tilde{K}} = \frac{r^B}{\alpha(1-\tau^K)} + \frac{\delta}{\alpha}, \tag{5.38e}$$

$$\tilde{Y} = \tilde{C} + \tilde{I} + \tilde{G}, \tag{5.38f}$$

$$\tilde{I} = (\psi - 1 + \delta)\tilde{K}. \tag{5.38g}$$

In the economy, the real return on bonds in the long run is determined by the Euler condition:

$$1+r^B = \frac{\psi^\sigma}{\beta}.$$

As a consequence, the real interest rate r and, hence, the capital-labor ratio $\tilde{K}/L$ depend on the capital income tax rate τ^K but not on the labor income tax rate τ^L according to[42]:

$$1 + (1-\tau^K)(r-\delta) = \frac{\psi^\sigma}{\beta}.$$

Therefore, an increase in the labor income tax rate τ^L does not change the steady-state wage $\tilde{w}$ before taxes, and the total incidence of the labor income tax change is borne by the workers.

---

[41] You are asked to derive these equations in Problem 5.4. Notice that the stationary value of the Lagrange multiplier λ_t is represented by $\tilde{\lambda}_t = \lambda_t \left(\psi^t\right)^\sigma$.

[42] This observation does not hold in the OLG model. Why is this the case? In Sect. 7.5, we embed the model of Trabandt and Uhlig (2011) in an OLG framework to study the dynamics of debt and the real interest rate.

5.5.1.5 Calibration

We follow Trabandt and Uhlig (2011), who calibrate their model for the US economy. In particular, the real interest rate on bonds is set equal to 4% annually, $r^B = 0.04$. The empirical estimates for the Frisch labor supply elasticity, which measures the (absolute) percentage change in the labor supply if the wage increases by 1%, vary considerably.[43] We use a value $\nu_1 = 1.0$ that is in the upper range of the estimates. The intertemporal elasticity of substitution is equal to $1/\sigma = 1/2.0$. The tax rates on labor income, capital income, and consumption are identical to those values observed in the US economy: $\tau^L = 0.28$, $\tau^K = 0.36$, and $\tau^C = 0.05$. In addition, the growth factor of the economy is calibrated as $\psi = 1.02$, the debt-GDP ratio of the US economy is set to $B/Y = 63\%$, and government consumption as a share of GDP amounts to $G/Y = 18\%$. Finally, the production parameter values for the capital income share, $\alpha = 0.38$, and the depreciation rate, $\delta = 0.07$, are close to the values that we used in previous chapters. The value of $\nu_0 = 3.732$ is calibrated such that steady-state labor supply is equal to $\bar{L} = 0.25$, as in Trabandt and Uhlig (2011). The solution is computed with the help of the GAUSS program *Ch5_laffer.g*.

5.5.2 Results

We separately study the effects of a change in the two tax rates, τ^L and τ^K, on revenues, holding the other tax rates constant. Figure 5.14 presents the Laffer curve for the labor income tax rate τ^L (holding the other taxes τ^C and τ^K constant). At the benchmark with $\tau^L = 28\%$, labor income tax revenues (the solid red line) are equal to 0.08278, which amounts to 17.4% of GDP. Labor income tax revenues peak at $\tau^L = 71\%$ and can be increased by 67%. Accordingly, there is still considerable leeway for the US government to increase taxes.[44]

As the labor income tax rate τ^L increases, labor supply L decreases, while the capital-labor ratio $\tilde{K}/L$, as argued above, remains constant. As a consequence, capital and labor decrease by the same percentage rate, and capital income taxes decline with a higher τ^L. The effect of τ^L on total revenues T_t is illustrated by the broken green line in Fig. 5.14. In this case, the revenue-maximizing labor income tax rate τ^L is lower and amounts to only 65%. Total tax revenue increases by only 31.5% because of the decline in the capital stock and, hence, tax revenue from capital income and consumption. Nevertheless, the government can raise its revenues by

[43] See also the discussion of these values in Sect. 4.4.5.

[44] Bear in mind that in all tax scenarios that we consider in this subsection, we only compare steady states and neglect transition dynamics.

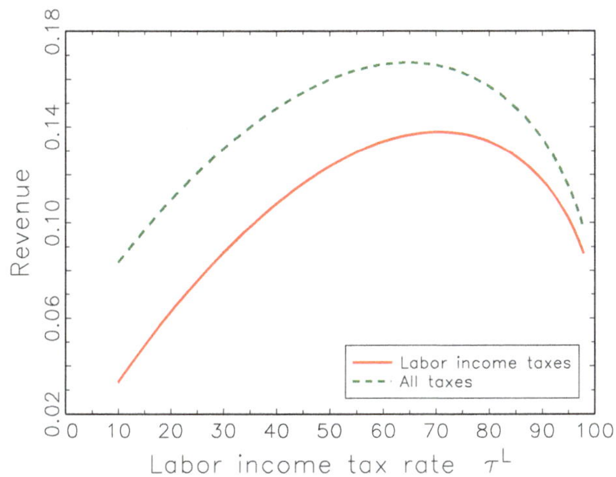

Fig. 5.14 US Laffer curve: labor income tax rate τ^L

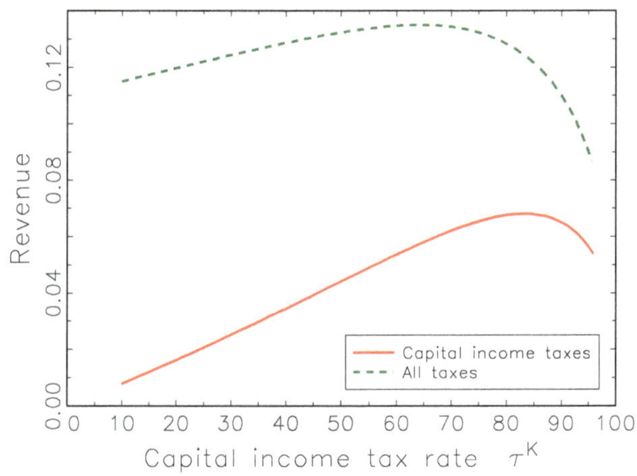

Fig. 5.15 US Laffer curve: capital income tax rate τ^K

8.4% of present GDP. Accordingly, government consumption can be increased to approximately 26.4% of GDP.[45]

The Laffer curve for the capital income tax rate τ^K (holding the other taxes τ^C and τ^L constant) is presented in Fig. 5.15. Capital income tax revenue peaks

[45]Trabandt and Uhlig (2011) also consider 14 EU countries and there ability to generate additional revenues with the help of income taxation. They find that all Scandinavian EU countries Denmark, Finland, and Sweden, and some other European countries, e.g., Austria, Italy, France, and Belgium cannot raise their labor income tax revenues by more than 5% because they are already so close to the peak of the Laffer curve.

at $\tau^K = 83\%$. Total tax revenue, however, already fall beyond the tax rate $\tau^K = 64\%$. At this high rate of capital income taxation, the decline in the capital stock and, hence, the marginal product of labor also reduces revenue from labor income taxation to such an extent that it offsets the increase in capital income taxes.[46] Since capital income taxes are a small share of total tax revenue, the capital income tax only exhibits limited potential for increasing tax revenue. For capital income to make a more significant contribution to total tax revenues, one possible solution would be to no longer make depreciation tax-deductible.[47]

5.6 Growth Effects of Taxes

Tax policy can affect economic growth through various channels. First, income taxes might alter factor inputs. For example, lower labor income taxes could increase labor supply, while lower capital income taxes could increase savings and investments. Second, lower income taxes might increase incentives to invest in human capital. For example, lower labor income taxes increase the benefit from both education and working. Third, lower taxes on or even subsidies for research activities may increase total factor productivity growth. In addition, how taxes are spent will affect economic growth. For example, if taxes are mainly used for redistributive purposes, the effect on economic growth may be less pronounced than in the cases in which revenues are spent on infrastructure, education, or health.

The results of cross-country growth regressions offer little support for a strong effect of taxes on economic growth.[48] For example, Mendoza, Milesi-Ferretti, and Asea (1997) find that tax policies do not affect long-run growth. In particular, the effect of tax rates on growth rates is weak once additional control variables such as initial GDP or other conditioning variables are included, as noted by Levine and Renelt (1992). However, these growth regressions are subject to various estimation problems. First, it is difficult to empirically identify the correct tax variable. For example, should the researcher use average or marginal tax rates? Second, the estimation suffers from the endogeneity problem. Taxes and government expenditures are highly correlated. Higher government expenditures temporarily increase output via the multiplier effect, as argued in the previous section. However, higher income may also increase government spending and, hence, taxes, especially if the tax code is progressive. More recent studies have considered the importance of the tax

[46] Furthermore, labor supply attains a minimum at $\tau^K = 73\%$ and increases for higher capital income tax rates beyond this threshold. For these high capital income tax rates and corresponding low wage rates, the income effect dominates the substitution effect, and lower wages imply higher labor supply.

[47] You are also asked to estimate the Laffer curves for this case in Problem 5.4.

[48] For an overview of these studies, see Chapter 12 of Barro and Sala-i-Martin (2003).

structure rather than the level of taxes. In his OECD panel study, Arnold (2008) finds that the distortions and the forgone economic growth from taxation increase in the following order[49]:

1. Property taxes
2. Consumption tax
3. Personal income tax
4. Corporate income tax.

He also finds a negative effect of the progressivity of the personal income tax system on economic growth.

In addition to these regression studies, one can gauge the effects of different tax policies on economic growth with the help of both qualitative and quantitative analysis of dynamic general equilibrium models. We will pursue this approach in the following. In the first section, we present a model in which income taxes are used to finance public expenditures, such as public investments in infrastructure, that increase aggregate productivity. We find that the growth-maximizing tax policy is the one where the income tax rate is set equal to the production elasticity of the public input. In the second section, we analyze the Lucas (1990) supply-side model in which growth is driven by investment in human capital. In this model, we find that an increase in capital relative to labor income taxation fosters growth because the latter has a stronger disincentive effect on human capital accumulation.

5.6.1 Endogenous Growth with Government Expenditures

In Chap. 4, we analyzed the effects of public expenditures on consumption. However, many government expenditures on public infrastructure such as roads, education, an efficient administration, or health also have significant effects on productivity. The seminal article by Aschauer (1989) presents empirical evidence on how an increase in productive government expenditures raises output. Moreover, Easterly and Rebelo (1993) show that public investment in infrastructure increases the long-run growth rate.

Hereinafter, we will account for the productivity-enhancing nature of (some) public expenditures and present a derivation of the optimal amount of government services based on the analysis of Barro (1990).[50] Government expenditures will enter the production function. In addition to the production sector, we will also

[49]Di Sanzo, Bella, and Graziano (2017) also study the empirical effects of the tax structure on economic growth. In a panel cointegrated VAR analysis, they find that a property tax has the least harmful effects on growth, while they cannot verify a significant difference between the growth effects of the income and the consumption tax when the total tax burden (relative to GDP) exceeds a threshold of 30%.

[50]However, we formulate the model in discrete time to comply with the approach used in the rest of the book. In addition, we consider exogenous labor supply in our model.

specify the behavior of the household sector, public finances, and the competitive equilibrium for our model in the following.

5.6.1.1 Production

The government provides services G_t free of charge in period t that are used as a production input. When we model productive government spending, we need to decide whether to consider a flow variable, a stock variable, or both. For simplicity, we will assume government expenditures to be a flow variable, but you will be asked in Problem 5.6 to derive results for government expenditures as a stock variable. In addition, we have to make an assumption regarding the congestion effect that is caused by increased use of the public good. For example, excessive use of public infrastructure such as roads or the legislative system may result in lower productivity for individual users. In the model, we exclude congestion effects, but you will be asked to analyze them in Problem 5.7.

There is a unit mass of firms that can be studied by means of a representative firm. Production $F(K_t, L_t, G_t)$ in period t uses capital K_t, labor L_t, and public services G_t as inputs according to:

$$Y_t = F(K_t, L_t, G_t) = AL_t^{1-\alpha} K_t^{\alpha} G_t^{1-\alpha}. \tag{5.39}$$

Notice that the production function $F(.)$ is characterized by constant returns to scale in the private inputs K_t and L_t. As a consequence, Euler's theorem (on linearly homogeneous functions) holds in markets with perfect competition, meaning that profits are equal to zero. In addition, we assume constant returns to scale in the two production factors K_t and G_t, which are able to grow at strictly positive rates γ_K and γ_G, respectively, in the long run, while the labor factor cannot grow without bound.

The assumptions concerning the production function are not innocuous. For a more general production function, e.g., $\tilde{F}(K, L, G) = K^{\alpha} L^{\beta} G^{\epsilon}$, there would be no endogenous growth if $\beta + \epsilon < 1$. In this case, the marginal product of capital would diminish in the long run. In steady state, to maintain growth, $\beta + \epsilon = 1$ must hold for capital and government expenditures to grow at the same rate, $\gamma_K = \gamma_G$. Moreover, if $\beta + \epsilon > 1$, the possibility of multiple equilibria may arise. Therefore, we restrict our consideration to the knife-edge assumption of constant returns to scale in K and G.[51]

We assume perfect competition in goods and factor markets. The representative firm maximizes profits Π_t in period t

$$\Pi_t = [1-\tau] AL_t^{1-\alpha} K_t^{\alpha} G_t^{1-\alpha} - r_t K_t - w_t L_t, \tag{5.40}$$

[51]For this reason, Irmen and Kuehnel (2009) suggest considering the growth effects of productive government expenditures in models with Schumpeterian innovation instead of the simple 'Ak'-model.

where w and r denote the real wage and interest rate, respectively. Production is taxed at rate τ. The necessary first-order conditions are given by

$$w_t = (1-\tau)F_{L_t} = (1-\tau)(1-\alpha)AL_t^{-\alpha}K_t^{\alpha}G_t^{1-\alpha}, \tag{5.41a}$$

$$r_t = (1-\tau)F_{K_t} = (1-\tau)\alpha AL_t^{1-\alpha}K_t^{\alpha-1}G_t^{1-\alpha}. \tag{5.41b}$$

In equilibrium, profits are zero.

5.6.1.2 Government

The government finances the public input by a tax on production.[52] In equilibrium, the budget is balanced in each period t:

$$G_t = \tau Y_t. \tag{5.42}$$

5.6.1.3 Households

The representative household is infinitely-lived and maximizes its intertemporal utility

$$U = \sum_{t=0}^{\infty} \beta^t u(C_t), \tag{5.43}$$

where $\beta < 1$ denotes its discount factor.

Instantaneous utility is presented by

$$u(C_t) = \frac{C_t^{1-\sigma} - 1}{1-\sigma}, \tag{5.44}$$

where $1/\sigma$ denotes the intertemporal elasticity of substitution.

Household labor supply is exogenous and given by $L_t = L$. The household owns the capital stock K_t and receives interest income from capital $r_t K_t$ and wage income from labor $w_t L$. Capital depreciates at rate δ. Accordingly, the household budget constraint is represented by

$$C_t + K_{t+1} = w_t L + (1 + r_t - \delta) K_t. \tag{5.45}$$

The optimization problem can be studied with the help of the Lagrangian

$$\mathscr{L} = \sum_{t=0}^{\infty} \beta^t \left\{ u(C_t) + \lambda_t \left[w_t L + (1 + r_t - \delta) K_t - C_t - K_{t+1} \right] \right\},$$

[52] Alternatively, we could consider an income tax on both labor and capital income.

that implies the following first-order conditions:

$$\lambda_t = \frac{\partial u}{\partial C_t} = u_{C_t}, \tag{5.46a}$$

$$\lambda_t = \beta\lambda_{t+1}\left(1 + r_{t+1} - \delta\right), \tag{5.46b}$$

where we can eliminate the Lagrange multiplier to obtain the Euler equation:

$$\left(\frac{C_{t+1}}{C_t}\right)^{\sigma} = \beta\left(1 + r_{t+1} - \delta\right). \tag{5.47}$$

5.6.1.4 Competitive Equilibrium

To derive the properties of the competitive equilibrium, we express consumption growth as a function of the tax rate. Therefore, we first rewrite the real interest rate r_t as a function of τ. Notice that

$$G_t = \tau Y_t = \tau A L^{1-\alpha} K_t^{\alpha} G_t^{1-\alpha},$$

or, noticing that $L_t = L$,

$$G_t = (\tau A)^{\frac{1}{\alpha}} L^{\frac{1-\alpha}{\alpha}} K_t.$$

Insert this expression for public consumption G_t into (5.41b) to derive

$$r_t = (1-\tau)\alpha A^{\frac{1}{\alpha}}\left[L\tau\right]^{\frac{1-\alpha}{\alpha}}. \tag{5.48}$$

Next, insert the expression for the real interest rate into (5.47):

$$\begin{aligned}\frac{C_{t+1} - C_t}{C_t} &= \left\{\beta\left[1 + r_{t+1} - \delta\right]\right\}^{\frac{1}{\sigma}} - 1 \\ &= \beta^{\frac{1}{\sigma}}\left[1 + (1-\tau)\alpha A^{\frac{1}{\alpha}}\left[L\tau\right]^{\frac{1-\alpha}{\alpha}} - \delta\right]^{\frac{1}{\sigma}} - 1.\end{aligned} \tag{5.49}$$

Notice that the growth rate of consumption

$$\gamma_t^C \equiv \frac{C_{t+1} - C_t}{C_t}$$

is constant over time and attains its maximum if the factor

$$(1-\tau)\tau^{\frac{1-\alpha}{\alpha}}$$

is maximized. To find the maximum, take the logarithm of this expression

$$\ln(1-\tau)+\frac{1-\alpha}{\alpha}\ln\tau$$

and differentiate it with respect to τ, implying:

$$\frac{1}{1-\tau}=\frac{1-\alpha}{\alpha}\frac{1}{\tau},$$

or

$$\tau=1-\alpha.$$

Accordingly, the maximum consumption growth rate is attained for a production tax $\tau=1-\alpha$:

$$\gamma^C=\beta^{\frac{1}{\sigma}}\left[1+\alpha^2(1-\alpha)^{\frac{1-\alpha}{\alpha}}A^{\frac{1}{\alpha}}L^{\frac{1-\alpha}{\alpha}}-\delta\right]^{\frac{1}{\sigma}}-1. \tag{5.50}$$

For this value of τ, the marginal product of the public input is equal to one and, therefore, equal to its marginal costs:

$$\frac{\partial F(K_t,G_t,L)}{\partial G_t}=(1-\alpha)\frac{Y_t}{G_t}=(1-\alpha)\frac{Y_t}{\tau Y_t}=\frac{1-\alpha}{1-\alpha}=1.$$

If $\tau>1-\alpha$, we observe that a reduction in taxes increases consumption growth.

In a competitive equilibrium, all variables G, K, and C grow at the same rate γ. To see this, let us begin with the growth rate of consumption (5.49). For a constant γ^C, the interest rate r is also constant. A constant interest rate, however, implies a constant ratio of government expenditures G to capital K according to (5.41b). Therefore, the growth rates of the two inputs G and K must be equal, $\gamma^K=\gamma^G$. Finally, if we divide the resource constraint, $Y_t=C_t+G_t+K_{t+1}-(1-\delta)K_t$, by K_t, we obtain

$$\frac{C_t}{K_t}=(1-\tau)AL^{1-\alpha}\left(\frac{G_t}{K_t}\right)^{1-\alpha}+1-\delta-\underbrace{\frac{K_{t+1}}{K_t}}_{:=\gamma_t^K}.$$

From this equation, we notice that C_t/K_t must also be constant in steady state. Therefore, $\gamma^C = \gamma^K$ must hold.

In the present model setup, an additional consumption tax $\tau^C > 0$ helps to increase the growth rate γ. The growth rate of consumption γ^C can be derived as above. Only the individual budget constraint (5.51)

$$(1+\tau^C)C_t + K_{t+1} = w_t L + (1 + r_t - \delta)\, K_t \tag{5.51}$$

and the government budget constraint (5.42)

$$G_t = \tau Y_t + \tau^C C_t \tag{5.52}$$

need to be adjusted.[53]

5.6.1.5 Pareto Efficiency

How does the solution in the competitive equilibrium compare with that in the command optimum? To determine this, we consider the case in which the central planer maximizes (5.43) subject to the resource constraint

$$C_t + G_t + K_{t+1} = AL^{1-\alpha} K_t^{\alpha} G_t^{1-\alpha} + (1-\delta)K_t. \tag{5.53}$$

The first-order conditions are given by

$$\lambda_t = C_t^{-\sigma}, \tag{5.54a}$$

$$\lambda_t = \beta\lambda_{t+1}\left(1 + \alpha AL^{1-\alpha} K_t^{\alpha-1} G_t^{1-\alpha} - \delta\right), \tag{5.54b}$$

$$1 = (1-\alpha)AL^{1-\alpha} K_t^{\alpha} G_t^{-\alpha}. \tag{5.54c}$$

With the help of (5.54c), we derive the Pareto-efficient level of government expenditures:

$$G_t = [A(1-\alpha)]^{\frac{1}{\alpha}} L^{\frac{1-\alpha}{\alpha}} K_t.$$

If we plug this expression for G_t into the Euler condition, we derive

$$\frac{C_{t+1} - C_t}{C_t} \equiv \tilde{\gamma}^C = \left[\beta\left(1 + \alpha(1-\alpha)^{\frac{1-\alpha}{\alpha}} A^{\frac{1}{\alpha}} L^{\frac{1-\alpha}{\alpha}} - \delta\right)\right]^{\frac{1}{\sigma}} - 1.$$

Notice that, for $0 < \alpha < 1$, the Pareto-efficient growth rate $\tilde{\gamma}^C$ is larger than the growth-maximizing rate γ^C in Eq. (5.50) for the case of a decentralized economy!

[53] You are asked to solve this case in Problem 5.5.

5.6.2 Capital Taxation in a Growth Model with Human Capital

In the following, we present the analysis of the capital income tax rate in the Lucas (1990) supply-side model. The central new element in Lucas' analysis is the consideration of human capital accumulation, which drives economic growth. The more time people spend on education, the higher the economic growth rate will be. Since the benefit from better education consists of higher wages and, hence, higher labor income in future periods, the opportunity costs of learning are inversely related with labor income tax levels. Accordingly, in the Lucas supply side model, higher capital income taxes are associated with higher growth because the government is able to reduce labor income taxation. We will present this mechanism in greater detail in the following.

The model that we specify is a discrete-time version of the Lucas (1990) model. In this way, we are able to maintain the same model setting as in the rest of the chapter. In addition, we consider a constant population rather than a growing one as in the model of Lucas (1990). The rest of the model is basically identical to that of Lucas.

5.6.2.1 Production

Firms use both physical and human capital K_t and H_t as production inputs. Let u_t denote the working time of the household, and production is described by the following constant elasticity of substitution (CES) function:

$$Y_t = A_0 \left(\alpha K_t^{\rho_p} + (1-\alpha)(u_t H_t)^{\rho_p} \right)^{\frac{1}{\rho_p}}, \tag{5.55}$$

where $\sigma_p = 1/(1-\rho_p)$ denotes the CES in production. In factor market equilibrium, the factor prices are equal to their marginal products:

$$w_t = A_0(1-\alpha) \left(\alpha K_t^{\rho_p} + (1-\alpha)(u_t H_t)^{\rho_p} \right)^{\frac{1}{\rho_p}-1} (u_t H_t)^{\rho_p - 1}, \tag{5.56a}$$

$$r_t = A_0 \alpha \left(\alpha K_t^{\rho_p} + (1-\alpha)(u_t H_t)^{\rho_p} \right)^{\frac{1}{\rho_p}-1} (K_t)^{\rho_p - 1}. \tag{5.56b}$$

Notice that w denotes the wage per efficiency unit uH, which is the product of working time and human capital.

5.6.2.2 Human Capital Accumulation

Individuals spend v_t units of time on education, and, hence, human capital h_t accumulates according to:

$$h_{t+1} = h_t + D v_t^{\varsigma} \bar{h}_t, \tag{5.57}$$

where $\bar{h}_t$ and ς denote the average human capital in the economy and the elasticity of human capital growth with respect to learning v, respectively. The household takes $\bar{h}_t$ as given.

5.6.2.3 Households

The number of households is equal to one, which allows us to study the behavior of the individual and aggregate households by means of the representative household. The household allocates its time endowment, which is normalized to one, to working, u_t, learning, v_t, and leisure, $1 - u_t - v_t$. The household derives utility from consumption C_t and leisure $1 - u_t - v_t$.

The representative household maximizes intertemporal utility:

$$\sum_{t=0}^{\infty} \beta^t \frac{[C_t(1 - u_t - v_t)^{\iota}]^{1-\sigma}}{1 - \sigma} \tag{5.58}$$

subject to the budget constraint in period t

$$(1 - \tau_t^L) w_t u_t h_t + \left(1 + (1 - \tau_t^K) r_t - \delta\right) K_t + tr_t = C_t + K_{t+1}. \tag{5.59}$$

The household pays taxes on labor and capital income at the rates τ_t^L and τ_t^K, respectively. In accordance with Lucas (1990), we assume that depreciation cannot be deducted from capital income taxes.[54] In addition, the household receives transfers tr_t from the government.

The Lagrangian of the household is presented by

$$\begin{aligned}
\mathscr{L} = \sum_{t=0}^{\infty} \beta^t \Bigg\{ & \frac{[C_t(1 - u_t - v_t)^{\iota}]^{1-\sigma}}{1 - \sigma} \\
& + \lambda_t \left[(1 - \tau_t^L) w_t u_t h_t + \left(1 + (1 - \tau_t^K) r_t - \delta\right) K_t + tr_t - C_t - K_{t+1} \right] \\
& + \mu_t \left[h_t + D v_t^{\varsigma} \bar{h}_t - h_{t+1} \right] \Bigg\}.
\end{aligned}$$

Since average human capital $\bar{h}_t$ is taken as exogenous by the individual, the first-order conditions with respect to C_t, K_{t+1}, u_t, v_t, and h_{t+1} are presented by:

$$\lambda_t = C_t^{-\sigma} (1 - u_t - v_t)^{\iota(1-\sigma)}, \tag{5.60a}$$

$$\lambda_t = \beta \lambda_{t+1} \left(1 + (1 - \tau_{t+1}^K) r_{t+1} - \delta\right), \tag{5.60b}$$

[54] You are asked to consider this change in assumptions in Problem 5.8.

$$\lambda_t(1-\tau_t^L)w_t h_t = \iota C_t^{1-\sigma}(1-u_t-v_t)^{\iota(1-\sigma)-1}, \tag{5.60c}$$

$$\mu_t \varsigma D v_t^{\varsigma-1}\bar{h}_t = \iota C_t^{1-\sigma}(1-u_t-v_t)^{\iota(1-\sigma)-1}, \tag{5.60d}$$

$$\mu_t = \beta\left[\mu_{t+1} + \lambda_{t+1}(1-\tau_{t+1}^L)w_{t+1}u_{t+1}\right]. \tag{5.60e}$$

5.6.2.4 Government

The government finances exogenous government consumption G_t and transfers Tr_t with the help of labor and capital income taxes such that its fiscal budget is balanced:

$$G_t + Tr_t = \tau_t^L w_t u_t H_t + \tau_t^K r_t K_t. \tag{5.61}$$

We assume that government consumption G_t and transfers Tr_t grow at the economic growth rate γ of the economy.

5.6.2.5 Equilibrium

In equilibrium, the following resource constraint holds:

$$Y_t = K_{t+1} - (1-\delta)K_t + C_t + G_t. \tag{5.62}$$

In addition, both aggregate human capital H_t and average human capital $\bar{h}_t$ are equal to individual human capital h_t:

$$H_t = \bar{h}_t = h_t. \tag{5.63}$$

Moreover, aggregate transfers are equal to individual transfers:

$$Tr_t = tr_t. \tag{5.64}$$

5.6.2.6 Steady State

In steady state, the aggregate variables Y_t, H_t, K_t, and C_t all grow at the endogenous growth rate γ, and the time allocation of the household is constant, $u_t = u$ and $v_t = v$. To derive these properties, assume that consumption grows at a constant rate γ. As a consequence, substitution of (5.60a) into (5.60b) implies that the interest rate $r_t = r$ is also constant:

$$\frac{1}{\beta}\left(\frac{C_{t+1}}{C_t}\right)^\sigma = \frac{(1+\gamma)^\sigma}{\beta} = \left(1+(1-\tau^K)r-\delta\right).$$

With the help of (5.56b), we can express the steady-state interest rate r as a function of the stationary variable $\tilde{K} \equiv K/H$:

$$r = A_0\alpha\left(\alpha\tilde{K}^{\rho_p} + (1-\alpha)u^{\rho_p}\right)^{\frac{1}{\rho_p}-1}\tilde{K}^{\rho_p-1}. \tag{5.65}$$

Since r is constant in steady state, K/H must also be constant, meaning that K and H grow at the same rate. Similarly, the wage rate per efficiency unit uH is also constant:

$$w = A_0(1-\alpha)\left(\alpha\tilde{K}^{\rho_p} + (1-\alpha)u^{\rho_p}\right)^{\frac{1}{\rho_p}-1} u^{\rho_p-1}. \tag{5.66}$$

Since both H and K grow at the same rate $\gamma^H = \gamma^K$, Y also grows at the same rate $\gamma^Y = \gamma^H = \gamma^K$. Since exogenous government consumption G_t is also assumed to grow at the rate of output $\gamma^G = \gamma^Y$, we derive from the resource constraint that all aggregate variables must grow at the same rate (after dividing (5.62) by H_t):

$$\tilde{Y}_t \equiv \frac{Y_t}{H_t} = (1+\gamma^H)\tilde{K}_{t+1} - (1-\delta)\tilde{K}_t + \tilde{C}_t + \tilde{G}_t,$$

or, in steady state,

$$\tilde{Y} = (\gamma+\delta)\tilde{K} + \tilde{C} + \tilde{G}. \tag{5.67}$$

In sum, the steady state is described by the following nine equations in the nine endogenous variables $\tilde{K}$, $\tilde{Y}$, u, v, $\tilde{C}$, γ, r, w, and τ^L[55]:

$$\frac{(1+\gamma)^\sigma}{\beta} = 1 + (1-\tau^K)r - \delta, \tag{5.68a}$$

$$\gamma = Dv^\varsigma, \tag{5.68b}$$

$$\frac{(1+\gamma)^{\sigma-1}}{\beta} = 1 + u\varsigma Dv^{\varsigma-1}, \tag{5.68c}$$

$$(1-\tau^L)w = \iota\frac{\tilde{C}}{1-u-v}, \tag{5.68d}$$

$$r = A_0\alpha\left(\alpha\tilde{K}^{\rho_p} + (1-\alpha)u^{\rho_p}\right)^{\frac{1}{\rho_p}-1}\tilde{K}^{\rho_p-1}, \tag{5.68e}$$

$$w = A_0(1-\alpha)\left(\alpha\tilde{K}^{\rho_p} + (1-\alpha)u^{\rho_p}\right)^{\frac{1}{\rho_p}-1} u^{\rho_p-1}, \tag{5.68f}$$

$$\tilde{Y} = (\gamma+\delta)\tilde{K} + \tilde{C} + \tilde{G}, \tag{5.68g}$$

$$\tilde{Y} = A_0\left(\alpha\tilde{K}^{\rho_p} + (1-\alpha)u^{\rho_p}\right)^{\frac{1}{\rho_p}}, \tag{5.68h}$$

$$\tilde{G} + \tilde{T}r = \tau^L wu + \tau^K rK. \tag{5.68i}$$

[55]The Lagrange multipliers λ_t and μ_t are transformed into stationary variables by the division by $H_t^{-\sigma}$, $\tilde{\lambda}_t = \lambda_t/H_t^{-\sigma}$ and $\tilde{\mu}_t = \mu_t/H_t^{-\sigma}$.

5.6.2.7 Calibration

We follow Lucas (1990) in selecting most parameters. For the production parameters, we set $\rho_p = -2/3$ (implying a substitution elasticity $\sigma_p = 0.6$) and $\alpha = 0.361$. The steady-state growth rate is set equal to $\gamma = 1.5\%$, and the elasticity of human capital growth with respect to the time spent learning is set equal to $\varsigma = 0.8$. Annual depreciation amounts to $\delta = 8.0\%$. The productivity parameter A_0 is normalized to one.

For the preference parameters, we choose $\beta = 0.96$, and thus, the annual discount rate is equal to 4%. The intertemporal elasticity of substitution is set equal to $1/\sigma = 1/2$. We also assume that the working time amounts to $u = 30\%$. Finally, we set the government share equal to $G/Y = 19\%$, and the two income tax rates are set as $\tau^K = 41\%$ and $\tau^L = 28\%$ as above.

The remaining parameters are calibrated using the equilibrium conditions of the model in steady state. Equation (5.68a) implies the steady-state value of the real interest rate r, which allows us to solve (5.68e) for $\tilde{K}$. The wage per efficiency unit w is implied by (5.68f), while production $\tilde{Y}$ is given by (5.68i). From the resource constraint (5.68g), we obtain $\tilde{C}$. Next, we have to solve (5.68b) and (5.68c) to compute $v = 0.0628$ and $D = 0.137$. From (5.68d), we compute the preference parameter $\iota = 5.327$.[56]

For our calibration, government transfers are equal to 13.1% of GDP, with $\tilde{T}r = 0.0435$, $\tilde{G} = 0.0631$, and $\tilde{Y} = 0.322$ in the benchmark of our economy.

5.6.2.8 Growth Effects

We consider the effect of a change in the capital income tax rate τ^K on the economic growth rate γ. For this purpose, we follow Lucas (1990) and assume that government expenditures relative to human capital are constant across the different tax policy scenarios, $\tilde{G} = 0.0631$ and $\tilde{tr} = 0.0435$.[57] In addition, τ^L adjusts such that the government budget is balanced.

The effects of a change in the capital income tax rate τ^K on the economic growth rate γ are illustrated in Fig. 5.16. For a reduction in the capital income tax rate τ^K from 41% to 0%, the economic growth rate falls by 4.6%, from 1.50% to 1.43%.

What explains this decline in economic growth? To answer this question, consider Fig. 5.17, which presents the endogenous variables labor income tax rate τ^L, working hours u, learning v, capital $\tilde{K}$, output $\tilde{Y}$, and consumption $\tilde{C}$ as functions of the capital income tax rate τ^K. To finance government expenditures

---

[56]The calibration and the computation of the steady states are implemented in the Gauss program *Ch5_lucas.g*.

[57]Grüner and Heer (2000) note that this assumption is not innocuous and favors a policy that does not tax capital. As capital income taxes decrease, labor income taxes must increase, and consequently, human capital declines relative to physical capital and output. Therefore, if G_t is held constant relative to H_t, government expenditures decline relative to GDP, and we compare economies with different sizes of the government sector. In particular, Grüner and Heer (2000) derive that the welfare-maximizing flat rate of capital τ^K increases from 9% to 32% if G/Y is held constant instead. Their result explicitly accounts for transitional dynamics of tax policies where a once-and-for-all change in τ_t^K is announced in period $t = 0$ and the tax rate $\tau_t^K = \tau^K$ is held constant during the transition and in steady state.

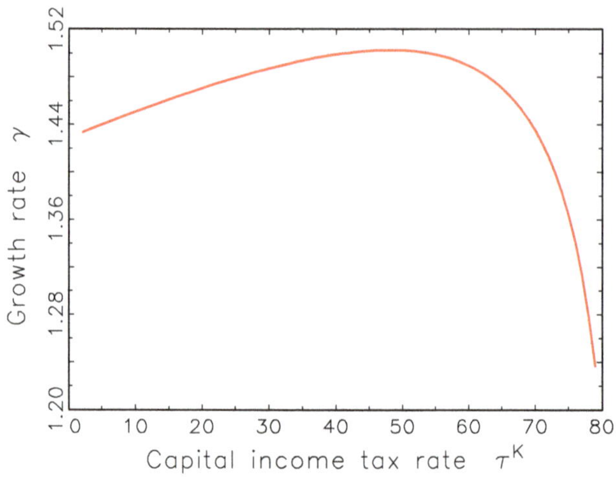

Fig. 5.16 Growth rate effects of capital income taxation τ^K

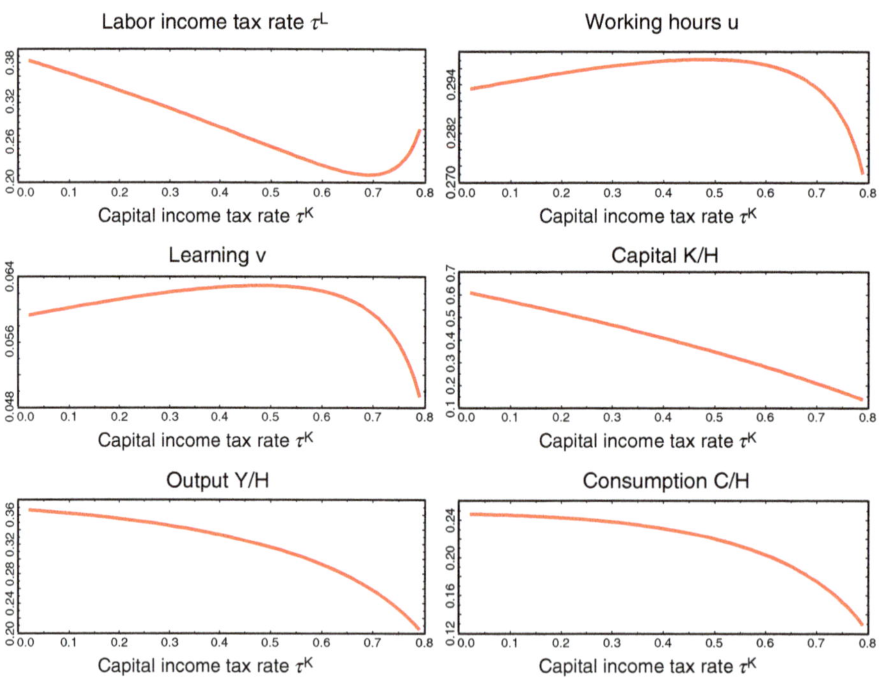

Fig. 5.17 Steady-state effects of capital income taxation τ^K

$\tilde{G}+\tilde{T}r$, the government has to increase labor income taxes from 28.0% to 38.4%. As a consequence, the opportunity costs of leisure decline, and the household increases leisure, $1-u-v$. Therefore, the household reduces both working hours u and learning time v. The opportunity costs of reduced learning time are the forgone wage income in the future due to the lower human capital h_{t+1}. As a consequence, the economic growth rate that is determined by v declines.

In addition, we observe that working hours are hump-shaped and decline again for values of τ^K above 50%. At high levels of the capital income taxes, the decline in savings and, hence, the marginal product of labor and wages is so strong that is offsets the effects of lower labor income tax rates. Net wages decline beyond a capital income tax rate of 50% and, consequently, workers reduce their labor supply. Eventually, the government has to increase labor income taxes at capital income taxes in excess of 70% in order to finance government expenditures. Therefore, growth declines again for high values of the capital income tax rate and the growth rate is a hump-shaped function of the capital income tax rate τ^K, too.

Notice that our steady-state analysis does not allow for a welfare analysis of capital income taxes. In steady state, consumption C_t grows at different rates for the particular tax policies $\{\tau^K, \tau^L\}$, meaning that we cannot simply compare instantaneous utilities. To make the welfare comparison meaningful, one has to compare the entire time paths of consumption and leisure and, hence, the utility, associated with different tax policies. In this vein, Grüner and Heer (2000) consider the benchmark case of the model as the initial state prior to period $t = 0$ and compare different tax policies that announce a once-and-for-all change in the capital income tax rate $\tau_t^K = \tau^K$ starting in period $t = 0$.[58] They compare the discounted lifetime utility associated with these different paths and find that the optimal capital income tax rate is equal to $\tau^K = 9\%$.

D'Erasmo, Mendoza, and Zhang (2016) also consider endogenous utilization of capital in their model. In this case, it is not optimal to fully tax the inelastic source (the initial capital stock) in the presence of endogenous utilization. The distortion in the utilization rate makes capital income taxes more welfare-reducing. Limited depreciation allowances (e.g., for residential capital) actually increase the distortionary effects of capital income taxation, and it is optimal to further decrease capital income taxes.

5.7 The Real Business Cycle Model and Stochastic Taxes

In the following, we introduce a stochastic capital and labor income tax into our RBC models from Sect. 4.4. We will find that stochastic taxes significantly help to improve the modeling of empirical business cycle effects, e.g., the low correlation of wages and employment, which was one of the main puzzles for early RBC studies.

[58] One reason to restrict the analysis to constant capital income tax rates τ_t^K is the time-consistency problem associated with capital income taxation. See also Footnote 33 in this chapter.

5.7.1 Literature

Various articles report that stochastic taxes help to improve the business-cycle properties of RBC models with respect to both the second moments of the time series and the dynamic effects of fiscal policy. In this vein, McGrattan (1994) shows that stochastic taxes help to explain the volatility of output, investment, and hours of work.

Burnside, Eichenbaum, and Fisher (2004) find that the simple stochastic neoclassical model with stochastic government consumption and taxes can qualitatively account for the empirical effects of fiscal policy shocks. To replicate the quantitative properties, they suggest introducing habit formation and capital adjustment costs.[59]

In a more recent contribution, Gomme, Ravikumar, and Rupert (2011) show that the standard RBC model produces a volatility of the return to capital relative to output that is too low and only 50% of the values observed empirically. One of the most promising ingredients of the RBC model to align its second moments of the return to capital data with the asset returns computed from the S&P 500 index is the consideration of stochastic taxes on capital and labor income. They show that the model with a joint stochastic process for total factor productivity, the capital income tax, and the labor income tax can explain nearly 80% of the volatility of the return to capital.

With the emergence of news shock in the more recent literature on business cycle models, shocks affecting future productivity, monetary policy or exogenous stochastic technological progress have become more important. In this regard, Mertens and Ravn (2011) also consider the anticipated effects of tax policy shocks and show that tax cuts can provide stimulus to the economy prior to the tax policy implementation.

5.7.2 The Model

In the following, we extend the RBC model with stochastic government in Sect. 4.4 to include stochastic taxes.

5.7.2.1 Households

We study the behavior of a representative household. For this reason, we assume that households are identical and of measure one. Households are infinitely lived and maximize expected value of intertemporal utility

$$\mathbb{E}_0 \sum_{t=0}^{\infty} \beta^t u(C_t, 1 - L_t), \quad 0 < \beta < 1, \tag{5.69}$$

[59]Recall that we presented a New Keynesian model that specifies habits and capital adjustment costs in Sect. 4.5.2.

where instantaneous utility $u(.,.)$ is discounted by the factor β and described by a function of effective consumption C_t and leisure $1 - L_t$ according to:

$$u(C_t, 1 - L_t) = \frac{\left(C_t^{\iota}(1 - L_t)^{1-\iota}\right)^{1-\sigma}}{1 - \sigma}. \tag{5.70}$$

Effective consumption is presented by the CES aggregator (4.21) of private consumption C_t^p and government consumption G_t:

$$C_t = \left[\phi \left(C_t^p\right)^{1-1/\rho_c} + (1 - \phi)G_t^{1-1/\rho_c}\right]^{\frac{1}{1-1/\rho_c}}.$$

We will also consider the special case with $\phi = 1$ in which government consumption does not affect utility, meaning that $C_t = C_t^p$.

The household owns the capital stock, which evolves according to

$$K_{t+1} = (1 - \delta)K_t + I_t. \tag{5.71}$$

Capital depreciates at rate δ. The household receives income from labor $w_t L_t$ and capital $r_t K_t$, which are taxed at the rates τ_t^L and τ_t^K, respectively, and lump-sum transfers tr_t. Depreciation δK is tax-deductible. The household spends its income on private consumption C_t^p and investment I_t. The budget constraint of the individual is presented by

$$C_t^p + I_t = (1 - \tau_t^L)w_t L_t + (1 - \tau_t^K)r_t K_t + \tau_t^K \delta K_t + tr_t. \tag{5.72}$$

The household maximizes intertemporal utility (5.69) subject to (5.72), resulting in the following first-order conditions:

$$\lambda_t = \iota\phi C_t^{\iota(1-\sigma)-1}(1 - L_t)^{(1-\iota)(1-\sigma)} (\Xi_t)^{\frac{1}{1-1/\rho_c}-1} \left(C_t^p\right)^{-\frac{1}{\rho_c}}, \tag{5.73a}$$

$$\lambda_t(1 - \tau_t^L)w_t = (1 - \iota)C_t^{\iota(1-\sigma)}(1 - L_t)^{(1-\iota)(1-\sigma)-1}, \tag{5.73b}$$

$$\lambda_t = \beta\mathbb{E}_t\left\{\lambda_{t+1}\left[1 + (1 - \tau_{t+1}^K)(r_{t+1} - \delta)\right]\right\}, \tag{5.73c}$$

with Ξ_t as defined in (4.25):

$$\Xi_t \equiv \phi \left(C_t^p\right)^{1-1/\rho_c} + (1 - \phi)G_t^{1-1/\rho_c}.$$

5.7.2.2 Production

Firms are owned by the households and maximize profits with respect to their labor and capital demand. Production Y_t is characterized by constant returns to scale in labor L_t and capital K_t:

$$Y_t = Z_t K_t^{\alpha} L_t^{1-\alpha}. \tag{5.74}$$

Production is also subject to a technology shock Z_t that is governed by the following AR(1) process:

$$\ln Z_t = \rho^Z \ln Z_{t-1} + \epsilon_t^Z, \quad \epsilon_t^Z \sim N\left(0, \sigma^Z\right), \tag{5.75}$$

The individual firm takes Z_t as exogenous.

In a factor market equilibrium, factors are compensated by their marginal products:

$$w_t = (1-\alpha) Z_t K_t^{\alpha} L_t^{-\alpha}, \tag{5.76a}$$

$$r_t = \alpha Z_t K_t^{\alpha-1} L_t^{1-\alpha}. \tag{5.76b}$$

5.7.2.3 Government

The government purchases an amount G_t of the final good in each period t. G_t follows a first-order autoregressive process:

$$\ln G_t = (1-\rho^G)\ln G + \rho^G G_{t-1} + \epsilon_t^G, \quad \epsilon_t^G \sim N(0, \sigma^G), \tag{5.77}$$

where G denotes steady-state government consumption.

The labor and capital income taxes also follow stochastic processes. We follow Gomme, Ravikumar, and Rupert (2011) and assume that the capital income tax rate follows an AR(1) process[60]:

$$\ln \tau_t^K = (1-\rho^K)\ln \tau^K + \rho^K \ln \tau_{t-1}^K + \epsilon_t^K, \quad \epsilon_t^K \sim N(0, \sigma^K), \tag{5.78}$$

while the labor income tax rate can be described by an AR(2) process:

$$\ln \tau_t^L = (1-\rho^{L1}-\rho^{L2})\ln \tau^L + \rho^{L1}\ln \tau_{t-1}^L + \rho^{L2}\ln \tau_{t-2}^L + \epsilon_t^L, \quad \epsilon_t^L \sim N(0, \sigma^L). \tag{5.79}$$

Government expenditures are financed with taxes on labor and capital income. The residual tax revenues that are not spent on government consumption are transferred lump-sum to the households in the amount Tr_t such that the government budget is balanced in each period t:

$$Tr_t = \tau_t^L w_t L_t + \tau_t^K (r_t - \delta) K_t - G_t. \tag{5.80}$$

[60]As one of the first articles in this literature, McGrattan (1994) assumed a VAR process of order 2 in the variables Z_t, G_t, τ_t^K, and τ_t^L. Burnside, Eichenbaum, and Fisher (2004) even use lags of order 50 and 16 for government consumption and the two tax rates, respectively.

5.7.2.4 Competitive Equilibrium

In a competitive equilibrium, (1) households maximize their intertemporal utility, (2) firms maximize profits, (3) the government balances its budget, (4) the sum of individual transfers equals aggregate transfers, and (5) the goods market clears:

$$Y_t = C_t^p + G_t + I_t. \tag{5.81}$$

The last equation can be derived by inserting (5.80) into the individual budget constraint (5.72) and noticing that production is subject to constant returns to scale such that $Y_t = w_t L_t + r_t K_t$.

To summarize, the equilibrium of the economy can be characterized by the following eight equations in the eight variables Y_t,C_t^p,C_t,I_t, L_t, w_t,r_t,λ_t:

$$\lambda_t = \iota\phi C_t^{\iota(1-\sigma)-1}(1-L_t)^{(1-\iota)(1-\sigma)}\,(\Xi_t)^{\frac{1}{1-1/\rho_C}-1}\left(C_t^p\right)^{-\frac{1}{\rho_C}}, \tag{5.82a}$$

$$\lambda_t(1-\tau_t^L)w_t = (1-\iota)C_t^{\iota(1-\sigma)}(1-L_t)^{(1-\iota)(1-\sigma)-1}, \tag{5.82b}$$

$$\lambda_t = \beta\mathbb{E}_t\left\{\lambda_{t+1}\left[1+(1-\tau_{t+1}^K)(r_{t+1}-\delta)\right]\right\}, \tag{5.82c}$$

$$C_t = \left[\phi\left(C_t^p\right)^{1-1/\rho_C} + (1-\phi)G_t^{1-1/\rho_C}\right]^{\frac{1}{1-1/\rho_C}}, \tag{5.82d}$$

$$Y_t = C_t^p + I_t + G_t, \tag{5.82e}$$

$$K_{t+1} = (1-\delta)K_t + I_t, \tag{5.82f}$$

$$w_t = (1-\alpha)Z_t K_t^{\alpha} L_t^{-\alpha}, \tag{5.82g}$$

$$r_t = \alpha Z_t K_t^{\alpha-1} L_t^{1-\alpha}. \tag{5.82h}$$

The four exogenous variables $\{Z_t, G_t, \tau_t^K, \tau_t^L\}$ follow the AR-processes (5.75) and (5.77)–(5.79).

5.7.2.5 Calibration

To compute the model, we need to calibrate the parameters α, δ, β, ι, ϕ, σ, ρ_C, ρ^Z, σ^Z, ρ^G, σ^G, τ^K, ρ^K, σ^K, τ^L, ρ^{L1}, ρ^{L2}, and σ^L. The parameters α, δ, β, σ, ρ^Z, σ^Z, ρ^G, and σ^G are taken from Table 4.3. The preference parameters for the composite consumption good are set to $\phi = 3/4$ and $\rho_C = 0.5$. $\tau^K = 32\%$ and $\tau^L = 28\%$ are chosen as the values of the capital and labor income tax rates for 2008.iv reported in Gomme, Ravikumar, and Rupert (2011). From these authors, we also take the values $\rho^K = 0.9725$, $\rho^{L1} = 0.7841$, and $\rho^{L2} = 0.2047$. Notice that both autoregressive processes for the tax rates are highly persistent.

ι is calibrated such that $L = 0.3$ in steady state. For this reason, notice that in steady state, $Z_t = Z_{t+1} = Z = 1.0$ and

$$r = \frac{1/\beta - 1}{1 - \tau^K} + \delta = 3.99\%.$$

From (5.76b), we can compute the steady-state capital intensity K/L:

$$\frac{K}{L} = \left(\frac{\alpha}{r}\right)^{\frac{1}{1-\alpha}} = 31.15,$$

which implies $K = \frac{K}{L}L = 9.345$ for $L = 0.30$. Therefore, production amounts to $Y = ZK^{\alpha}L^{1-\alpha} = 1.03$. Government consumption is equal to 20% of production, or $G = 0.207$. Steady-state investment follows from (5.71) for $K_{t+1} = K_t = K$ such that $I = \delta K = 0.234$. From the goods market equilibrium (5.81), we obtain $C^p = Y - G - I = 0.594$. Given the values of ρ_c and ϕ, we can compute $C = 0.405$ with the help of (4.21). The equilibrium wage follows with the help of (5.76a), $w = 2.21$. Dividing (5.73b) by (5.73a), we find that

$$(1 - \tau^L)w = \frac{1-\iota}{\iota}\frac{C}{1-L}\phi \Xi^{\frac{1}{1-1/\rho_c}-1}\left(C^p\right)^{-\frac{1}{\rho_c}},$$

which we can solve for $\iota = 0.511$. The solution is implemented in the GAUSS program *Ch5_RBC_stoch_tax.g*, which computes the solution in the form of policy functions for the next-period capital stock $K'(K, Z, G, \tau^K, \tau^L)$, consumption $C^p(K, Z, G, \tau^K, \tau^L)$, labor supply $L(K, Z, G, \tau^K, \tau^L)$, and investment $I(K, Z, G, \tau^K, \tau^L)$ as functions of the state variables K, Z, G, τ^K, and τ^L.

Autoregressions of the HP-filtered components of the (logarithmic) two tax rates τ^K and τ^L, of order 1 and 2, respectively, results in standard deviations of the residual equal to $\sigma^K = 0.027$ and $\sigma^L = 0.028$.

5.7.3 Results

5.7.3.1 Impulse Responses

The impulse responses for the technology shock and the government consumption shock are identical to those in the case with deterministic taxes, as presented in Figs. 4.12 and 4.13, respectively. Therefore, the model is able to replicate the empirical evidence from VAR studies that (1) a positive technology shock increases output, investment, consumption, wages, and the interest rate, and that (2) a positive shock to government consumption increases output, private consumption, employment, and interest rates, while it crowds out investment.

The response of the equilibrium variables to a one-standard-deviation shock (2.7%) to the capital income tax rate τ_t^K (increasing it from 32.00% to 32.86%)

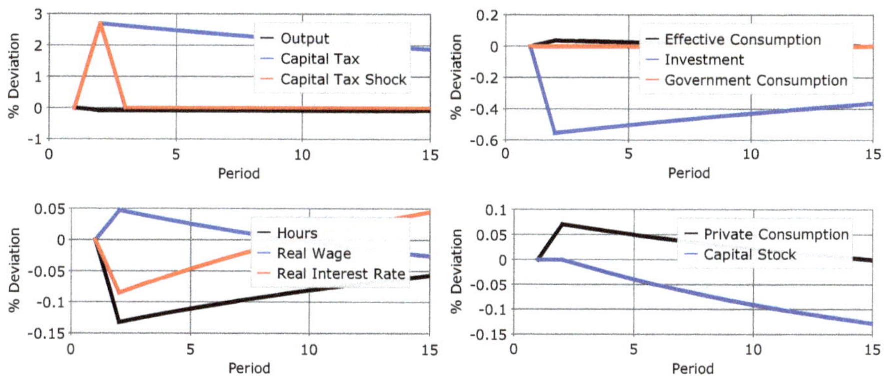

Fig. 5.18 Impulse responses to a capital income tax shock

is presented in Fig. 5.18. Higher capital income taxes reduce investment (see the upper-right panel in Fig. 5.18), and thus, income is shifted from savings into consumption. The total effect on demand is negative, and output declines. The household also reduces its labor supply because its incentives to generate income and save it for future periods is reduced.[61] Since the effect of reduced labor supply dominates the effect of a smaller capital stock, the marginal product of capital (labor) decreases (increases), and therefore, the interest rate (wage) falls (rises).

Figure 5.19 presents the impulse responses of equilibrium variables to a one-standard-deviation shock (2.8%) to the labor income tax rate τ_t^L, increasing it from 28.00% to 28.78%. Since the opportunity costs of leisure are decreased, households work less, and hours decline by 0.7% (the black line in the lower-left panel in Fig. 5.19). As a consequence, both output and income fall, and the household has to reduce both consumption and savings. Therefore, investments also decline. As the quantitative effect of higher labor income taxes on labor is more pronounced than that on capital, the marginal product of labor (equal to the wage w) rises, while the real interest rate r declines.

In summary, the impulse responses of the model variables are in accordance with economic intuition from the AS-AD framework. Labor income taxes primarily reduce the labor supply, while capital income taxes reduce investment. In both cases, output, investment and employment all decline. Private consumption increases in the case of higher capital taxation, while it declines in the case of higher labor

[61] Recall that one basic mechanism in the standard RBC model is the intertemporal substitution of labor. If wages rise, the household increases labor supply in the present period. If the real interest rate increases, the household shifts its working hours from future periods into the present period because the discounted income in future periods from a marginal increase in its labor supply is reduced.

Fig. 5.19 Impulse responses to a labor income tax shock

taxation.[62] In both cases, we also find that the effect on labor supply relative to that on the capital stock dominates and wages rise, while the real interest rate decreases.

5.7.3.2 Explanation of Second Moments

Table 5.5 presents the second moments from the simulation of the model with stochastic taxes. In comparison with the case of deterministic taxes presented in Table 4.4, output, investment and consumption are more volatile. This observation is not surprising because both stochastic taxes introduce more volatility in these variables. In addition, investment is also more volatile relative to output, which is in better accordance with empirical observations. Therefore, our results confirm those found by McGrattan (1994).

Given the impulse responses of labor to shocks to the tax rates, it is not surprising that both tax rates τ^K and τ^L are negatively correlated with labor supply L, exhibiting correlation coefficients equal to -0.15 and -0.63, respectively, while they are positively correlated with the wage rate (not presented). As a consequence, the positive correlation of wages with employment falls to 0.02 (from 0.55 in the case without stochastic taxes) and is in much better accordance with empirical evidence. As presented in Table 2.1, the correlation of wages with output and employment amounted to -0.27 and -0.26 in the US economy during the period 1953–2014.[63] In sum, the introduction of stochastic taxes helps to improve both the volatility and correlation behavior of the model variables relative to empirical observations.

---

[62]The reader is invited to determine how the response of consumption to higher capital taxes depends on the elasticity of substitution between private and public consumption.

[63]In the RBC literature, many studies have analyzed and attempted to replicate the fact that wages and labor productivity are uncorrelated or even negatively correlated with working hours. One of the early studies is by Burnside, Eichenbaum, and Rebelo (1993), who introduces labor hoarding into the standard RBC model.

Table 5.5 Second moments for the RBC model with stochastic taxes

Variable	s_x	s_x/s_Y	r_{xY}	r_{xL}	r_{xG}
Output Y	1.30	1.00	1.00	0.78	0.22
Private consumption C^p	0.58	0.45	0.84	0.85	0.27
Investment I	4.50	3.46	0.95	0.61	−0.05
Hours L	1.00	0.77	0.78	1.00	0.45
Real wage w	0.81	0.62	0.63	0.02	−0.20
Real interest rate r	1.35	1.04	0.96	0.79	0.21
Public consumption G	1.23	0.94	0.22	0.45	1.00
Capital tax τ^K	3.38	2.60	−0.06	−0.15	0.01
Labor tax τ^L	3.06	2.35	−0.35	−0.63	0.00

Notes: s_x: = Standard deviation of the HP-filtered simulated time series x, s_x/s_Y: = standard deviation of the variable x relative to the standard deviation of output Y, r_{xY}: = Cross-correlation of the variable x with output Y, r_{xL}: = Cross-correlation of the variable x with labor L, r_{xG}: = Cross-correlation of the variable x with government consumption G

Appendix 5.1: Derivation of the Chamley-Judd Result

The optimality conditions for the derivation of (5.22) with respect to τ_t^K and τ_t^L are given by (after the replacement of the factor prices r and w by their marginal products F_K and F_L, respectively):

$$\psi_t K_t - \mu_{2t-1} u_{C_t} = 0, \tag{5.83}$$

and

$$\psi_t L_t - \mu_{1t} u_{C_t} = 0. \tag{5.84}$$

The optimality condition with respect to K_t is represented by:

$$\psi_t \left[\tau_t^K \left(F_{K_t} + K_t F_{K_t K_t}\right) + \tau_t^L F_{L_t K_t} L_t\right] + \theta_t \left[F_{K_t} + 1 - \delta\right] - \frac{1}{\beta}\theta_{t-1}$$
$$+ \mu_{1t} u_{C_t} \left[1 - \tau_t^L\right] F_{L_t K_t} + \mu_{2t-1} u_{C_t} \left[1 - \tau_t^K\right] F_{K_t K_t} = 0.$$

Inserting Eqs. (5.83) and (5.84) into the above equation and noticing that for a constant-returns to scale production function, $F_{KK}K + F_{LK}L = 0$, holds,[64] we derive

$$\psi_t \tau_t^K F_{K_t} + \theta_t \left[F_{K_t} + 1 - \delta\right] - \frac{1}{\beta}\theta_{t-1} = 0.$$

In steady state, all variables are constant: $K_t = K$, $C_t = C$, $L_t = L$, $\theta_t = \theta$, and $\psi_t = \psi$:

$$\psi \tau^K F_K + \theta \left[F_K + 1 - \delta\right] - \frac{1}{\beta}\theta = 0. \tag{5.85}$$

In addition, the Euler equation of the household

$$\beta = \frac{1}{1 - \delta + \left(1 - \tau^K\right) F_K}$$

can be substituted into (5.85), yielding:

$$\left[\psi + \theta\right] \tau^K F_K = 0$$

This equation is only fulfilled if $\tau^K = 0$.

Appendix 5.2: Data Sources

In addition to the macroeconomic data presented in Appendices 2.4 and 4.6, we introduce the following variables in our empirical analysis:

- **Tax Revenue** The data for Fig. 5.1 are retrieved from OECD Revenue Statistics 1965–2015: OECD (2018), Tax revenue (indicator). doi: 10.1787/d98b8cf5-en (Accessed on 26 March 2018).
  https://data.oecd.org/tax/tax-revenue.htm.
  The data for Table 5.1 are retrieved from OECD Revenue Statistics 1965–2016 (2017). and OECD (2018), Tax on corporate profits (indicator). doi: 10.1787/d30cc412-en (Accessed on 26 March 2018).

[64] $F_{KK}K + F_{LK}L = 0$ follows from Euler's theorem and the derivation of

$$F(K_t, L_t) = F_{K_t}(K_t, L_t)K_t + F_{L_t}(K_t, L_t)L_t$$

with respect to K_t.

The shares of Italian taxes on goods and services in GDP and total revenues are retrieved from the OECD: OECD (2018), Tax on goods and services (indicator). doi: 10.1787/40b85101-en (Accessed on 26 March 2018).
https://data.oecd.org/tax/tax-on-goods-and-services.htm#indicator-chart.

- **Income Tax Rates** The US income tax schedule presented in Fig. 5.3 is generated with data from the Tax Foundation (Accessed on 26 March 2018).
  http://taxfoundation.org/article/2016-tax-brackets.
  For Germany, the respective data (as presented in Figs. 5.4 and 5.5) are retrieved from the web page of the *Bundesministerium für Finanzen* (Ministry of Finance) (Accessed on 26 March 2018).
  https://www.bmf-steuerrechner.de/ekst/.

Problems

5.1. Recompute the model of Sect. 5.3.2. Instead of (5.5), use the following additive utility function:

$$u(C_t, L_t) = \frac{C_t^{1-\sigma}}{1-\sigma} - \nu_0 \frac{L_t^{1+\frac{1}{\nu_1}}}{1+\frac{1}{\nu_1}},$$

where ν_1 denotes the Frisch elasticity of labor supply. Set the Frisch labor supply elasticity equal to 0.2, $\nu_1 = 0.2$. All other parameters are set as in Sect. 5.3.2.

1. Calibrate ν_0 such that the steady-state labor supply is equal to $L = 0.3$.
2. Compute the welfare effects of an increase in τ^L from 23% to 24% in partial equilibrium and in general equilibrium (for both the steady state and the transition). Are the consumption equivalent changes close to one another and insensitive to the Frisch labor supply elasticity?

5.2. This problem follows Prescott (2004). Assume that instantaneous utility is logarithmic. Lifetime utility is given by

$$U = \sum_{t=0}^{\infty} \beta^t \left[\ln C_t + \iota \ln(1 - L_t)\right].$$

The capital stock accumulates according to

$$K_{t+1} = (1-\delta)K_t + I_t,$$

and production is Cobb-Douglas:

$$Y_t = A_t K_t^\alpha L_t^{1-\alpha}.$$

In goods market equilibrium,

$$Y_t = C_t + G_t + I_t.$$

The household's budget constraint is represented by

$$(1+\tau^C)C_t + (1+\tau^I)I_t = (1-\tau^L)w_t L_t + (1-\tau^K)(r_t-\delta)K_t + \delta K_t + Tr_t,$$

where τ^C, τ^I, and τ^K denote the tax rates on consumption, investment, and capital income, respectively. Capital depreciation is tax-exempt. The government spends tax revenues, $\tau^C C_t + \tau^I I_t + \tau^L w_t L_t + \tau^K(r_t - \delta)K_t$, on government consumption G_t and transfers Tr_t. Assume that $\alpha = 0.36$, $\delta = 0.10$, and $\beta = 0.96$.

1. Derive the first-order conditions of the household and the firm. Notice that the *wedge* on labor income also depends on the consumption tax rate τ^C.
2. Calibrate the parameter ι such that the steady-state labor supply is equal to $L = 0.3$ for $\tau^I = 0$, $\tau^K = 0.42$, $\tau^L = 0.23$, and $\tau^C = 0.26$. τ^C is set such that $\frac{\tau^C+\tau^L}{1+\tau^L} = 0.40$ as given in Table 2 of Prescott (2004) for the US economy during the periods 1970–1974 and 1993–1996. Compare the values of the endogenous variables C, L, K, and Y with those in the the case in which τ^L and τ^C increase to the Italian values of the tax rates, $\tau^C = 0.485$ and $\tau^L = 0.429$.[65] How large is the change in the steady-state labor supply? Does this account for the observation that Americans worked 56% more hours than Italians during this period?

5.3. Show that the Chamley-Judd result also holds for the wealth tax. Use the budget constraint (5.25) to derive the first-order conditions of the household optimization problem:

$$\lambda_t(1-\tau_t^L)w_t = -\frac{\partial u}{\partial L_t} = -u_{L_t}, \tag{5.86a}$$

$$\lambda_t = \frac{\partial u}{\partial C_t} = u_{C_t}, \tag{5.86b}$$

$$\lambda_t = \beta\lambda_{t+1}\left(1 + r_{t+1} - \delta - \tau_{t+1}^V\right). \tag{5.86c}$$

[65] According to Table 2 in Prescott (2004), the tax wedge $\frac{\tau^C+\tau^L}{1+\tau^L}$ amounted to 64% in Italy during the period 1993–1996.

Show that these conditions are equivalent to those presented in (5.19) if

$$\tau_t^K r_t = \tau_t^V,$$

implying that the tax revenues are the same in both cases:

$$\tau^K r K = \tau^V K.$$

5.4. Consider the Laffer curve in Sect. 5.5.

1. Show that the equilibrium conditions (5.38) hold.
2. Recompute the Laffer curves in Fig. 5.14.
3. Compute the sensitivity of the results with respect to a Frisch labor supply elasticity $\nu_l = 0.3$.
4. Recompute Fig. 5.15 for the case in which depreciation is not tax-deductible.

5.5. Show that a replacement of the income tax τ with a consumption tax τ^C increases the economic growth rate in the decentralized economy of the endogenous growth model with the public input good presented in Sect. 5.6.1.

5.6. Productive Government Expenditures as a Stock Variable In contrast to (5.39), assume that production uses public capital K_t^G as an input:

$$Y_t = F(K_t, L_t, G_t) = A L_t^{1-\alpha} K_t^{\alpha} \left(K_t^G\right)^{1-\alpha}. \tag{5.87}$$

Public capital accumulates according to:

$$K_{t+1}^G = (1-\delta) K_t^G + I_t^G.$$

Public investment I_t^G is financed by a production tax:

$$I_t^G = \tau Y_t.$$

Compute (1) the maximum growth rate in the decentralized economy and (2) the Pareto-efficient growth rate.

5.7. Congestion Effects and Productive Government Expenditures (follows Turnovsky 1996) To introduce congestion effects, we distinguish between the capital that is used in the production of the individual firm k and the aggregate capital K. The ratio k/K measures the size of the individual firm relative to the economy. Accordingly, the public expenditures G imply the following service g to the individual firm:

$$g_t = G_t \left(\frac{k_t}{K_t} \right)^{1-\sigma^G},$$

where $\sigma^G \in [0, 1]$ denotes the degree of congestion.

Assume that the production of the individual firm is represented by

$$y_t = A l_t^{1-\alpha} k_t^{\alpha} g_t^{1-\alpha}$$

Derive the equilibrium growth rate in the decentralized economy in which $K_t = k_t$, $l_t = L$, and $g_t = G_t$. In addition, assume that the individual firm does not consider the impact of its investment decision on G_t and K_t.

5.8. Analyze the effects of capital income taxation on the growth rate in the Lucas supply-side model if depreciation can be deducted from capital income taxation. Derive the growth-rate effects of capital income taxation and compare it to the case presented in Sect. 5.6.2.

5.9. Introduce adjustment costs of capital in the RBC model with stochastic taxes presented in Sect. 5.7. Use the specification of the adjustment cost function (4.54) with the parameterization $\zeta = 3.0$. How do adjustment costs affect the results presented in Table 5.5?

References

Arnold, J. (2008). *Do tax structures affect aggregate economic growth? Empirical evidence from a panel of OECD countries* (OECD Economics Department working papers, 643).

Aschauer, D. A. (1989). Is public expenditure productive? *Journal of Monetary Economics, 23*, 177–200.

Barro, R. J. (1990). Government spending in a simple model of endogenous growth. *The Journal of Political Economy, 98*, 103–125.

Barro, R. J., & Sala-i-Martin, X. (2003). *Economic growth* (2nd ed.). Cambridge: MIT Press.

Budría Rodriguez, S., Díaz-Giménez, J., & Quadrini, V. (2002). Updated facts on the U.S. distributions of earnings, income, and wealth. *Federal Reserve Bank of Minneapolis Quarterly Review, 26*, 2–35.

Burnside, C., Eichenbaum, M., & Rebelo, S. (1993). Labor hoarding and the business cycle. *Journal of Political Economy, 110*, 245–273.

Burnside, C., Eichenbaum, M., & Fisher, J. D. M. (2004). Fiscal shocks and their consequences. *Journal of Economic Theory, 115*, 89–117.

Chakraborty, I., Holter, H. A., & Stepanchuk, S. (2015). Marriage stability, taxation and aggregate labor supply in U.S. vs. Europe. *Journal of Monetary Economics, 72*(C), 1–20.

Chamley, C. (1986). Optimal taxation of capital income in general equilibrium with infinite lives. *Econometrica, 54*, 607–622.

Chetty, R., Guren, A., Manoli, D., & Weber, A. (2011). Are micro and macro labor supply elasticities consistent? A review of evidence on the intensive and extensive margins. *American Economic Review, 101*(1), 471–475.

Conesa, J. C., & Krueger, D. (2006). On the optimal progressivity of the income tax code. *Journal of Monetary Economics, 53*, 1425–1450.
D'Erasmo, P., Mendoza, E. G., & Zhang, J. (2016). What is a sustainable public debt? In J. B. Taylor & H. Uhlig (Eds.), *Handbook of the macroeconomics* (pp. 2493–2597). Amsterdam: Elsevier.
Di Sanzo, S., Bella, M., & Graziano, G. (2017). Tax structure and economic growth: A panel cointegrated VAR analysis. *Italian Economic Review, 3*, 239–253.
Domeij, D., & Heathcote, J. (2004). On the distribution effects of reducing capital taxes. *International Economic Review, 45*(2), 523–554.
Easterly, W., & Rebelo, S. (1993). Fiscal policy and economic growth. *Journal of Monetary Economics, 32*, 417–458.
Fischer, S. (1980). Dynamic inconsistency, cooperation and the benevolent dissembling government. *Journal of Economic Dynamics and Control, 2*, 93–107.
Gomme, P., Ravikumar, B., & Rupert, P. (2011). The return to capital and the business cycle. *Review of Economic Dynamics, 14*, 262–278.
Grüner, H.-P., & Heer, B. (2000). Optimal flat-rate taxes on capital – A re-examination of Lucas' supply-side model. *Oxford Economic Papers, 52*, 289–305.
Heer, B., & Trede, M. (2003). Efficiency and distribution effects of a revenue-neutral income tax reform. *Journal of Macroeconomics, 25*, 87–107.
Irmen, A., & Kuehnel, J. (2009). Productive government expenditure and economic growth. *Journal of Economic Surveys, 23*(4), 692–733.
Judd, K. L. (1985). Redistributive taxation in a simple perfect-foresight model. *Journal of Public Economics, 28*, 59–83.
Keuschnigg, C. (2005). *Öffentliche Finanzen: Einnahmenpolitik*. Tübingen: Mohr Siebeck.
Kocherlakota, N. R. (2010). *The new dynamic public finance*. Princeton/Oxford: Princeton University Press.
Kydland, F. E., & Prescott, E. C. (1977). Rules rather than discretion: The inconsistency of optimal plans. *Journal of Political Economy, 85*, 185–200.
Levine, R., & Renelt, D. (1992). A sensitivity analysis of cross-country growth models. *American Economic Review, 82*, 942–963.
Lucas, R. E. (1990). Supply-side economics: An analytical review. *Oxford Economic Papers, 42*, 3–42.
McGrattan, E. R. (1994). The macroeconomic effects of distortionary taxation. *Journal of Monetary Economics, 33*, 573–601.
Mendoza, E. G., Milesi-Ferretti, G. M., & Asea, P. (1997). On the ineffectiveness of tax policy in altering long-run growth: Harberger's superneutrality conjecture. *Journal of Public Economics, 66*, 99–126.
Mertens, K., & Ravn, M. O. (2011). Understanding the aggregate effects of anticipated and unanticipated tax policy shocks. *Review of Economic Dynamics, 14*, 27–54.
Prescott, E. C. (2004). Why do Americans work so much more than Europeans? *Federal Reserve Bank of Minneapolis Quarterly Review, 28*, 2–13.
Trabandt, M., & Uhlig, H. (2011). The Laffer curve revisited. *Journal of Monetary Economics, 58*, 305–327.
Turnovsky, S. J. (1996). Optimal tax, debt, and expenditures policies in a growing economy. *Journal of Public Economics, 60*, 21–44.
Wickens, M. R. (2011). *Macroeconomic theory – A dynamic general equilibrium approach* (2nd ed.). Princeton/Oxford: Princeton University Press.

Part III

Social Security, Demographics, and Debt

Pensions 6

6.1 Introduction

In this chapter, we first review empirical facts of public pension systems in OECD countries. Subsequently, we introduce a public pension system in the standard two-period overlapping generations (OLG) model of Chap. 2. We consider two different social security systems, pay-as-you-go (PAYG) versus fully funded. While a fully funded pension system does not have any effect on aggregate savings if capital markets are perfect, aggregate savings fall significantly in a PAYG system. Since public pensions are likely to distort household labor supply decisions, we endogenize labor supply below. In addition, we extend the two-period model to a more realistic 70-period model in which the retirement period is smaller than the working period. Next, we derive the optimal amount of pensions in a PAYG system and study how the demographic transition and aging of the population affect the sustainability of social security. We also discuss the findings of the literature on quantitative pension studies in detail. Finally, we introduce the concept of fiscal space and point out its sensitivity with respect to the aging that takes place in many industrialized countries at present.

6.2 Empirical Regularities

The modern state public pension system began in Germany. In 1889, the national parliament (*the Reichstag*) enacted Otto von Bismarck's social legislation in the form of the *Old Age and Disability Insurance Bill*. The pension system was devised

B. Heer, *Public Economics*, Springer Texts in Business and Economics,
https://doi.org/10.1007/978-3-030-00989-2_6

as a PAYG system in which the contributions of the young are used to finance the pension payments to the old.[1]

At the time of its introduction, workers in Germany could retire at age 70.[2] Average life expectancy at this time amounted to 45 years. However, we have to take care in interpreting this number and using it as a parameter for evaluating the burden on the pension system because infant mortality was high at the end of the nineteenth century. Therefore, the life expectancy of a worker entering insurance, which amounted to 70 years, is more useful for understanding the financing needs of the public pension system. The contribution rate to the public pension system in Germany in 1889 amounted to a modest 1.7% of the gross wage income.

A critical number for the solvency and viability of a public pension system is the old-age dependency ratio (sometimes also called the aged dependency ratio), which is defined as the ratio of the number of people aged 65 and over to the number of working-age individuals aged 25–64.[3] In other words, the ratio expresses the number of young people (contributors) who have to finance the pension payments to the number of elderly people (pensioners).[4] The evolution of the old-age dependency ratio over time and its forecast for selected countries are presented in Fig. 6.1.[5] While the old-age dependency ratio approximately doubles for the US between 2000 and 2050, the number of retirees to workers increases even more significantly for Japan, Italy, Germany, and China. For instance, in the case of Japan, the ratio amounted to 30% in 2000 and 50% in 2015 and is forecasted to increase to 86% by 2050.

When inspecting Fig. 6.1, one needs to bear in mind that the accuracy of the dependency ratio forecasts declines with the forecast horizon. In Fig. 6.1, we used the numbers of the old-age dependency ratio (OADR) that are associated with the medium-fertility scenario of the UN projection. The United Nations uses Bayesian hierarchical models for their official population projections, which are described in greater detail in Alkema, Raftery, Gerland, Clark, Pelletier, Buettner, and Heilig (2011), Raftery, Li, Sevčíková, Gerland, and Heilig (2012), Raftery, Chunn, Gerland, and Sevčíková (2013), Gerland, Raftery, Sevčíková, Li, Gu, Spoorenberg, Alkema, Fosdick, Chunn, Lalic, Bay, Buettner, Heilig, and Wilmoth (2014) and UN

[1]In the United Kingdom, the modern pension system was introduced in 1908 with the help of the *Old Age Pensions Act*. The United States initially only provided pensions for federal employees under the *Civil Service Retirement System* in 1920.

[2]In 1916, the retirement age was reduced to 65 years.

[3]Sometimes, the working age is defined over a larger age range, e.g., 15–64, 20–64, or 20–69 years. Accordingly, take care when comparing old-age dependency ratios from different sources.

[4]In this book, we will sometimes distinguish the two old-age dependency ratios OADR2 and OADR3, which refer to the retirees aged 65 (and above) and those aged 70 (and above). While we primarily use OADR2 as our reference value, it may sometimes be important to consider the evolution of OADR3, e.g., when we study pension policies that extend the retirement age to 70 years.

[5]The data are taken from the UN population division and refer to the medium-fertility variant. See Appendix 6.3 for a more detailed description of the data.

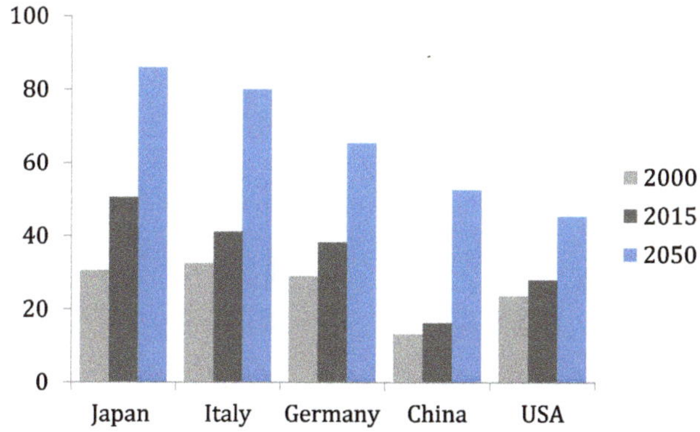

Fig. 6.1 Old-age dependency ratios in 2000, 2015, and 2050

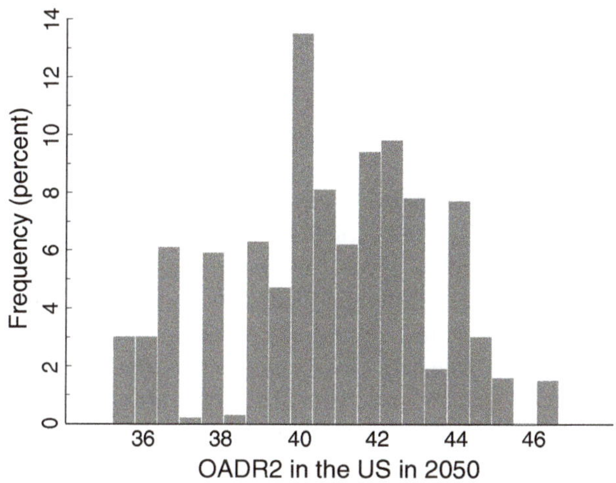

Fig. 6.2 Frequency distribution of the OADR2 in the US in 2050

(2015). In Fig. 6.2, we present the frequency distribution of the old-age dependency ratio (OADR2) for the United States in the year 2050 that results from this Bayesian model.[6] Notice that the distribution of the old-age dependency ratio has a standard deviation of 2.5 percentage points and is not shaped like the normal distribution.[7] A 90% confidence interval of the OADR2 for the year 2050 lies in the range 36.3%–

[6]Thanks to Hana Ševčíková for providing the data on dependency ratio forecasts in the United States and the EU14 countries.

[7]For the US economy in the year 2050, standard empirical distribution tests such as the Lilliefors, Cramer-von Mises, Anderson-Darling, and Watson tests reject the normality assumption of the OADR2 distribution at the 1% or 5% level of significance. This observation does not hold uniquely for all countries in our data sample (the US and 14 EU-countries).

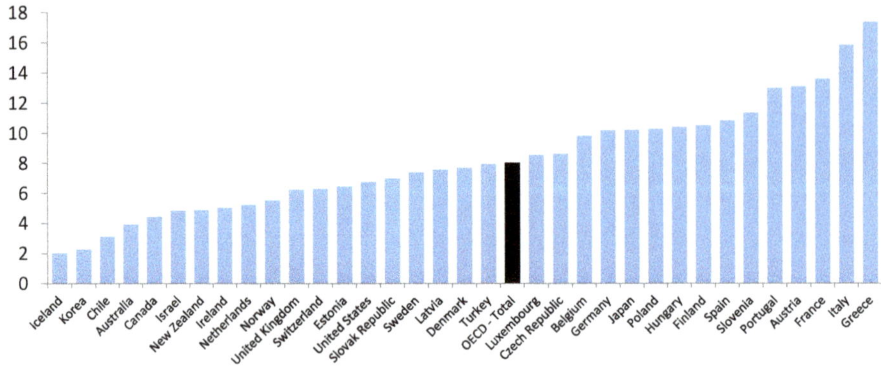

Fig. 6.3 Pension spending, OECD, 2012, as a percentage of GDP

44.5%. By the year 2100, the standard deviation of the old-age dependency ratio in the US increases to 7.1 percentage points.

Population aging, i.e., longer lifetimes and lower fertility rates, poses the most serious challenge to modern public pension systems.[8] Public pension expenditures already account for a large share of GDP, averaging 7.8% for the OECD countries in 2012 (see Fig. 6.3). Public pension payments vary considerably across OECD countries. In 2012, while they only constituted 4.5% and 6.8% of GDP in Canada and the US, they amounted to 13.6% in France and 15.9% in Italy. The large share of social security expenditures in GDP reflects the high generosity of most modern public pension systems. Figure 6.4 presents the gross replacement rate of pensions for men in selected OECD countries, which is defined as gross pension entitlement divided by gross pre-retirement earnings. The OECD average replacement rate amounts to 52.9% and ranges between 22.1% in the UK and 96.9% in the Netherlands.

When the old-age dependency ratio increases, pension expenditures will continue to rise, ceteris paribus. There are various ways to reform the public pension system to meet the challenges of aging societies: The government can (1) increase the contribution rates, (2) reduce pension payments, and/or (3) increase the retirement age. In addition, the social security system may use debt financing during the transition when adjustment costs for the respective cohorts are largest.[9]

In recent decades, the public pension systems of individual countries have predominantly responded to the demographic challenge by increasing contribution

[8]An excellent non-technical survey of pension designs to meet the demographic challenge is provided by Barr and Diamond (2008).

[9]There are many other public policies that may help to alleviate the pressure on public pension systems during the demographic transition that we do not study in here. Among others, the government could attempt to increase the labor force participation rate, encourage higher immigration (of young individuals), or to use family policies to increase the fertility rate.

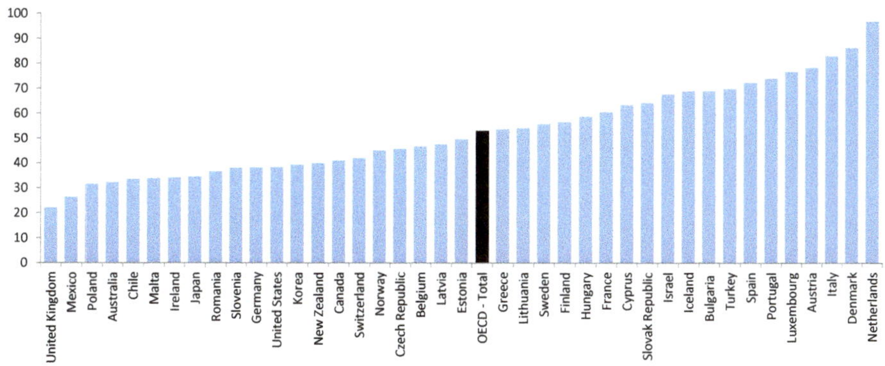

Fig. 6.4 Gross pension replacement rates, men, percentage of pre-retirement earnings, OECD, 2016

rates.[10] For example, the contribution rate to the German public pension system increased from 11.0% in 1955 to 17.0% in 1970 and 18.9% in 2013.[11] If pension benefit levels remain at their present levels, contribution rates are estimated to exceed 40% in Germany by 2030.[12] Similar developments are expected to take place in many other industrialized countries, as presented in Table 6.1. For example, the dynamics of pension payments relative to GDP in France are comparable to those in Germany, while in Italy, the contribution rate will have to increase to levels of 60% and above to finance pensions in 2030 (assuming that the levels of pension entitlements are not reduced and financed by higher contributions rather than debt). For the US economy, De Nardi, Imrohoroğlu, and Sargent (1999) find that, if the government were to use a tax on labor income to finance the additional fiscal burden from the pension system, the tax rate would have to increase by 29.8 percentage points between 2000 and 2060.[13]

[10]However, in recent years, as the baby boomers have begun to retire, many countries have also started to reduce pension benefits, e.g., Italy in 1995 through the Dini reform.

[11]In Germany, the contribution was evenly split between the employer and employee in 2013. Each side pays a contribution rate of 9.45%. In a Walrasian labor market, the legal incidence, i.e., who pays the contribution rate, does not affect the economic incidence, i.e., who actually bears the burden of the tax. If, however, the labor market is subject to a distortion such as a minimum gross wage, the legal incidence does affect the economic incidence.

[12]Börsch-Sopan and Winter (2001) estimate that the pension payments in Germany will increase from 10.0% of GDP in 1995 to 18.4% of GDP in 2030. To finance these higher expenditures resulting from an aging population, the contribution rate will have to increase to 41.1% by 2030.

[13]The authors study a variety of pension reform proposals for the US economy in a general equilibrium model with labor-augmenting technical progress, endogenous bequests, and endogenous labor supply similar to the model we will study in this chapter. They consider different scenarios for the reform of the public pension system. The number reported in the text above corresponds to their experiment 1.

Table 6.1 Pension payments and contribution rates

Country	1995 (%)	2030 (%)
France		
Pensions[a]	12.5	19.4
Contribution rates[b]	24.3	37.7
Germany		
Pensions[a]	10.0	18.4
Contribution rates[b]	22.6	41.1
Italy		
Pensions[a]	16.0	23.3
Contribution rates[b]	4.6	61.9

Notes: Source: Börsch-Sopan and Winter (2001)
[a]Pension payments relative to GDP
[b]If additional pensions are exclusively financed by higher contributions

In addition to changes affecting revenues and expenditures, national social security systems have also adjusted to the demographic transition by reforming the incentives of contributors. Many systems, e.g., the Italian pension system through the 1995 Dini reform, have transformed their benefit entitlements from defined benefits to defined contributions. In a defined benefit system, the pension payment is usually tied to the wage income during the last working period, while in a defined contribution system, the pension payment is related to the accumulated pension contributions over the working life. In Germany, for example, the two systems coexist. Public employees belong to a separate defined benefit social security system. Their pensions depend on the number of contribution years and the wage income in the last year. For example, if you have contributed to the pension system for 40 years, you receive a pension that amounts to 71% of your last gross wage income (independent of your income history). All other (non-public) employees in Germany belong to a separate defined contribution public pension system. For these employees, the pension depends on the accumulated sum of all contributions (without any discounting of contributions in later years), and the pension entitlement is computed as a (non-linear) function of the contributions.[14] Of course, a defined benefit system has much stronger labor supply disincentives than a defined contribution system, and we will quantitatively explore this issue in this chapter.

[14]There are other institutional details of the German social insurance system that we will neglect in the following, e.g., there is a social security contribution ceiling, meaning that, in 2013, contributions were only paid on a gross annual income up to an amount of €69.600 (West Germany) or €58.800 (East Germany). In addition, the number of contribution years enters the formula used to compute the pensions. For further details on the German pension system, see Fehr (1999).

Besides the PAYG pension system outlined above, in which current expenditures are financed by current revenues, some countries such as Switzerland and Chile have (additional) mandatory individual accounts where the savings are invested in financial markets and paid out in full as they come due.[15] These systems are called *fully funded*. The accumulated savings are usually paid out as an annuity during retirement. The advantages of fully funded versus PAYG systems will be considered analytically and quantitatively in the next Sect. 6.3. In particular, we will find that the PAYG systems dramatically reduce aggregate savings (public and private), while fully funded systems do not. In addition, the two systems are subject to two different types of risk, financial market risk on the one hand and demographic risk on the other hand.[16] Moreover, the transition from a PAYG system, which is now the most common in OECD countries, to a fully funded system implies a huge burden on the current young generation, which has to finance both the pensions of the current elderly and its own pensions.

In Sect. 6.4, we study PAYG systems and the optimal contribution rate (or, equivalently, level of pensions). For this reason, we will consider the extension of the standard two-period model introduced in Sect. 3.2 to a multi-period framework. In particular, we will present the model of Auerbach and Kotlikoff (1987) with 70 overlapping generations. As a consequence, we are much better able to model the demographic structure of the population and, thus, derive much more accurate predictions of the efficiency and distributional effects of pension reform proposals.

6.3 Fully Funded Versus Pay-As-You-Go Public Pension Systems

In the following, we will contrast a fully funded public pension system with a PAYG system. For this reason, we extend the two-period OLG model of Sect. 3.2 for a government that collects lump-sum contributions d_t. The government either invests d_t in the capital market or pays these contributions out to current pensioners. Again, the young population consists of N_t households and grows at rate n, $N_{t+1} = (1+n)N_t$. A special focus of our analysis will be the effects of the two systems on savings.

---

[15]In Switzerland, for example, the fully funded pension system is part of a three-part pension system. The other two parts are a mandatory PAYG public pension and an employer-based pension.

[16]From the theory of finance, we know that it is optimal to diversify risk, which implies that pension systems should combine the two different systems as is done, for example, in the three-part public pension system in Switzerland.

6.3.1 OLG Model with Fully Funded Pensions

In a fully funded system, the young generation pays the contribution d_t to the government, and the government invests d_t in the capital market. In old age, the government repays the contribution plus interest, $(1 + r_{t+1})d_t$, to the households.[17] Accordingly, the budget constraints of the household in young age and old age are represented by

$$w_t = c_t^1 + s_t + d_t, \tag{6.1a}$$

$$c_{t+1}^2 = (1 + r_{t+1})(d_t + s_t), \tag{6.1b}$$

where, again, c_t^i denotes consumption in young ($i = 1$) and old ($i = 2$) age, s_t denotes savings, w_t the wage rate, and r_t the real interest rate in period t. Notice that the government and the household are assumed to receive the same rate of return r_t in the capital market. Labor supply is assumed to be inelastic and is normalized to 1, meaning that the wage rate w_t is also equal to the wage income of the young household.

Household preferences are described as in Sect. 3.2 by the time-separable lifetime utility function (3.3), which we restate for the reader's convenience:

$$U_t = u(c_t^1) + \beta u(c_{t+1}^2), \tag{6.2}$$

where β denotes the discount factor, and the standard assumptions with respect to the utility function apply; in particular, instantaneous utility $u(.)$ is a concave function.

The first-order condition, the so-called *Euler equation*, is identical to (3.7) in the OLG model without public pensions:

$$u'(c_t^1) = \beta(1 + r_{t+1})u'(c_{t+1}^2). \tag{6.3}$$

Substituting the household budget constraint (6.1) into the Euler equation (6.3) and noticing that, in the capital market equilibrium, total private and public savings, $(s_t + d_t)N_t$, at the end of period t are equal to the beginning-of-period capital stock K_{t+1} we derive the following equilibrium conditions:

$$u'\left(w_t - (s_t + d_t)\right) = \beta(1 + r_{t+1})u'\left((1 + r_{t+1})(s_t + d_t)\right), \tag{6.4a}$$

$$s_t + d_t = (1 + n)k_{t+1}. \tag{6.4b}$$

The production sector of the economy is modeled as in Sect. 3.2.3. In particular, production is characterized by constant returns to scale, and firms are competitive,

[17] All payments are made at the end of the period.

meaning that the factor prices are equal to their marginal products, as presented in Eq. (3.12).

Therefore, to compare the economy without and with a fully funded public pension system, we need only consider (6.4) for the two cases $d_t = 0$ and $d_t > 0$. All other equilibrium conditions are identical. As a consequence, the young household chooses exactly the same intertemporal consumption profile $\{c_t^1, c_{t+1}^2\}$ in the two cases. If $\tilde{s}_t$ denotes the household's optimal private savings in the case without a public pension system, it simply saves $s_t = \tilde{s}_t - d_t$ in the case of a fully funded system with $d_t > 0$. As a consequence, aggregate savings (private plus public) are identical in the two models, and therefore, (6.4b) implies that the dynamics of the capital stock per capita k_t are also identical. As a result, we find that a fully funded social security system does not have any effect on total savings and capital accumulation, equilibrium factor prices, or the lifetime utility of the household. In the following, we discuss the sensitivity of this result to two assumptions, perfect capital markets and lump-sum contributions.

6.3.1.1 Imperfect Capital Markets

To derive the previous result, we assumed that capital markets are perfect. This assumption might not be very realistic. First, we assumed that debtors and creditors pay or receive the same interest rate r_t. Usually, this is not the case, and banks demand a default premium from the debtors. Assume that without a public pension system, $d_t = 0$, the private households are creditors, $s_t > 0$. If a fully funded pension system with $d_t > s_t$ is introduced, private savings might fall below zero, and the household then becomes a debtor. If the interest rates were different for creditors and debtors, the interest rate in (6.3) would change, and therefore, the optimal intertemporal consumption profile $\{c_t^1, c_{t+1}^2\}$. Things get even uglier if the household were credit-constrained $s_t \geq 0$, making it unable to receive a loan. If the constraint is binding, $s_t = 0$, a higher contribution d_t to the public pension system increases aggregate savings. Of course, the lifetime utility of the household would decrease.

6.3.1.2 Social Security Tax on Wage Income and Elastic Labor

In modern pension systems, contributions are mostly levied upon the individual (wage) income.[18] To study the possible distortionary effects of a social security tax on wage income, let us consider the case of elastic labor supply. For this reason, assume that lifetime utility is given by the following time-separable function:

$$U(c_t^1, c_{t+1}^2, l_t) = u^y(c_t^1, l_t) + u^o(c_{t+1}^2), \tag{6.5}$$

where u^y and u^o denote the instantaneous utility when young and old. The young household is working l_t hours. Total endowment over time is normalized to one such that leisure is given by $1 - l_t$. Social security contributions d_t are levied on wage

[18] As one exception, the first pillar of the Swiss pension system is also financed by sources other than wage income, e.g., by a levy on the income of the self-employed.

income $w_t l_t$ at rate τ, and thus, the budget constraints in young and old age are given by

$$(1-\tau)w_t l_t = c_t^1 + s_t, \tag{6.6a}$$

$$c_{t+1}^2 = (1+r_{t+1})(d_t + s_t), \tag{6.6b}$$

$$d_t = \tau w_t l_t. \tag{6.6c}$$

Inserting (6.6b) and (6.6c) into (6.6a), we derive the intertemporal budget constraint of the household:

$$w_t l_t = c_t^1 + \frac{c_{t+1}^2}{1+r_{t+1}}, \tag{6.7}$$

which does not depend on τ. As a consequence, maximizing (6.5) with respect to the intertemporal budget constraint (6.7) implies the Euler equation and the first-order condition with respect to labor supply

$$\frac{\partial u^y(c_t^1, l_t)}{\partial c_t^1} = \beta(1+r_{t+1})\frac{\partial u^o(c_{t+1}^2)}{\partial c_{t+1}^2}, \tag{6.8a}$$

$$-\frac{\partial u^y(c_t^1, l_t)}{\partial l_t} = w_t, \tag{6.8b}$$

that also do not depend on τ. All other equilibrium conditions, i.e., the first-order conditions of the firms and the capital market equilibrium condition (see Sect. 3.2) are unchanged and do not depend on τ either:

$$w_t = \frac{\partial F(K_t, L_t)}{\partial L_t} = f(k_t) - k_t f'(k_t), \tag{6.9a}$$

$$r_t = \frac{\partial F(K_t, L_t)}{\partial K_t} = f'(k_t), \tag{6.9b}$$

$$(1+n)k_{t+1} = s_t + d_t = (1-\tau)w_t l_t - c_t^1 + \tau w_t l_t = w_t l_t - c_t^1, \tag{6.9c}$$

with $L_t \equiv N_t l_t$. Since all equilibrium conditions, (6.7), (6.8), and (6.9), are independent of τ, we have shown that a social security contribution that is levied proportional to wage income does not affect the equilibrium allocation in a fully funded public pension system either.

6.3.2 OLG Model with a PAYG Pension

In this section, we introduce a PAYG public pension system in the two-period OLG model. We will study the quantitative effects of the PAYG system on capital and output and will find that public pensions dramatically reduce aggregate savings and output. For example, in the case of inelastic labor supply, the capital stock falls by 64% if a contribution rate of 30% is imposed on wage income!

We will study the sensitivity of this result with respect to various modifications in the next sections. In addition to endogenous labor, we will also consider the effects of pension savings accounts where future pensions are related to past contributions, and we will introduce growth, which might help to improve the return from a PAYG system. If future generations have higher wage income due to productivity growth, a transfer from the young to the old in the future will imply higher pensions relative to one own's contributions. In addition, we also derive the welfare implications of the introduction of a PAYG system. We will find that substantial welfare losses are associated with a public pension system irrespective of whether we assume elastic or inelastic labor or a defined contribution pension system. Only growth helps, to some extent, to reduce the welfare losses from a PAYG system.

6.3.2.1 The Model

In the PAYG system, each of the N_t young households pays a contribution d_t to the social security system, while each of the N_{t-1} old households receives a pension pen_t from it. We assume that the budget of the social security system is balanced:

$$N_t d_t = N_{t-1} pen_t.$$

If the (young) population grows at rate n, $N_t = (1+n)N_{t-1}$, the social security budget constraint can be expressed in stationary variables:

$$pen_t = N_t d_t / N_{t-1} = (1+n) d_t. \tag{6.10}$$

Social security is a pure transfer scheme, and the 'rate of return' is related to the population growth rate n. For constant contributions, $d_{t-1} = d_t = d$, pensions are given by $pen_t = (1+n)d$, meaning that $n = \frac{pen_t - d_{t-1}}{d_{t-1}}$ is also the rate of return for the individual who 'invests' the amount d into the PAYG system.

The budget constraints of the young and old households are given by:

$$w_t = c_t^1 + s_t + d_t, \tag{6.11a}$$

$$c_{t+1}^2 = (1+r_{t+1}) s_t + pen_{t+1}, \tag{6.11b}$$

implying the following intertemporal budget constraint (after solving the old household's budget constraint for s_t and substituting it into the young household's budget constraint):

$$c_t^1 + \frac{c_{t+1}^2}{1+r_{t+1}} = w_t - d_t + \frac{pen_{t+1}}{1+r_{t+1}}. \tag{6.12}$$

Households maximize lifetime utility (6.2) subject to (6.12). Differentiating the Lagrange function

$$\mathscr{L} = u(c_t^1) + \beta u(c_{t+1}^2) + \lambda \left[w_t - d_t + \frac{pen_{t+1}}{1+r_{t+1}} - c_t^1 - \frac{c_{t+1}^2}{1+r_{t+1}} \right]$$

with respect to c_t^1 and c_{t+1}^2 results in the following first-order condition:

$$u'\left(c_t^1\right) = \beta(1+r_{t+1})u'\left(c_{t+1}^2\right). \tag{6.13}$$

If we substitute the intertemporal budget constraint (6.12) and the social security budget (6.10) constraint into the Euler equation (6.13), we derive the equilibrium condition:

$$u'\left(w_t - s_t - d_t\right) = \beta(1+r_{t+1})u'\Big((1+r_{t+1})s_t + (1+n)d_{t+1}\Big). \tag{6.14}$$

To understand how the PAYG pension system affects private savings, let us assume that the wage and the interest rates are given (implying $r_{t+1} = r$) and that social security contributions are changed by $dd_t = dd_{t+1}$ in period t and $t+1$ (changes are expressed with respect to the steady state value d so that $d_t = d + dd_t$ is also equal to $d_{t+1} = d + dd_{t+1}$). The total differential of the above equation results in[19]:

$$\frac{\partial s_t}{\partial d_t} = -\frac{u_1'' + \beta(1+n)(1+r)u_2''}{u_1'' + \beta(1+r)^2 u_2''} < 0,$$

with $u_1'' \equiv \frac{\partial^2 u(c_t^1)}{\left(\partial c_t^1\right)^2}$ and $u_2'' \equiv \frac{\partial^2 u(c_{t+1}^2)}{\left(\partial c_{t+1}^2\right)^2}$. From this equation, we derive that, in general, savings are reduced for higher contributions and pensions because $u'' < 0$.

Moreover, we observe in most industrialized countries that the population growth rate n is low and smaller than the real interest rate r, $n < r$. Therefore, lifetime income, as represented by the right-hand side of (6.12), reduces to the following term (assuming constant social security contributions, $d_t = d_{t+1} = d$, and dropping time indices for convenience):

$$w + \frac{n-r}{1+r}d.$$

Therefore, lifetime income declines with higher contributions d for $n < r$, and savings decrease unanimously if consumption in old age c_{t+1}^2 is a normal good.[20]

[19]This argument is taken from Chapter 3.2 in Blanchard and Fischer (1989).

[20]Recall that $(1+r)s_t = c_{t+1}^2 - (1+n)d$.

Of course, in the derivation of this partial equilibrium effect, we have ignored general equilibrium effects. The decrease in savings s_t reduces the capital stock k_{t+1} in the capital market equilibrium, meaning that wages fall and interest rates rise. The fall in wages decreases savings further because lifetime income is reduced.[21] The effect of higher interest rates on savings, however, is ambiguous. If the substitution effect (savings are more attractive because of higher interest rates) dominates the income effect (discounted lifetime income decreases), savings will actually increase. The total effect, therefore, can only be derived by considering the effect of higher contributions d in general equilibrium. For this reason, consider the capital market equilibrium:

$$(1+n)k_{t+1} = s\left[w_t(k_t), r_{t+1}(k_{t+1}), d_t\right].$$

Differentiating this equation with respect to d_t for a given k_t, we derive for $s_r > 0$:

$$\frac{dk_{t+1}}{dd_t} = \frac{\frac{\partial s_t}{\partial d_t}}{1+n-s_r f''(k_{t+1})} < 0. \tag{6.15}$$

Recall that, in Sect. 3.2.5, we showed that the two-period OLG model is stable if $s_r > 0$.[22] As a consequence, the derivative is smaller than zero because $\frac{\partial s_t}{\partial d_t} < 0$.

As you learned in Sect. 3.3.2, the market equilibrium in the OLG model can be Pareto-inefficient if $f'(k) < n$. In this case, the households accumulate excessive capital, and a PAYG system that reduces savings and capital accumulation can improve welfare. As we argued in Sect. 3.3.2, the condition $f'(k) < n$ is rarely fulfilled in practice. In addition, we need to add a word of caution at this point. Thus far, we have assumed that labor supply is inelastic and that contributions are lump-sum. If labor supply is endogenous and contributions are levied on wage income, the PAYG system distorts labor supply decisions and reduces welfare. Therefore, with endogenous labor supply, a PAYG system is not dynamically efficient, even for $r < n$. Before we demonstrate this result in an OLG model with elastic labor supply, we first present a numerical example to develop an idea of the magnitude of the quantitative effects.

[21] This effect is absent in the derivation of (6.15) where we only consider an unexpected change of d_t and d_{t+1} on the savings of the generation born in period t. Since the capital stock k_t is predetermined, the wage w_t does not change for the present (working) generation, but only for future generations.

[22] The intuition for this result given in Sect. 3.2.5 was that increasing the capital stock results in lower interest rates r and, with $s_r > 0$, in lower savings s, such that capital cannot grow without bound.

6.3.2.2 Numerical Example: PAYG System with Inelastic Labor Supply

Assume utility to be logarithmic in consumption and additively separately according to

$$U_t = \ln c_t^1 + \beta \ln c_{t+1}^2 - \nu_0 \frac{l_t^{1+\frac{1}{\nu_1}}}{1+\frac{1}{\nu_1}}. \tag{6.16}$$

Labor $l_t = \bar{l}$ is exogenous and causes disutility $\nu_0 l_t^{1+1/\nu_1}/(1+1/\nu_1)$ to the household. ν_1 denotes the Frisch labor supply elasticity and measures the percentage change in the labor supply if the net wage increases by 1%.[23] In our example, we set $\bar{l} = 0.3$ to ensure that, subsequently, we are able to compare the results for the case of exogenous labor with those for the case of endogenous labor (where, in the initial steady state, we will assume that the household works 30% of its time endowment and ν_0 is calibrated accordingly).

As above, the contributions to the PAYG pension system are constant, $d_t = d$, meaning that the equilibrium condition of the social security system, $N_t d_t = N_{t-1} pen_t$, implies the intertemporal budget constraint

$$w_t l_t + \frac{n - r_{t+1}}{1 + r_{t+1}} d = c_t^1 + \frac{c_{t+1}^2}{1 + r_{t+1}}. \tag{6.17}$$

Maximizing lifetime utility (6.16) subject to the above budget constraint with respect to c_t^1 and c_{t+1}^2 implies:

$$\frac{1}{c_t^1} = \lambda_t, \tag{6.18a}$$

$$\frac{\beta}{c_{t+1}^2} = \frac{\lambda_t}{1 + r_{t+1}}, \tag{6.18b}$$

where λ_t denotes the Lagrange multiplier of the intertemporal budget constraint. The first-order conditions can be solved for c_{t+1}^2:

$$c_{t+1}^2 = \beta c_t^1 (1 + r_{t+1}).$$

Substitution of this expression into the intertemporal budget constraint results in:

$$c_t^1 = \frac{1}{1+\beta} \left[w_t \bar{l} + \frac{n - r_{t+1}}{1 + r_{t+1}} d \right],$$

[23] See Appendix 4.2 for the derivation of the Frisch labor supply elasticity in the case of a Cobb-Douglas utility function.

which yields private savings

$$\begin{aligned} s_t &= w_t \bar{l} - c_t^1 - d_t \\ &= \frac{\beta}{1+\beta} w_t \bar{l} - \frac{1+\beta+\beta r_{t+1}+n}{(1+\beta)(1+r_{t+1})} d. \end{aligned}$$

In capital market equilibrium, aggregate savings at the end of period t are equal to the beginning-of-period $t+1$ capital stock, $s_t = (1+n)k_{t+1}$, and the dynamics of the capital stock per capita, $k_t = K_t/N_t$, for given initial capital stock k_0 are described by the following first-order difference equation:

$$(1+n)k_{t+1} = \frac{\beta}{1+\beta} w_t \bar{l} - \frac{1+\beta+\beta r_{t+1}+n}{(1+\beta)(1+r_{t+1})} d. \tag{6.19}$$

We, again, assume that production is described by a Cobb-Douglas technology:

$$Y_t = K_t^{\alpha} (N_t l_t)^{1-\alpha}. \tag{6.20}$$

Therefore, wages and the real interest rate are given by:

$$w_t = (1-\alpha)K_t^{\alpha} (N_t l_t)^{-\alpha} = (1-\alpha)k_t^{\alpha} l_t^{-\alpha}, \tag{6.21a}$$

$$r_t = \alpha K_t^{\alpha-1} (N_t l_t)^{1-\alpha} = \alpha k_t^{\alpha-1} l_t^{1-\alpha}. \tag{6.21b}$$

We numerically solve the problem described by Eqs. (6.19) and (6.21). The parameter values of α, β, and n are taken from Sect. 3.2, with $\alpha = 0.36$, $\beta = 0.40$, and $n = 0.1$. The steady-state labor supply is set equal to $l \equiv 0.3$, and the Frisch labor supply elasticity is set equal to $\nu_l = 0.3$.[24] Pensions initially amount to 0% of the steady-state wage. For comparative dynamic analysis, we consider pension benefits that are equal to 30% of the (new) steady-state wage. Notice that this gross pension replacement rate of 30% is approximately equal to the lower limit among the OECD countries presented in Fig. 6.4 and is close to that observed for Poland or Australia.

First, we compute the steady state for $d = 0$ with the help of Eq. (6.19) for $k_t = k_{t+1} = k$. The numerical problem is one of solving a non-linear equation for which we use the Gauss program `Ch6_social_security1.g`. The solution for the steady-state capital stock is given by $k^0 = 0.0182$. Second, we compute the steady-state k^d for $d_t = d = 0.3wl = (1-\alpha)\left(k^d\right)^{\alpha} l^{1-\alpha}$, implying the value $k^d = 0.00687$. As a result, we find that the introduction of pensions equal to 30% of the wage reduces the steady-state capital stock by almost 62%! As presented in Table 6.2, output per capita, $y = Y/N = k^{\alpha}$, falls by 29%, from 0.109 to 0.0770.

[24] See also Sect. 4.4.5 for a discussion of the empirical evidence on ν_l.

Table 6.2 Efficiency and welfare effects of PAYG

	$\tau = 0$	$\tau = 30\%$		
	Inelastic labor	Inelastic labor	Elastic labor	CD pensions
Capital stock k	0.0182	0.00687	0.00672	0.00679
Labor l	0.300	0.300	0.294	0.297
Output y	0.109	0.0770	0.0754	0.0761
Welfare Δ	0.0%	−38.4%	−38.5%	−38.4%

Notes: Values in column $\tau = 0$ report the numbers for the case without PAYG pensions; values in the three columns with $\tau = 30\%$ report the values for a gross replacement rate of pensions equal to 30%; CD pensions: defined contribution pensions, Δ: consumption equivalent change

The change in the consumption profile, denoted by $\{c^1, c^2\}$ and $\{\tilde{c}^1, \tilde{c}^2\}$ in the old and the new steady state, affects lifetime utility (6.16). The associated values of (6.16) are −4.42 and −5.10 for $\tau = 0\%$ and $\tau = 30\%$, respectively.[25] However, these numbers are difficult to interpret, and thus, we use a more convenient measure, the *consumption equivalent change* Δ that you learned about in Sect. 5.3.2. To apply this concept in the present context, let utility in the old steady state with c^1, c^2, and $\bar{l}$ be given by

$$U = \ln c^1 + \beta \ln c^2 - \nu_0 \frac{\bar{l}^{1+\frac{1}{\nu_1}}}{1 + \frac{1}{\nu_1}}.$$

Utility in the new steady state with $\tilde{c}^1$, $\tilde{c}^2$, and $\tilde{l}$ is represented by

$$\tilde{U} = \ln \tilde{c}^1 + \beta \ln \tilde{c}^2 - \nu_0 \frac{\tilde{l}^{1+\frac{1}{\nu_1}}}{1 + \frac{1}{\nu_1}}.$$

The consumption equivalent change Δ is defined implicitly by the following equation:

$$\tilde{U} = \ln(1 + \Delta)c^1 + \beta \ln(1 + \Delta)c^2 - \nu_0 \frac{\bar{l}^{1+\frac{1}{\nu_1}}}{1 + \frac{1}{\nu_1}}.$$

Accordingly, Δ measures the percentage change by which we need to raise the initial consumption levels $\{c^1, c^2\}$ in the old steady state to have the same utility as in the

[25] You should attempt to compute these values. In the computation, we set $\nu_0 = 257.15$ as implied by the calibration in the case of elastic labor supply (see the following section).

new steady state.[26] For our logarithmic utility with exogenous labor supply $\bar{l} = \tilde{l}$, Δ can be computed with the help of

$$(1+\beta)\ln(1+\Delta) = \tilde{U} - U.$$

In our example, an increase in the pension replacement rate τ from 0% to 30% is associated with a welfare change Δ equal to a decline in total consumption by 38.4% (see Table 6.2).

Finally, we will consider the dynamics following a change in the pension system. To consider a realistic scenario that is likely to prevail in modern industrialized countries in the coming decades, we consider the effects of a reduction in the level of pension benefits. As an illustration, we consider the transition from a replacement rate of 30% to 0%, which is announced in period $t = 1$ and effective in period $t = 2$. As a consequence, the capital stock increases from an initial $k_0 = 0.00687$ to $k_{20} = 0.0182$ in the long run. We also assume that the transition is complete after 20 periods (which corresponds to 600 (!) years if each period is equal to 30 years).

In period $t = 0$, we are in the old steady state. The utility of the present (young) generation at $t = 0$ is 38.4% (as measured by the consumption equivalent change Δ) below that of the generation at the end of the transition at $t = 20$. Therefore, one might have the impression that all generations may benefit from such a change. Unfortunately, this is not the case. The reason is simple: The young generation in period $t = 1$ still has to finance the pension of the old in period $t = 1$; however, they will not receive a pension when they are old in period $t = 2$. We, therefore, have to reformulate the above equilibrium conditions for this generation. In particular, consumption by the young generation in period $t = 1$ is represented by

$$c_1^1 = \frac{1}{1+\beta}\left[w_1\bar{l} - d\right],$$

which yields private savings (using the household first-order condition)

$$\begin{aligned} s_1 &= w_1\bar{l} - c_1^1 - d \\ &= \frac{\beta}{1+\beta}\left(w_t\bar{l} - d\right). \end{aligned}$$

However, the generation born in period $t = 1$ does not receive a pension when old. Therefore, consumption by the old generation in period $t = 2$ is given by

$$c_2^2 = (1+r_2)s_1.$$

[26]As an alternative measure of the welfare change, some authors use output equivalent change. In this case, Δ is computed as the percentage of output by which the consumption levels in the old steady state need to be raised to obtain the same utility as in the new steady state.

In capital market equilibrium, $(1+n)k_2 = s_1$, the dynamics of the capital stock per capita k_2 are described by the following first-order difference equation:

$$(1+n)k_2 = \frac{\beta}{1+\beta}\left(w_1\bar{l} - d\right). \tag{6.22}$$

Notice that (6.22) differs from (6.19) because the pension payments d_t are no longer constant but change between periods $t = 1$ and $t = 2$.

More generally, we can restate the dynamics of the model as follows:

$$(1+n)k_{t+1} = \frac{\beta}{1+\beta}\left(w_t\bar{l} - d_t\right) - \frac{1}{1+\beta}\frac{1+n}{1+r_{t+1}}d_{t+1}, \tag{6.23}$$

with

$$d_t = \tau_t w_t \bar{l},$$

and $\tau_t = 30\%$ for $t = 0, 1$ and $\tau_t = 0\%$ for $t = 2, \ldots$.

The dynamics are also computed with the help of the Gauss program `Ch6_social_security1.g`. For a given initial capital stock k_0, we can compute k_1 with the help of (6.19). To obtain the value k_2, we use (6.22). For $t < 2$, we, again, use (6.19) to compute k_{t+1} for a given k_t in periods $t = 2, \ldots, 19$. Once we have computed the final capital stock k_{20}, we compare it to the value of the capital stock in the new steady state. If the values are close (which is indeed the case), we are done. Otherwise we would have to increase the number of transition periods, which we set equal to 20. We find that the new steady state of the capital stock is stable and the convergence is smooth.

The dynamics of the (per capita) capital stock k_t and the consumption equivalent change Δ_t are presented in Figs. 6.5 and 6.6. Δ is evaluated with respect to the utility level in the final steady state with $\tau = 0\%$. The capital stock converges to the new steady state within 4–5 periods (corresponding to 120–150 years). Although utility increases in the long run, the transition generation (born in period $t = 1$) experiences a decline in welfare because it still has to finance the pensions of the old but does not receive pensions. The consumption equivalent change Δ with respect to the new steady state is equal to -43.7%. If there had been no change in the pension policy, Δ_1 would only have amounted to -38.4%. Consequently, the generation born in $t = 1$ would have been better off by an equivalent of 5.3% of total consumption (in units in the new steady state).

To alleviate the pressure on the generations during the transition to new pension systems, most industrialized countries, therefore, have opted to gradually reform their social security systems. For example, in Germany, the parliament (Bundestag) decided in 2007 to increase the retirement age gradually from age 65–67. Between

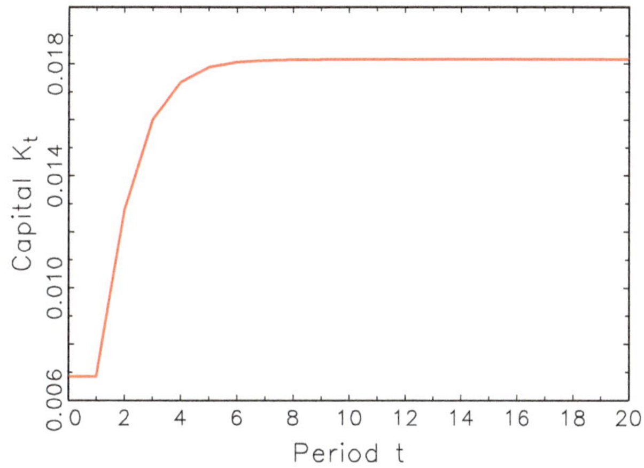

Fig. 6.5 Dynamics of the capital stock: abolition of PAYG pensions

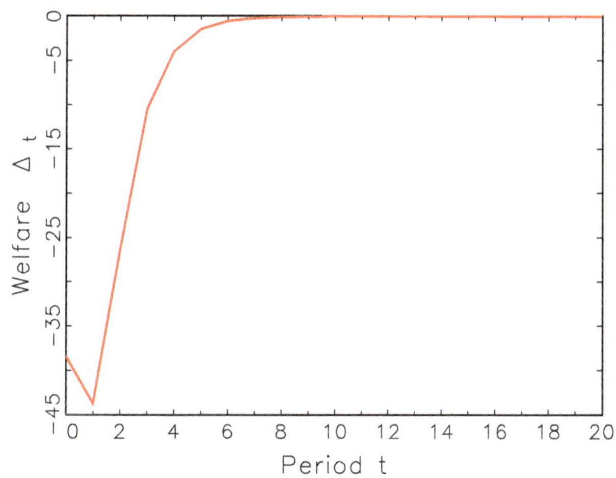

Fig. 6.6 Dynamics of the consumption equivalent change Δ (in percent): abolition of PAYG pensions

2012 and 2023, the retirement age increases by 1 month each year. Starting in 2024, the retirement age increases by 2 months each year until it reaches its final value of 67 years in 2029. Therefore, all workers born in 1964 or after need to work until age 67 to receive the full benefit.[27]

[27]We will consider the effects of a gradual policy change in Sect. 6.4.3.

6.3.3 Public Pensions and Elastic Labor Supply

Thus far, we have assumed that the pension system has no distortionary effect on the labor supply. However, pensions are financed by worker contributions, which are raised at the rate τ on wage income. In essence, contributions act like a tax on labor supply, and consequently, the marginal rate of substitution, $MRS = \frac{dc_t}{dl_t}$, is no longer equal to the marginal product of labor, $MPL_t = (1-\alpha)k_t^{\alpha}\bar{l}^{-\alpha} = w_t$, but rather equal to $(1-\tau)w_t$. Depending on the strengths of the income and substitution effects, this reduces (increases) the labor supply if the latter is stronger (smaller) than the former. For our calibration with standard preference parameters for the US economy, the substitution effect will dominate, and therefore, higher pensions will decrease aggregate labor supply.

In the following, we assume that pensions are financed by a tax τ on wage income:

$$d_t = \tau w_t l_t.$$

The budget constraints of the young and old generations are now given by:

$$(1-\tau)w_t l_t = c_t^1 + s_t, \tag{6.24}$$

$$c_{t+1}^2 = (1+r_{t+1})s_t + pen_{t+1}, \tag{6.25}$$

implying the following intertemporal budget constraint (and imposing a balanced social security budget, $pen_t = (1+n)d_t$):

$$(1-\tau)w_t l_t + \frac{1+n}{1+r_{t+1}} d_{t+1} = c_t^1 + \frac{c_{t+1}^2}{1+r_{t+1}}. \tag{6.26}$$

Maximizing (6.16) subject to (6.26) with respect to c_t^1, c_{t+1}^2, and l_t implies the first-order conditions:

$$\frac{1}{c_t^1} = \lambda_t, \tag{6.27a}$$

$$\frac{\beta}{c_{t+1}^2} = \frac{\lambda_t}{1+r_{t+1}}, \tag{6.27b}$$

$$\nu_0 l_t^{\frac{1}{\nu_1}} = \lambda_t (1-\tau) w_t, \tag{6.27c}$$

where λ_t denotes the Lagrange multiplier of the intertemporal budget constraint. Notice that for constant λ, (6.27c) implies that[28]

$$\frac{\partial l_t}{\partial\left((1-\tau)w_t\right)}\frac{(1-\tau)w_t}{l_t}=\nu_1,$$

so that ν_1 is equal to the Frisch labor supply elasticity. The first-order conditions can be solved for c_{t+1}^2:

$$c_{t+1}^2=\beta c_t^1\left(1+r_{t+1}\right).$$

Substitution of this expression in the intertemporal budget constraint results in:

$$c_t^1=\frac{1}{1+\beta}\left[(1-\tau)w_t l_t+\frac{1+n}{1+r_{t+1}}d_{t+1}\right].$$

Combining this equation with the first-order condition (6.27c), we derive a non-linear equation in l_t that can only be solved numerically:

$$\frac{(1-\tau)w_t}{\nu_0 l_t^{\frac{1}{\nu_1}}}=\frac{1}{1+\beta}\left[(1-\tau)w_t l_t+\frac{1+n}{1+r_{t+1}}d_{t+1}\right]. \tag{6.28}$$

Savings follow from

$$\begin{aligned} s_t &= (1-\tau)w_t l_t - c_t^1 \\ &= \frac{\beta}{1+\beta}(1-\tau)w_t l_t - \frac{1}{1+\beta}\frac{1+n}{1+r_{t+1}}d_{t+1}. \end{aligned} \tag{6.29}$$

In capital market equilibrium with $(1+n)k_{t+1}=s_t$, the dynamics of k_t are described by:

$$(1+n)k_{t+1}=\frac{\beta}{1+\beta}(1-\tau)w_t l_t-\frac{1}{1+\beta}\frac{1+n}{1+r_{t+1}}d_{t+1}. \tag{6.30}$$

---

[28] To derive the elasticity, simply take the logarithm on both sides of (6.27c) implying

$$\ln l_t=\nu_1\left[\ln\left((1-\tau)w_t\right)+\ln\lambda_t-\ln\nu_0\right]$$

and notice that

$$\frac{\partial\ln l_t}{\partial\ln\left((1-\tau)w_t\right)}=\frac{\partial l_t}{\partial\left((1-\tau)w_t\right)}\frac{(1-\tau)w_t}{l_t}=\nu_1.$$

The equilibrium dynamics are therefore described by the four equations (6.21), (6.28), and (6.30). In steady state, these four equations imply the steady state values for $k_t = k_{t+1} = k$, w, r, and l. We calibrate ν_0 such that steady state labor supply amounts to 30% of the time endowment, $l = 0.3$, for the case in which $\tau = 0$ to make our results comparable with the numerical example of a pension reform in an economy with inelastic labor supply presented in the previous section. For this reason, we solve the 4 equations (6.21), (6.28), and (6.30) for the variables k, w, r, and ν_0 for $l = 0.3$, implying $\nu_0 = 257.15$. The Gauss program `Ch6_social_security2.g` computes the calibration of ν_0.

Our main results are presented in the third entry column of Table 6.2. We find that the declines in the capital stock, output, and welfare following the introduction of a PAYG pension system are even more pronounced than in the case of inelastic labor. Since labor supply falls by 2%, the steady-state output falls by 30.8%. The impact of endogenous labor on steady-state lifetime utility is of approximately the same magnitude. The introduction of a pension that is financed by a contribution rate of 30% on wages also implies welfare costs that are equivalent to reducing total consumption by 38.5%.

The Gauss program `Ch6_social_security2.g` also computes the dynamics following a decrease in τ from 30% to 0% in period $t = 1$, where the economy is in steady state in $t = 0$ for $\tau = 30\%$. Since the dynamics are similar to those displayed in Figs. 6.5 and 6.6, we refrain from presenting them. However, it will be instructive to describe the numerical computation of the dynamics because we encounter a new problem. In the case of inelastic labor supply, we could simply iterate over the dynamic equation (6.19) given an initial value k_0. In the case of elastic labor supply, we cannot proceed in the same way. For this reason, assume that we would like to solve Eqs. (6.21), (6.28) and (6.30) in $t = 0$ for given k_0 and l_0. Since $d_{t+1} = d_1 = \tau w_1 l_1$ and $r_{t+1} = r_1 = \alpha k_1^{\alpha-1} l_1^{1-\alpha}$, the system of five equations has six unknowns k_1, w_0, w_1, r_1, l_0, l_1.[29] Households need to know the labor supply of the next generation to compute their pensions. The next generation, which is born in period $t = 1$, however, needs to know the labor supply of the generation in period $t = 2$ to optimally choose its labor supply, and so forth. As a consequence, the generation born in period t needs to project the labor supply and the capital stock over the complete transition period until the new steady state is reached. The method that starts the iteration over k_t with a guess of k_1 is called *forward shooting* and, evidently, does not work well in the present case.

Instead, we have to use a method called *reverse shooting*.[30] We start with the final steady state. We assume that the final steady state is reached in period $t = 6$ (we took the number of transition periods from the case with inelastic labor supply).[31]

---

[29]More specifically, we need to solve (6.21a) for w_0 and w_1 in periods 0 and 1, (6.21b) for r_1 in period 1, and (6.28) and (6.30) in period 0.

[30]See also Appendix 4.1.

[31]The algorithm used in our program is also very sensitive to the number of transition periods. In particular, for a larger number of transition periods, the solution does not converge.

Let $k_7 = 0.0182$ and $l_7 = 0.300$ denote final steady-state values associated with the contribution $\tau = 0\%$. Next, we provide a guess for k_6 that is close to k_7. Since savings are higher for $\tau = 0\%$ than for $\tau = 30\%$, we project that the capital stock approaches the final steady-state value from below. Therefore, we choose $\tilde{k}_6 = 0.99 \cdot k_7$ as an initial guess for the true k_6. With the help of Eqs. (6.21) and (6.30), we can solve for l_6, w_6, and r_6. In the next step, we iterate backward in time and compute the equilibrium values of k_t, l_t, w_t, and r_t at $t = 5$. Therefore, we solve four equations (6.21), (6.28), and (6.30) in $t = 5$ for the four endogenous values k_5, l_5, r_5, and w_5. Notice that this system in four equations is now a system in four unknowns, and we exploit its recursive nature. Continuing in this fashion, we can compute the sequence of all endogenous variables $\left\{\tilde{k}_t, \tilde{l}_t, \tilde{w}_t, \tilde{r}_t\right\}_{t=0}^{6}$. If $\tilde{k}_0$ is equal to the steady-state value of k_t associated with the case of no pensions, $\tau = 30\%$, we are done. If not, we need to adjust our guess $\tilde{k}_6$ and restart. In essence, we are computing the solution to a non-linear equation problem $f(\tilde{k}_6) = \tilde{k}_0 - k_0 = 0$ and use conventional procedures for the solution, e.g., the Gauss-Newton Algorithm as described in Appendix 2.2.

6.3.4 Contribution-Based Benefits

We extend our numerical example with elastic labor supply from the previous section and allow for defined contribution pensions. Let the contributions of the young amount to $d_t = \tau w_t l_t$. In old age, their pension now depends on the contributions according to[32]:

$$pen_{t+1} = pen_{min} + \rho_{pen} d_t. \tag{6.31}$$

Depending on the values of $\{pen_{min}, \rho_{pen}\}$, a pension system redistributes among retirees. In a pure flat-rate pension system with $\rho_{pen} = 0$, pensions are completely independent of the earnings history, while for the polar case of $pen_{min} = 0$, pensions are completely proportional to pre-retirement income. We will characterize a pension system as more *progressive* if it is more redistributive (higher pen_{min}). Two measures of the public pension system's progressivity that are used in the literature are the Gini coefficient of the pension income and the pension progressivity index. The latter is computed as 100 minus the ratio of the Gini coefficient of pension income and the Gini coefficient of earnings. Table 6.3 displays the values for selected countries in 2007. Canada and the UK have very progressive pension systems, while the pension systems of Germany and Italy are more earnings-based. The US system is an intermediate case.

---

[32]Of course, we are quite idealistic in assuming that the individual is able to determine the parameters ρ_{pen} and pen_{min}. Usually, the pension is a complicated function of contributions and the number of contribution years. In Germany, for example, years of university eduction increase your pension despite that students do not contribute to the PAYG system.

Table 6.3 Progressivity of pensions

Country	Gini pensions	Progressivity index
Canada	3.7	86.6
Germany	20.0	26.7
Italy	26.4	3.1
UK	5.1	81.1
US	16.1	40.9

Notes: The data are taken from the CESifo DICE report 4/2007 and is based upon OECD, *Pensions at a glance*, 2007, pages 44–45

With contribution-based benefits, households maximize intertemporal utility (6.16) subject to the budget constraint

$$(1-\tau)w_t l_t + \frac{pen_{t+1}}{1+r_{t+1}} = c_t^1 + \frac{c_{t+1}^2}{1+r_{t+1}}, \tag{6.32}$$

and (6.31). The differentiation of the Lagrange function

$$\begin{aligned}\mathscr{L} = & \ln c_t^1 + \beta \ln c_{t+1}^2 - \nu_0 \frac{l_t^{1+\frac{1}{\nu_1}}}{1+\frac{1}{\nu_1}} \\ & + \lambda_t \left[(1-\tau)w_t l_t + \frac{pen_{min} + \rho_{pen}\tau w_t l_t}{1+r_{t+1}} - c_t^1 - \frac{c_{t+1}^2}{1+r_{t+1}} \right]\end{aligned}$$

with respect to c_t^1, c_{t+1}^2, and l_t results in the first-order conditions of the household

$$\frac{1}{c_t^1} = \lambda_t, \tag{6.33a}$$

$$\frac{\beta}{c_{t+1}^2} = \frac{\lambda_t}{1+r_{t+1}}, \tag{6.33b}$$

$$\nu_0 l_t^{\frac{1}{\nu_1}} = \lambda_t(1-\tau)w_t + \lambda_t \frac{\rho_{pen}\tau w_t}{1+r_{t+1}}. \tag{6.33c}$$

Notice that the first-order condition for the optimal labor supply (6.33c) has an additional additive term on the right-hand side not present in Eq. (6.27c). Young households choose their labor supply while considering their increased entitlement to pension payments when old. For $\rho_{pen} = 0$, the economy is equivalent to that with elastic labor supply studied in the previous section.

In equilibrium, the social security budget is balanced:

$$N_t d_t = N_{t-1} pen_t. \tag{6.34}$$

The rest of the model is identical to that of the previous Sect. 6.3.3. The new parameters are chosen as $\tau = 0.3$ and $\rho_{pen} = 0.5$. Accordingly, half of the contributions are accounted for the pension in old age. The steady state is computed with help of the GAUSS program `Ch6_social_security3.g`, which solves the following system of seven non-linear equations in the seven endogenous variables c^1, c^2, k, l, w, r, and pen_{min}:

$$c^2 = c^1\beta(1+r), \tag{6.35a}$$

$$(1+n)\tau wl = pen_{min} + \rho_{pen}\tau wl, \tag{6.35b}$$

$$w = (1-\alpha)k^{\alpha}l^{-\alpha}, \tag{6.35c}$$

$$r = \alpha k^{\alpha-1}l^{1-\alpha}, \tag{6.35d}$$

$$\nu_0 l^{\frac{1}{\nu_1}} c^1 = (1-\tau)w + \frac{\rho_{pen}\tau w}{1+r}, \tag{6.35e}$$

$$c^1 + \frac{c^2}{1+r} = (1-\tau)wl + \frac{pen_{min} + \rho_{pen}\tau wl}{1+r}, \tag{6.35f}$$

$$(1+n)k = (1-\tau)wl - c^1. \tag{6.35g}$$

The results are displayed in the rightmost column of Table 6.2. Following an introduction of a defined contribution pension with $\tau = 30\%$, the changes in capital, output, and labor are an intermediate case between the two cases of inelastic and elastic labor supply. The decline in labor is reduced by half, from 2% to 1% in comparison with the elastic labor supply and $\rho_{pen} = 0$. Welfare still declines by 38.4% of total consumption (as measured by the consumption equivalent change Δ). Although the labor supply decisions are now less distorted than in the case with $\rho_{pen} = 0$ (because the individual considers the effect of working more hours on his pension), pensions are still welfare-reducing to a quantitatively significant extent. One reason for this result is that, with $\rho_{pen} = 50\%$, only part of the contributions helps to increase future pensions. In addition, the return on the contributions to the PAYG system is still determined by the population growth rate n, while the return on savings, the real interest rate r, is much larger. For our calibration, the annual return is approximately 5.5%, corresponding to a 30-year return of $r = 404\%$. Obviously, this return is much larger than the return from population growth with $n = 10\%$.

Thus far, however, we have neglected productivity growth. The return on the public pension increases if wages and, hence, pensions increase over time and old generations participate in this productivity growth.

6.3.5 Public Pensions and Growth

We introduce growth into the two-period OLG model as described in Sect. 3.4. In particular, production is given by:

$$Y_t = K_t^{\alpha}\left(A_t N_t l_t\right)^{1-\alpha}. \tag{6.36}$$

Labor-augmenting technology A_t grows at the exogenous rate γ:

$$A_t = (1+\gamma)A_{t-1}.$$

In the following, we set $\gamma = 0.80 = 80\%$, such that output per capita grows at approximately 2.0% per year (periods correspond to 30 years).[33] The effective labor supply of the household at a young age is equal to $A_t l_t$, and the household receives the wage rate w_t per *efficiency unit* $A_t l_t$, implying the following savings s_t, old-age consumption c_{t+1}^2, and intertemporal budget:

$$s_t = (1-\tau)w_t A_t l_t - c_t^1, \tag{6.37a}$$

$$c_{t+1}^2 = (1+r_{t+1})s_t + pen_{t+1}, \tag{6.37b}$$

$$c_t^1 + \frac{c_{t+1}^2}{1+r_{t+1}} = (1-\tau)w_t A_t l_t + \frac{pen_{t+1}}{1+r_{t+1}}. \tag{6.37c}$$

The first-order conditions implied by maximizing (6.16) subject to (6.37c) are given by:

$$c_{t+1}^2 = \beta(1+r_{t+1})c_t^1, \tag{6.38a}$$

$$\nu_0 l_t^{\frac{1}{\nu_1}} = \frac{(1-\tau)w_t A_t}{c_t^1}, \tag{6.38b}$$

and in stationary variables $\tilde{c}_t^1 \equiv c_t^1/A_t$, $\tilde{c}_{t+1}^2 \equiv c_{t+1}^2/A_{t+1}$:

$$(1+\gamma)\tilde{c}_{t+1}^2 = \beta(1+r_{t+1})\tilde{c}_t^1, \tag{6.39a}$$

$$\nu_0 l_t^{\frac{1}{\nu_1}} = \frac{(1-\tau)w_t}{\tilde{c}_t^1}, \tag{6.39b}$$

[33] In steady state, output and capital both grow at the rate γ. The US growth rate of real GDP per capita amounted to 2.00% during the period 1960–2011.

implying equilibrium stationary savings $\tilde{s}_t \equiv s_t/A_t$:

$$\tilde{s}_t = \frac{\beta}{1+\beta}(1-\tau)w_t l_t - \frac{1}{1+\beta}\frac{(1+\gamma)\tilde{pen}_{t+1}}{1+r_{t+1}},$$

where $\tilde{pen}_t \equiv pen_t/A_t$.

In equilibrium, the social security budget is balanced:

$$N_t pen_{t+1} = N_{t+1}\tau w_{t+1}A_{t+1}l_{t+1},$$

or

$$\tilde{pen}_{t+1} = (1+n)\tau w_{t+1}l_{t+1}. \tag{6.40}$$

Following the description of the production sector in Sect. 3.4, firms maximize profits:

$$\Pi_t = K_t^{\alpha}\left(A_t N_t l_t\right)^{1-\alpha} - w_t A_t N_t l_t - r_t K_t,$$

implying the first-order conditions

$$w_t = (1-\alpha)\left(\frac{K_t}{A_t N_t}\right)^{\alpha} l_t^{-\alpha} = (1-\alpha)k_t^{\alpha}l_t^{-\alpha}, \tag{6.41a}$$

$$r_t = \alpha k_t^{\alpha-1}l_t^{1-\alpha}, \tag{6.41b}$$

with $k \equiv K/(AN)$.

In capital market equilibrium,

$$N_t s_t = K_{t+1},$$

or after dividing by $A_t N_t$ on both sides,

$$\tilde{s}_t = k_{t+1}(1+\gamma)(1+n) = \frac{\beta}{1+\beta}(1-\tau)w_t l_t - \frac{1}{1+\beta}\frac{(1+\gamma)\tilde{pen}_{t+1}}{1+r_{t+1}}. \tag{6.42}$$

In steady state, all stationary variables $\tilde{k}_t$, $\tilde{c}^1$, $\tilde{c}^2$, l, w, r, and $\tilde{pen}$ are constant and can be computed as the solution of the following non-linear system of equations:

$$(1+\gamma)\tilde{c}^2 = \beta(1+r)\tilde{c}^1, \tag{6.43a}$$

$$\nu_0 l^{\frac{1}{\nu_1}}\tilde{c}^1 = (1-\tau)w, \tag{6.43b}$$

$$w = (1-\alpha)k^{\alpha}l^{-\alpha}, \tag{6.43c}$$

$$r = \alpha k^{\alpha-1}l^{1-\alpha}, \tag{6.43d}$$

Table 6.4 Output and welfare effects of PAYG pension with productivity growth

	$\tau = 0$	$\tau = 30\%$
$\tilde{k}$	0.00725	0.00259
l	0.300	0.293
$\tilde{y}$	0.0786	0.0534
Δ	0.0%	−27.5%

Notes: Δ denotes the consumption equivalent change. Entries are computed with the help of the Gauss program *Ch6_social_security4.g*

$$k(1+\gamma)(1+n) = \frac{\beta}{1+\beta}(1-\tau)wl - \frac{1}{1+\beta}\frac{(1+\gamma)\tilde{pen}}{1+r}, \tag{6.43e}$$

$$\tilde{pen} = (1+n)\tau wl, \tag{6.43f}$$

$$\tilde{c}^1 + \frac{(1+\gamma)\tilde{c}^2}{1+r} = (1-\tau)wl + \frac{(1+\gamma)\tilde{pen}}{1+r}. \tag{6.43g}$$

The introduction of the pension system results in a decline in the labor supply relative to the case without a pension system because the social security contribution reduces the net wage and distorts the individual labor supply decision, as presented by (6.43b). According to Table 6.4, labor supply falls by 2.2%, from 0.300 (for $\tau = 0\%$) to 0.293 (for $\tau = 30\%$). In the case of public pensions, savings decrease because the young have less need to provide for old age, and thus, the steady-state capital stock is reduced by 64.4%. Savings are also reduced because net income falls; the decline in savings in the presence of economic growth is comparable to that in the case without growth.[34]

For this economy, we cannot compute the welfare changes Δ by comparing it to the case with inelastic labor supply and no growth. How can we compare two economies when one eventually outgrows the other? To do so, we compute the consumption equivalent change in Table 6.4 by comparing lifetime utility in the growth economy for $\tau = 0$ and $\tau = 0.3$. Clearly, the welfare losses, to some extent, are reduced in the presence of growth. In fact, the consumption equivalent change still amounts to a loss of $\Delta = 27.5\%$. Although the return from the PAYG pension system has increased due to productivity growth γ, it still falls short of the return on savings as represented by the real interest rate r. Furthermore, the social security contribution rate τ still distorts the labor supply of the household, resulting in reduced lifetime utility.

[34]The results are computed with the help of the GAUSS program *Ch6_social_security4.g*, which solves the non-linear system of equations (6.43) with the Gauss-Newton Algorithm. In addition, we recalibrate ν_0 such that, for $\tau = 0$ and $\gamma = 80\%$, the steady-state labor supply is equal to $l = 0.30$.

6.4 Optimal Pensions

In the preceding two sections, we learned that public PAYG pension systems imply huge welfare costs and considerably decrease equilibrium savings and output. Nevertheless, we observe that most OECD countries have rather generous PAYG pension systems. One obvious reason for their prevalence is, of course, political. When a public pension system is introduced, the old generation benefits unanimously, and politicians may be inclined to capture their votes. In addition, most public pension systems, e.g., in the US or Germany, were introduced when the population growth rate was much higher than at present, while life expectancy was much lower. At the time, the return from a public pension system was much higher than at present.

In the following, we will focus on the economic rather than the political reasons for a PAYG public pension system and analyze whether a PAYG public pension may be welfare-increasing given the present demographic situation in the US economy. Welfare will be measured by the average lifetime utility of the newborn generation. There are basically five effects that will affect household utility:

1. Distortion of labor supply
2. Decrease in aggregate savings
3. Credit constraints
4. Insurance against negative income shocks
5. Uncertain lifetime: missing annuities markets.

The first two effects are detrimental to output and welfare. Households reduce their labor supply because the net wage is lower due to pension contributions. We saw in the previous section that the quantitative effect on labor supply is reduced but not completely offset if pensions are based on contributions rather than operated on a defined benefit basis. In addition, we observed substantial reductions in savings and, hence, the aggregate capital stock in the presence of public pensions.

In the model in this section, we introduce a third distortion. We will study heterogeneous agents that are subject to idiosyncratic productivity shocks. Therefore, some individuals will be confronted with very low wages and would like to indebt themselves. However, we will assume that financial markets do not provide access to credit for these households, meaning that these individuals are credit-constrained, $k_t \geq 0$. In the presence of public pensions, this constraint becomes more binding because households receive lower net wages and, hence, welfare is further reduced.

The remaining two effects (4) and (5) help to explain why public pensions may be welfare-improving. In our model (and in real life), households are subject to stochastic productivity shocks. Their hourly wage can vary for deterministic reasons, e.g., it increases initially with age, but also for stochastic reasons, e.g., due to health or unemployment risks. A public pension system that also redistributes from the income-rich to the income-poor might therefore increase average utility

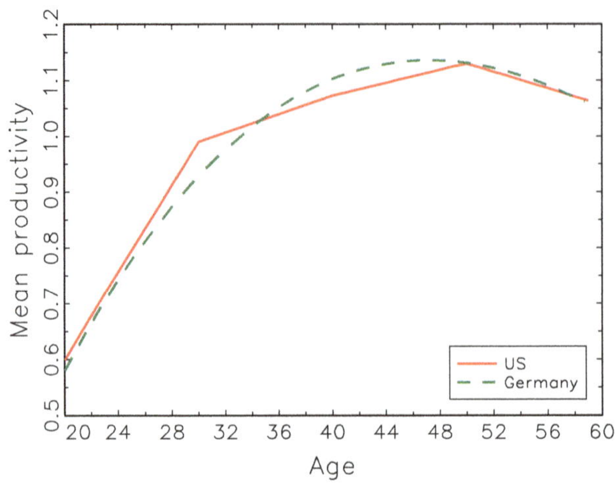

Fig. 6.7 Age-productivity profiles in the US and Germany

because those households with lower income and wealth are also the households with higher marginal utility of consumption.

Households in our model, in particular, will face wages that depend on an age component, a permanent efficiency type, and a stochastic individual productivity. The age component is hump-shaped and depicted in Fig. 6.7 for the US and German economies. In both countries, efficiency peaks at ages 46–48 and is much higher at the age of retirement than in youth.[35] As a second component, we distinguish two types of permanent productivity, $\epsilon \in \{\epsilon_1, \epsilon_2\}$, which may be interpreted as households with college (high school) education. The third component is a stochastic component, $\eta \in \{\eta_1, \eta_2\}$, which may simply reflect good or bad luck during employment. For example, technological change may favor some types of workers while detrimentally affecting others.

Our specification of the individual productivity process and, hence, the labor earnings is a simplification with respect to recent empirical evidence. Guvenen, Karahan, Ozkan, and Sang (2015), for example, study the dynamics of individual labor earnings over the life-cycle and find, among others, that (1) earnings shocks display substantial deviations from log-normality in the form of an extremely high kurtosis and that (2) the statistical properties of the labor earnings process vary over the life-cycle. In our analysis, we neglect age-specificity in the variance of earnings

[35]The data for the US economy are taken from Hansen (1993). The efficiency profile for Germany is computed with the help of the average hourly wages of s-year-olds during the period 1990–1997 following the method of Hansen (1993). Average productivity is normalized to one. We further interpolated the productivity-age profile with a polynomial function of order 3. We used data from the Cross National Data Files for West Germany during the period 1990–1997, which are extracted from the German Socio Economic Panel, GSOEP.

and do not model the top income percentiles.[36] With respect to the lower tail of the labor income distribution, our model is able to closely replicate the empirical income and wealth heterogeneity. In particular, we are also able to model the fact that a substantial fraction of households are credit-constrained. These households will be subject to considerable changes in lifetime utility if the public pension system is altered.

Finally, a public pension system may increase welfare because financial markets are imperfect (in addition to not providing credit). Specifically, we assume that annuities markets are absent, meaning that individuals cannot insure against early death. What would perfect financial markets look like? Assume that there are no transaction costs. In this case, a bank would collect savings k_t from the individuals at the end of period $t-1$, borrow the savings to firms and earn interest rate r_t. Between periods $t-1$ and t, only a fraction ϕ of households survive. At the end of period t, the bank (which faces no transaction costs) would be able repay $(1+r_t)k_t/\phi$ to those who survive and nothing to the deceased. As a consequence, the banks would end up with zero profits (which would be the outcome if we assumed free entry into financial markets and if the law of large numbers holds). In the absence of perfect annuity markets, a public pension may, in part, substitute for these missing annuity markets. Those who survive receive a payment from the young, while those who die early do not. Consequently, it redistributes from those who may die early to those who survive until old age. In this regard, Hubbard and Judd (1987) show that a fully funded social security system can increase welfare in the absence of liquidity constraints and annuity markets because it provides insurance against longevity.[37]

Next, we describe the large-scale OLG model that we use to study these welfare questions in greater detail. The basic policy question that we address is the optimal size of the public pension. What is the optimal replacement rate of pensions with respect to wages? The details of the model with respect to the demographics, the household optimization, and the production and government sectors are described in turn.

6.4.1 The Model

The model is a simplified version of that studied in Heer (2018).[38]

[36]The consideration of the highest income households would not significantly affect our welfare results on optimal social security. The welfare effect of social security is a second-order consideration for these households since the pension income from social security is a relatively small share of total savings for the top income earners.

[37]In a recent study, however, Caliendo, Guo, and Hosseini (2014) demonstrate that this result is sensitive to the assumption of whether (1) bequest income is fixed or endogenous and (2) bequest income is redistributed anonymously or through a direct linkage between deceased parents and surviving children.

[38]In addition, Heer (2018) models income uncertainty from unemployment and specifies a more general utility function with Epstein-Zin preferences that include the Cobb-Douglas utility function (6.45) as a special case.

6.4.1.1 Demographics and Timing

A period, t, corresponds to 1 year. In each t, a new generation of households is born. Newborns have a real-life age of 20, denoted by $s = 1$. All generations retire at age $s = R = 46$ (corresponding to real-life age 65) and live up to a maximum age of $s = J = 70$ (real-life age 89).

Let $N_t(s)$ denote the number of agents of age s in t. We denote total population in t by N_t which grows at the rate n_t. In t, all agents of age s survive until age $s+1$ with probability $\phi_{t,s}$ where $\phi_{t,0} = 1$ and $\phi_{t,J} = 0$.[39]

6.4.1.2 Households

Each household comprises one (possibly retired) worker. Households maximize expected intertemporal utility at the beginning of age 1 in period t[40]

$$\max \mathbb{E}_t \sum_{s=1}^{J} \beta^{s-1} \left(\prod_{j=1}^{s} \phi_{t+j-2,j-1} \right) u(c_{t+s-1}^s, l_{t+s-1}^s), \tag{6.44}$$

where $\beta > 0$ denotes the discount factor; c_t^s and l_t^s denote consumption and labor supply of the s-year old in period t. Per-period utility $u(c,l)$ is a function of consumption c and labor supply l

$$u(c,l) = \frac{\left(c^{\iota}(1-l)^{1-\iota}\right)^{1-\sigma}}{1-\sigma}, \quad \sigma > 0, \quad \iota \in (0,1). \tag{6.45}$$

Households are heterogeneous in age s, individual labor efficiency $\eta \epsilon_j \bar{y}_s$, and wealth k_t^s. We stipulate that an agent's efficiency depends on its age, $s \in \mathcal{S} \equiv \{1, 2, \ldots, 70\}$, and its efficiency type, $\epsilon_j \in \mathcal{E} \equiv \{\varepsilon_1, \varepsilon_2\}$. We choose the age-efficiency profile $\{\bar{y}_s\}$ in accordance with the US wage profile presented in Fig. 6.7. The permanent efficiency types ϵ_1 and ϵ_2 are meant to capture differences in education and ability. In addition, we follow Krueger and Ludwig (2007) and assume that a household's labor productivity is affected by an idiosyncratic shock,

[39] Be careful when you compare our equilibrium conditions to those in the literature. In some cases, the indexation of the survival probabilities is different and ϕ_s denotes the probability to survive up to age s conditional on surviving up to age $s-1$ as in İmrohoroğlu, İmrohoroğlu, and Joines (1995) or Huggett (1996), while our notation follows Conesa and Krueger (1999).

[40] In the literature, expected lifetime utility is either stated in the form of (6.44) (e.g., in İmrohoroğlu, İmrohoroğlu, and Joines 1995 or Huggett 1996) or the product of the cumulative survival probabilities, $\prod_{j=1}^{s} \phi_{t+j-2,j-1}$, is dropped from this expression (e.g., in Conesa and Krueger 1999). In the latter case, expectations are also formed with respect to (stochastic) survival and instantaneous utility of being dead is set equal to zero. We adhere to the former notation so that expectations $\mathbb{E}_t\{.\}$ are only formed with respect to stochastic idiosyncratic productivity. This notation will be useful in a model of Chap. 7.5 where we analyze the demographic transition. In this model, survival of the individuals is stochastic, while individual productivity is deterministic. As a consequence, the derivation of the Euler equation that contains the survival probability of the individual as an additional factor will become more evident.

$\eta \in \Gamma \equiv \{\eta_1, \eta_2\}$, that follows a time-invariant Markov chain with transition probabilities

$$\pi(\eta'|\eta) = \begin{pmatrix} \pi_{11} & \pi_{12} \\ \pi_{21} & \pi_{22} \end{pmatrix}. \tag{6.46}$$

The net wage income in period t of an s-year-old household with efficiency type $\eta\epsilon$ is given by $(1 - \tau^w - \tau^p)w_t A_t \eta\epsilon \bar{y}_s l_t^s$, where w_t denotes the wage rate per efficiency unit in period t, and A_t is labor productivity. The wage income is taxed at the constant rate τ^w. Furthermore, the worker has to pay contributions to the pension system at rate τ^p.[41] A retired worker receives a lump-sum pension pen_t.[42]

Households are born without assets at the beginning of age $s = 1$; hence, $k_t^1 = 0$. Parents do not leave bequests to their children, and all accidental bequests are confiscated by the government. The household earns interest r_t on its wealth $k_t^s \in \mathbb{R}^+$. Capital income is taxed at the constant rate τ^K. In addition, households receive lump-sum transfers tr_t from the government. As a result, the budget constraint of an s-year-old household with productivity type $\eta\epsilon$ and wealth k_t^s in period t is represented by:

$$\begin{aligned} & c_t^s + k_{t+1}^{s+1} \\ & = \begin{cases} (1 - \tau^w - \tau_t^p)w_t A_t \eta\epsilon \bar{y}_s l_t^s + \left[1 + (1 - \tau^K)r_t\right] k_t^s + tr_t, \\ \text{for } s \leq 45, \\ \\ pen_t + \left[1 + (1 - \tau^K)r_t\right] k_t^s + tr_t, \quad \text{for } s > 45. \end{cases} \end{aligned} \tag{6.47}$$

In addition, households face a borrowing constraint, $k_t^s \geq 0$.

6.4.1.3 Production

The production sector is modeled as in Sect. 3.4, where we introduced economic growth into the OLG model. Production is assumed to be described by a Cobb-Douglas function:

$$Y_t = K_t^\alpha (A_t L_t)^{1-\alpha},$$

where labor-augmenting technological progress A_t grows at the exogenous rate γ:

$$A_t = (1 + \gamma)A_{t-1}. \tag{6.48}$$

---

[41] In Chap. 5, we denoted the labor income tax rate by τ^L. Notice that the tax on labor income in this model is the sum of the wage income tax and the social security contribution rate, $\tau^L = \tau^w + \tau^p$.

[42] In contrast to Sect. 6.3.4, we do not assume pensions to be related to the individual's lifetime social security contributions. Our simplifying assumption is supported by the results of Fehr, Kallweit, and Kindermann (2013) and Heer (2018), who find in their studies with earnings-dependent pensions that pensions should optimally be provided lump-sum rather than earnings-dependent.

L_t denotes aggregate efficient labor that will be defined in greater detail below for the stationary equilibrium.

Firms maximize profits:

$$\Pi_t = K_t^{\alpha} (A_t L_t)^{1-\alpha} - w_t A_t L_t - r_t K_t - \delta K_t,$$

implying the first-order conditions

$$w_t = (1-\alpha)\left(\frac{K_t}{A_t L_t}\right)^{\alpha} = (1-\alpha)\tilde{k}_t^{\alpha}\tilde{L}_t^{-\alpha}, \tag{6.49a}$$

$$r_t = \alpha \tilde{k}_t^{\alpha-1}\tilde{L}_t^{1-\alpha} - \delta, \tag{6.49b}$$

where $\tilde{k} \equiv K/(AN)$ is defined as capital per efficiency population, and $\tilde{L} \equiv L/N$. For convenience, we will also refer to $\tilde{k}$ as capital.

6.4.1.4 Government

The government collects income taxes T_t to finance its expenditures on government consumption G_t and transfers Tr_t. In addition, it confiscates all accidental bequests Beq_t. The government budget is balanced in every period t, i.e.,

$$G_t + Tr_t = T_t + Beq_t. \tag{6.50}$$

In view of the tax rates τ^w and τ^K, the government's tax revenue is given by

$$T_t = \tau^w w_t A_t L_t + \tau^K r_t K_t. \tag{6.51}$$

Government spending is exogenous and grows at the rate of labor augmenting-technological progress γ and the population growth rate n_t:

$$G_t = G_{t-1}(1+\gamma)(1+n_t). \tag{6.52}$$

6.4.1.5 Social Security

The social security system is a PAYG system. The social security authority collects contributions from the workers to finance its pension payments to the retired agents. The pension is provided lump-sum, with the net replacement rate being denoted by θ_t^p:

$$pen_t = \theta_t^p (1 - \tau^w - \tau_t^p) w_t A_t \bar{l}, \tag{6.53}$$

where $\bar{l}$ denotes the average labor supply (working hours) of the workers.[43]

[43]The mean of the workers' efficiency $\eta \epsilon \bar{y}_s$ is normalized to one.

In equilibrium, the social security budget is balanced such that total expenditures on pensions Pen_t are equal to total contributions:

$$Pen_t = \tau_t^p w_t A_t L_t. \tag{6.54}$$

6.4.1.6 Stationary Equilibrium

In the stationary equilibrium, individual behavior is consistent with the aggregate behavior of the economy: firms maximize profits, households maximize intertemporal utility, and factor and goods markets clear. To express the equilibrium in terms of stationary variables only, we have to divide aggregate quantities by $A_t N_t$ (with the exception of aggregate labor supply L_t) and individual variables by A_t. Therefore, we define the following stationary aggregate variables:

$$\tilde{Beq}_t \equiv \frac{Beq_t}{A_t N_t}, \quad \tilde{T}_t \equiv \frac{T_t}{A_t N_t}, \quad \tilde{G}_t \equiv \frac{G_t}{A_t N_t}, \quad \tilde{Pen}_t \equiv \frac{Pen_t}{A_t N_t}, \quad \tilde{L}_t \equiv \frac{L_t}{N_t},$$
$$\tilde{C}_t \equiv \frac{C_t}{A_t N_t}, \quad \tilde{Y}_t \equiv \frac{Y_t}{A_t N_t},$$

and stationary individual variables:

$$\tilde{c}_t^s \equiv \frac{c_t^s}{A_t}, \quad \tilde{pen}_t \equiv \frac{pen_t}{A_t}, \quad \tilde{k}_t^s \equiv \frac{k_t^s}{A_t}, \quad \tilde{tr}_t \equiv \frac{tr_t}{A_t}.$$

Notice that we divide aggregate labor supply L_t by total population in period t, N_t, to obtain a stationary variable. The mass of all individuals in our economy, therefore, is normalized to one in every period t.

Let f_t denote the cross-sectional measure of households in period t. The household policy functions depend on an individual household's wealth $\tilde{k}_t^s$, age s, permanent efficiency type ϵ, and idiosyncratic productivity η.

A *stationary equilibrium* for a constant government policy $\{\tau^K, \tau^w, \tau^p, \theta^p, \tilde{G}, \tilde{tr}\}$ corresponds to a price system, an allocation, and a sequence of aggregate productivity indicators $\{A_t\}$ that satisfy the following conditions:

1. Population grows at the constant rate $n = \frac{N_{t+1}}{N_t} - 1$ and the survival probabilities are constant, $\phi_{t,s} = \phi_s$.
2. The aggregate productivity indicator A_t evolves according to (6.48).
3. The individual productivity shock η follows the Markov transition matrix (6.46).

4. Households maximize intertemporal utility (6.44) subject to the budget constraint (6.45), $l_t^s \in [0,1]$, and $\tilde{k}_t^s \geq 0$. In stationary variables, the budget constraint is represented by

$$\tilde{c}_t^s + (1+\gamma)\tilde{k}_{t+1}^{s+1} = \begin{cases} (1-\tau^w-\tau_t^p)w_t\eta\epsilon\bar{y}_s l_t^s + \left[1+(1-\tau^K)r_t\right]\tilde{k}_t^s + \tilde{tr}_t, \\ \text{for } s \leq 45, \\ \tilde{pen}_t + \left[1+(1-\tau^K)r_t\right]\tilde{k}_t^s + \tilde{tr}_t, \quad \text{for } s > 45. \end{cases} \tag{6.55}$$

   Moreover, there is a transversality condition requiring $\tilde{k}_t^{71} = 0$.
   As a result, for each period t, individual labor supply $l_t(\tilde{k}, s, \epsilon, \eta)$, consumption $\tilde{c}_t(\tilde{k}, s, \epsilon, \eta)$, and optimal next-period assets $\tilde{k}_t'(\tilde{k}, s, \epsilon, \eta)$ are functions of the individual state variables $\tilde{k} \in \mathscr{K}$, $s \in \mathscr{S}$, $\epsilon \in \mathscr{E}$, and $\eta \in \Gamma$ and are constant over time in the stationary equilibrium.
5. Firms maximize profits by satisfying (6.49a) and (6.49b). In equilibrium, firm profits are zero.
6. Aggregate variables are equal to the sum of the individual variables, the capital market is in equilibrium, and all accidental bequests are collected from the deceased:

$$\tilde{L}_t = \frac{1}{N_t}\int \eta\epsilon\bar{y}_s l_t(\tilde{k}, s, \epsilon, \eta)\; f_t(d\tilde{k} \times ds \times d\epsilon \times d\eta), \tag{6.56a}$$

$$\tilde{C}_t = \frac{1}{N_t}\int \tilde{c}_t(\tilde{k}, s, \epsilon, \eta)\; f_t(d\tilde{k} \times ds \times d\epsilon \times d\eta), \tag{6.56b}$$

$$\tilde{T}_t = \tau^w w_t \tilde{L}_t + \tau^K r_t \tilde{K}_t, \tag{6.56c}$$

$$\tilde{K}_{t+1} = \frac{1}{N_{t+1}}\int \tilde{k}_t'(\tilde{k}, s, \epsilon, \eta)\; f_t(d\tilde{k} \times ds \times d\epsilon \times d\eta), \tag{6.56d}$$

$$\tilde{Beq}_{t+1} = \frac{1}{N_{t+1}}\int (1-\phi_{t,s})(1+r_{t+1}(1-\tau^K))\;\tilde{k}_t'(\tilde{k}, s, \epsilon, \eta)\; f_t(d\tilde{k} \times ds \times d\epsilon \times d\eta). \tag{6.56e}$$

7. The government budget is balanced:

$$\tilde{G} + \tilde{tr}_t = \tilde{T}_t + \tilde{Beq}_t. \tag{6.57}$$

8. The budget of the social security system is balanced:

$$\tilde{Pen}_t = \tau_t^p w_t \tilde{L}_t. \tag{6.58}$$

9. The final goods market clears:

$$\tilde{Y}_t = \tilde{K}_t^{\alpha} \tilde{L}_t^{1-\alpha} = \tilde{C}_t + \tilde{G}_t + (1+\gamma)\tilde{K}_{t+1} - (1-\delta)\tilde{K}_t. \tag{6.59}$$

10. The cross-sectional measure f_t evolves as

$$f_{t+1}(\tilde{\mathscr{K}} \times \mathscr{S} \times \mathscr{E} \times \Gamma) = \int P_t\left((\tilde{k}, s, \epsilon, \eta), \tilde{\mathscr{K}} \times \mathscr{S} \times \mathscr{E} \times \Gamma\right) f_t(d\tilde{k} \times ds \times d\epsilon \times d\eta)$$

for all sets $\tilde{\mathscr{K}}$, $\mathscr{S}$, $\mathscr{E}$, and Γ, where the Markov transition function P_t is given by

$$P_t\left((\tilde{k}, s, \epsilon, \eta), \tilde{\mathscr{K}} \times \mathscr{S} \times \mathscr{E} \times \Gamma\right) = \begin{cases} \phi_s \pi(\eta'|\eta) \text{ if } \tilde{k}_t'(\tilde{k}, s, \epsilon, \eta) \in \tilde{\mathscr{K}}, \\ \qquad\qquad \text{for } \epsilon \in \mathscr{E}, s+1 \in \mathscr{S}, \eta' \in \Gamma, \\ 0 \qquad\qquad else, \end{cases}$$

and for the newborns

$$f_{t+1}(\tilde{\mathscr{K}} \times 1 \times \mathscr{E} \times \Gamma) = N_{t+1}(1) \cdot \begin{cases} \Upsilon_1 \text{ if } 0 \in \tilde{\mathscr{K}}, \\ 0 \quad else. \end{cases}$$

The initial distribution $\Upsilon_1(\epsilon, \eta)$ of $\epsilon \in \mathscr{E} = \{\epsilon_1, \epsilon_2\}$ and $\eta \in \Gamma = \{\eta_1, \eta_2\}$ among the one-year-olds is chosen to be uniform: $\Upsilon_1(\epsilon_1, \eta_1) = \Upsilon_1(\epsilon_1, \eta_2) = \Upsilon_1(\epsilon_2, \eta_1) = \Upsilon_1(\epsilon_2, \eta_2) = 1/4$.

Notice that, in (6.56d), we divide the integral over the next-period capital stock by population size N_{t+1}. The capital market equilibrium condition is formulated in an analogous way to the condition (3.15) in the two-period OLG in Sect. 3. Accordingly, total savings in efficiency units, K_{t+1}/A_{t+1}, is equal to the sum of the individual savings $k'(.)$ (in efficiency units A_{t+1}) of the population at the end of period t. After division by N_{t+1}, (6.56d) holds. Equation (6.56e) states the equilibrium condition for the accidental bequests. According to this equation, the net interest payments of the deceased are also collected by the government. The economic reason for inclusion of the term $(1-\tau^K)r_{t+1}\tilde{k}_{t+1}$ in the accidental bequests is as follows. The s-year old households saves the capital stock $\tilde{k}_{t+1}^{s+1}$ at the end of period t. The capital stock is used in production and at the end of period $t+1$, the firms pay the capital plus net interest to the households and capital income taxes to the government. The fraction ϕ_s of the s-year old households do not survive until the end of period $t+1$ and, therefore, the government also collects the net interest payments (plus the principal) from these households. The formal argument for the setup of (6.56e) is presented in Appendix 6.1. The numerical computation of the model applies value function iteration and is described in greater detail in Appendix 6.2.

Table 6.5 Calibration of parameters in the large-scale OLG model

Parameter	Value	Description
n	$\{0.8\%, 0.4\%\}$	Population growth rate
β	1.011	Subjective discount factor
$1/\sigma$	{1/2, 1/4}	Intertemporal elasticity of substitution
ι	0.31	Relative weight of leisure and consumption in utility
α	0.36	Share of capital income
δ	8.0%	Rate of capital depreciation
γ	2.0%	Growth rate
$\{\epsilon_1, \epsilon_2\}$	{0.57, 1.43}	Permanent productivity types
$\{\eta_1, \eta_2\}$	{0.727, 1.273}	Stochastic individual productivities
G/Y	19.5%	Share of government spending in steady-state production
τ^w	24.8%	Wage income tax
τ^K	42.9%	Capital income tax
θ^p	50%	Net pension replacement rate
$\pi_{11} = \pi_{22}$	0.98	Persistence of idiosyncratic productivity shock

6.4.1.7 Calibration

We calibrate the parameters of the model in accordance with the US economy. The population forecast for the US until 2050 is taken from UN (2015). We use the two sets of the survival probabilities, $\{\phi_{t,s}\}_{s=1}^{70}$, $t \in \{2015, 2050\}$, and the corresponding population growth rates, $n = 0.8\%$ and $n = 0.4\%$, to study the optimal public pension policy. For simplification, we assume that the economy is in stationary competitive equilibrium in 2015 and 2050, respectively.

The preference and production parameters are set as in the economies in Sect. 6.3.2. The intertemporal elasticity of substitution $1/\sigma$ is chosen with $\sigma = 2.0$ in accordance with İmrohoroğlu, İmrohoroğlu, and Joines (1995). The parameter ι, which reflects the relative weight of consumption and leisure in utility, is set equal to 0.31 such that the average working hours amount to approximately 0.30 in the benchmark equilibrium for the year 2015. We choose the discount factor $\beta = 1.011$ in accordance with the empirical estimates of Hurd (1989), who explicitly accounts for mortality risk.[44] This choice of the discount factor implies a real interest rate of 5.8%. By our choice of β, the value of the real interest rate is somewhat higher than observed empirically and usually applied in these types of models. The calibration of the parameters is summarized in Table 6.5.

The elasticity of production with respect to capital is set equal to $\alpha = 0.36$, and capital depreciates at a rate of $\delta = 8.0\%$ annually. We set $\gamma = 2.0\%$, corresponding

[44]Related research that uses such a value for β includes İmrohoroğlu, İmrohoroğlu, and Joines (1995) and Huggett (1996). With this value of β, the effective time discount factor of the newborn for utility at age s, $\beta^{s-1}\left(\prod_{j=1}^{s} \phi_{j-1}\right)$, displays an increasing weight to instantaneous utility until real lifetime age 63, before it declines again and even falls below one after the real lifetime age 82 (for the survival probabilities for the year 2015).

to the average growth rate of US GDP per capita during the period 1960–2011 (using data provided by the Federal Reserve Bank of St. Louis, available at 'http://research.stlouisfed.org/fred2').

A s-year-old household of temporary productivity type i and permanent productivity type j has productivity $\eta_i \epsilon_j \bar{y}_s$. The age-efficiency profile $\{\bar{y}_s\}_{s=1}^{45}$ is taken from Hansen (1993) as illustrated in Fig. 6.7. The set of the equally distributed productivity types $\{\epsilon_1, \epsilon_2\} = \{0.57, 1.43\}$ is taken from Storesletten, Telmer, and Yaron (2004). Our choice of the stochastic individual productivity component, $\eta \in \{\eta_1, \eta_2\}$, is also motivated by Storesletten, Telmer, and Yaron (2004). In particular, the two-state Markov chain is calibrated such that the annual persistence amounts to 0.98 with an implied conditional variance of 8%. Accordingly, $\{\eta_1, \eta_2\} = \{0.727, 1.273\}$ and

$$\pi(\eta'|\eta) = \begin{pmatrix} \pi_{11} & \pi_{12} \\ \pi_{21} & \pi_{22} \end{pmatrix} = \begin{pmatrix} 0.98 & 0.02 \\ 0.02 & 0.98 \end{pmatrix}.$$

With this calibration, we are able to approximately replicate the empirical distribution of US wages. In our model, the Gini coefficient of the wage income distribution is equal to 0.38, which is somewhat lower than the empirical values reported by, e.g., Budría Rodriguez, Díaz-Giménez, and Quadrini (2002). The primary reason for this is that we omit the top income percentiles from our model.

Government expenditures $\tilde{G}$ are set such that the government share G/Y is equal to the average ratio of government consumption to GDP, $G/Y = 19.5\%$, in the US economy during the period 1959–1993 according to the Economic Report of the President (1994). The tax rates $\tau^w = 24.8\%$ and $\tau^K = 42.9\%$ are computed as the average values of the effective US tax rates over the period 1965–1988 that are reported by Mendoza, Razin, and Tesar (1994). Government transfers Tr are computed using the equilibrium condition of the government budget (6.50) and amounted to 5.4% of GDP in 2015.

The social security contribution rate on wage income τ^p is set such that the net replacement rate of pensions relative to wage income θ^p is equal to 50%, following İmrohoroğlu, İmrohoroğlu, and Joines (1999). The value is close to the most recent one reported by the OECD in 2014, which amounts to 45%.

6.4.1.8 Measuring Welfare Effects

In the following, we wish to compare different public pension policies with respect to their effects on the stationary-state allocation. In particular, we are interested in how much the value of lifetime utility, as measured by the average value of the newborns' value function, changes if pensions, as measured by their replacement rate θ^p, are adjusted. Therefore, we compute the average value of the newborn:

$$W(\theta^p) = \frac{1}{4} \sum_{\epsilon,\eta} \tilde{V}(0, 1, \epsilon, \eta),$$

where $\tilde{V}(.)$ denotes the (stationary) value function of a one-year-old with zero assets and individual productivity $\epsilon\eta$.[45]

The aggregate capital stock $\tilde{K}$ and efficient labor $\tilde{L}$ and, hence, output $\tilde{Y}$ will vary with the public pension policy $\{\theta^p\}$. Notice that we calibrated the model such that the government share of GDP is equal to $G/Y = 19.5\%$. In the computation of the allocations for different public pension policies $\{\theta^p\}$, we do not hold G/Y constant because this would imply different values of exogenous government spending G. Since government spending G does not have any effect in our economy and is pure waste, it would make no sense to compare economies that are characterized by different sizes of the government sector $\tilde{G}$. We, therefore, hold $\tilde{G}$ constant and equal to the value in our calibration for all other computations.[46]

To compute the welfare change associated with a different policy $\{\theta^p\}$, we compute the consumption equivalent change Δ as before. As our comparison case, we choose the benchmark policy with $\theta^p = 50\%$. The consumption equivalent change is now simply computed as the percentage by which we need to increase (or reduce) the consumption in the benchmark case to obtain the same welfare as under the policy $\{\theta^p\}$. Noticing the functional form of our utility function, Δ can be computed with the help of:

$$(1+\Delta)^{\iota(1-\sigma)} W(50\%) = W(\theta^p). \tag{6.60}$$

6.4.2 Results

We will first discuss the allocation in the stationary equilibrium in 2015 and compare it to the case without social security, before the optimal pension policy is studied. We will establish that the optimal pension in the stationary equilibrium in 2015 is rather low and only amounts to 14% of net wages. Next, we will analyze the effect of aging and find that the quantitative welfare effects of lower pensions are even larger for a grayer population.

6.4.2.1 Stationary Equilibrium in 2015

In the following, we assume that the population is stationary. For this reason, we assume that the survival probabilities $\phi_{t,s}$ are constant and equal to those prevailing in the year 2015. Figure 6.8 presents the survival probabilities estimated by UN (2015) for the years 2015, 2050, and 2100. Clearly, the probability of surviving from age s to age $s+1$ declines with age s and increases over time. In addition, we assume that population growth is constant and equal to 0.80%. Consequently, the implied stationary old-age dependency ratio of the retired relative to the 20–65-year-olds, $OADR2$, amounts to 27%. Furthermore, the distribution of the capital

[45] The concept of the value function is introduced in Appendix 6.2.

[46] An alternative would be to either let government expenditures be a production input or let the government provide a public consumption good. See the applications in Chaps. 4 and 5.

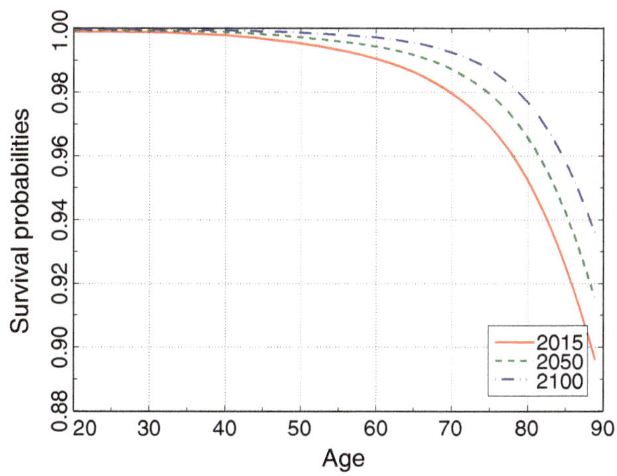

Fig. 6.8 (Projected) Survival probabilities in the years 2015, 2050, and 2100

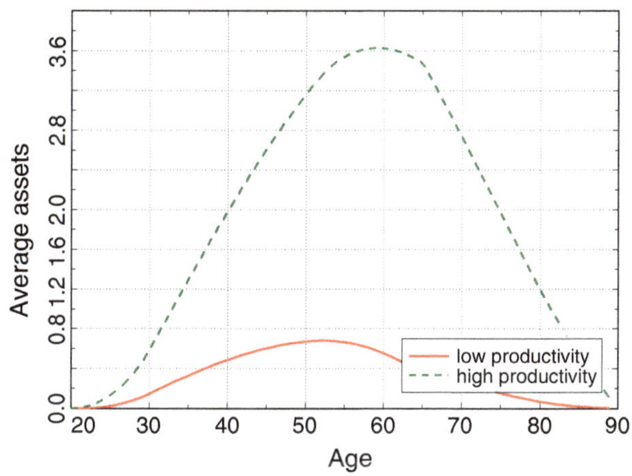

Fig. 6.9 Wealth-age profile

stock per (effective) capita, $\tilde{k} = K/(AN)$, is constant across age s and productivity types η and ϵ.

The average wealth $\tilde{k}^s$ of the s-year-old cohort with high (broken green line) and low (solid red line) permanent productivity over the life-cycle is graphed in Fig. 6.9. High-income (low-income) households accumulate savings until age 59 (53) before they start to dissave. In their effort to smooth consumption over their lifetime, households start to consume part of their savings as their income declines. The decline in wage income is caused by the decrease in age-dependent efficiency $\bar{y}_s$, which peaks at age 50 (see Fig. 6.7). The decline in wealth is accelerated as soon as the households retire because pensions are below the former wage income.

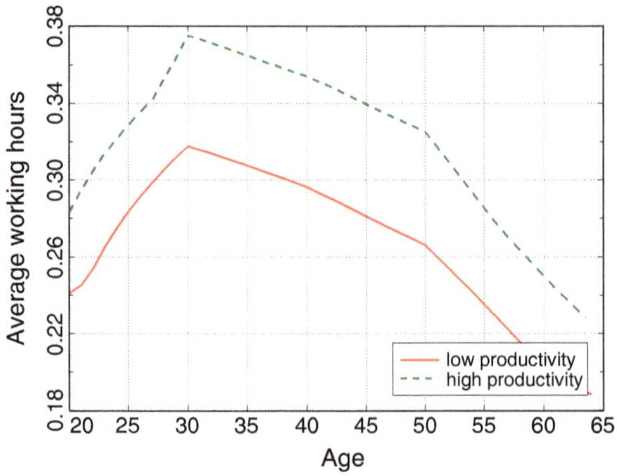

Fig. 6.10 Labor-supply-age profile

The profile of average working hours of the s-year-old over the life-cycle in Fig. 6.10 mirrors the age-productivity profile in Fig. 6.7 because the substitution effect of higher wages dominates the income effect. However, the peak of working hours (at age 30) takes place prior to the peak in age-dependent efficiency $\bar{y}_s$ because of increasing wealth over the lifetime, which reduces labor supply.

Labor supply and wealth also depend on the permanent and temporary productivity types $\{\epsilon, \eta\}$. Both increase with higher productivity, $\epsilon = \epsilon_2$ and $\eta = \eta_2$. The household with $\epsilon = \epsilon_1$ that experiences a negative productivity shock, $\eta = \eta_1$, is also liquidity-constrained, $\tilde{k} = 0$, if it has not accumulated sufficient savings in prior periods. In fact, the percentage of households without savings amounts to 36.5% in our benchmark calibration. Empirically, a large fraction of households are also credit-constrained. Budría Rodriguez, Díaz-Giménez, and Quadrini (2002) report that 2.5% of households have zero wealth and 7.4% actually have negative wealth in the 1998 Survey of Consumer Finances.

The average consumption-age profile displayed in Fig. 6.11 attains its maximum one period prior to retirement and declines in old age. These observations are in accordance with empirical evidence for the US economy reported by Fernández-Villaverde and Krueger (2007). If we had used the assumption of perfect annuities markets to address accidental bequests (rather than the case in which the government collects the accidental bequests and redistributes them lump-sum), the consumption-age profile would be increasing over the entire life-cycle, as demonstrated by Hansen and İmrohoroğlu (2008). This behavior, however, would be at odds with the empirically observed hump-shaped consumption-age profile. Notice further that consumption declines as the households enter retirement. This consumption behavior results from the household's effort to smooth utility over its lifetime.

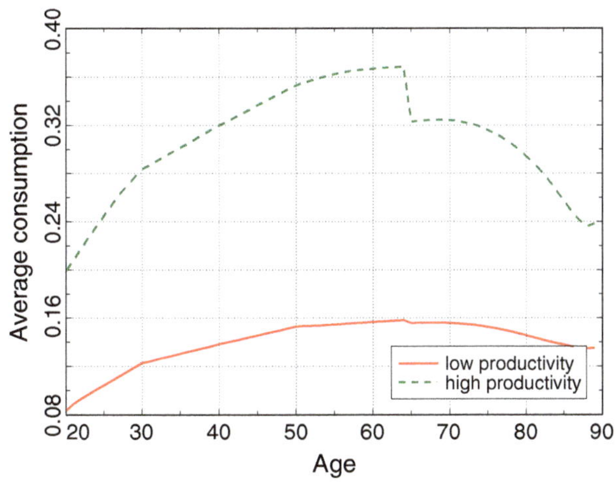

Fig. 6.11 Consumption-age profile

Since consumption and leisure are substitutes and leisure increases to 100% during retirement, consumption is reduced.[47]

The heterogeneity in individual productivity, $\epsilon\eta\bar{y}_s$, results in inequality in wages, income, and wealth. The Gini coefficient of wage income amounts to 0.379 and implies inequality in gross income and wealth, which are characterized by Gini coefficients of 0.390 and 0.669, respectively. The results are summarized in the second column of Table 6.6. Notice that the OLG model is able to generate much more inequality in wealth than in income as observed empirically.[48] However, all our inequality measures fall short of the values observed empirically. For example, Budría Rodriguez, Díaz-Giménez, and Quadrini (2002) report Gini coefficients of (gross) income and wealth equal to 0.553 and 0.803. Our model's inequality measures underestimate the empirical measures for two main reasons[49]: (1) We do not consider self-employed workers and entrepreneurs. Quadrini (2000) presents empirical evidence that the concentration of income and wealth is higher among entrepreneurs and that the introduction of entrepreneurs into a model similar to ours helps to reconcile the inequality in the model with that in the US economy. (2) We omit bequests. De Nardi and Yang (2016) develop a model that considers both bequests of wealth and inheritance of abilities from parents and is able to match the skewness of the distribution of income, wealth, and bequests.

[47]In Problem 6.6, you are asked to test whether consumption habits help to improve the modeling of consumption-age behavior in a standard Auerbach-Kotlikoff model that implies a downward jump in consumption at the age of retirement.

[48]One of the first studies to highlight the role of the OLG model in accounting for observed wealth heterogeneity was Huggett (1996).

[49]See De Nardi (2015) for a survey of modeling wealth heterogeneity in quantitative general equilibrium models.

Table 6.6 Allocative effects of social security

	2015			2050		
θ^p	50%	0%	14%	50%	0%	12%
$\tilde{Y}$	0.425	0.501	0.479	0.420	0.514	0.481
$\tilde{K}$	1.110	1.551	1.369	1.145	1.725	1.504
$\tilde{L}$	0.248	0.265	0.259	0.239	0.261	0.253
$\bar{l}$	0.298	0.319	0.310	0.307	0.336	0.324
τ^p	8.53%	0%	2.59%	10.92%	0%	2.94%
$\tilde{Tr}$	0.0229	0.0390	0.329	0.0179	0.0340	0.0286
Gini coefficients						
Wage income	0.379	0.367	0.373	0.374	0.360	0.366
Gross income	0.390	0.418	0.409	0.395	0.447	0.430
Wealth	0.669	0.604	0.631	0.665	0.588	0.616
Consumption	0.276	0.287	0.282	0.270	0.289	0.283
Liquidity-constrained	36.5%	26.6%	29.2%	35.8%	24.5%	27.0%
Δ	0%	1.30%	1.80%	0%	2.27%	2.91%

Note: Welfare is measured by the average lifetime utility of the newborn generation in stationary state. The welfare change Δ is computed as the consumption equivalent change relative to the benchmark case ($\theta^p = 50\%$) in the years 2015 and 2050

Our model is also able to replicate the fact that consumption inequality is considerably smaller than income inequality. Using US data from the Consumer Expenditure Survey, Krueger and Perri (2006) present evidence that the Gini coefficient of consumption was 0.26 in 2003, while it is equal to 0.276 in the model.

6.4.2.2 Abolition of Social Security

In the second and third columns of Table 6.6, the stationary-state allocation of the benchmark (with a net pension replacement rate θ^p of 50%) is compared with the case without social security ($\theta^p = 0\%$). The abolition of social security increases savings for old age considerably, and thus, the aggregate capital stock $\tilde{K}$ rises by 39.7%, from $\tilde{K} = 1.110$ to $\tilde{K} = 1.551$. In addition, the abolition of distortionary pension contributions τ^p increases the labor supply (which is also augmented because of the rise in the marginal product of labor) such that the average working hours increase by 7.0%, from $\bar{l} = 0.298$ to $\bar{l} = 0.319$. As a consequence, equilibrium output $\tilde{Y}$ increases by 17.9% in response to the abolition of social security, and tax revenue rises. Therefore, the government is able to increase the lump-sum transfers $\tilde{Tr}$ by 70%, from 0.0229 to 0.0390. The share of transfers relative to GDP, $\tilde{Tr}/\tilde{Y}$, also rises, from 5.4% to 7.8%.

Without social security, wage income is less concentrated because the substitution effect of a higher net wage rate affects the labor supply of the low-efficiency workers to a larger extent than that of the high-efficiency workers. Therefore, the Gini coefficient of wage income falls from 0.379 to 0.367. However, gross income is nevertheless more concentrated than in the case with social security because retired households with only interest income do not receive any income from pension payments in the no-social-security case. For this reason, the Gini coefficient of gross income increases from 0.390 to 0.418 if pensions are abolished. The inequality of the wealth distribution decreases without pensions because, in this case, many low-income workers have to save to provide for old age, and the number of households without any savings decreases from 36.5% to 26.6%. Accordingly, the Gini coefficient of wealth decreases considerably, from 0.660 to 0.604, if social security is abolished. Although the social security system redistributes from the income-rich to the income-poor, the distortionary effect of public pensions dominates, and welfare increases significantly by a consumption equivalent of 1.3% in the case without social security.

6.4.2.3 Optimal Pension Policy

The welfare effects of social security are ambiguous. On the one hand, public PAYG pensions increase the distortions affecting labor supply and savings because contributions are levied on labor income. On the other hand, social security insures against negative income shocks and against the risk of longevity if, in particular, annuity markets are missing. The overall effect can only be computed numerically.

The welfare changes associated with different net pension replacement rates θ^p are presented in Fig. 6.12. We find that the optimal benefit level in 2015 (solid red line) amounts to a replacement rate of 14% (if we assume the population to be stationary with the demographics, survival probabilities and population growth rate prevailing in 2015 and the economy to be in steady state). In this case, welfare increases by 1.80% of total consumption relative to the benchmark. This number likely represents a lower limit of the steady-state welfare effects. In particular, we assume in our computation that additional tax revenue is re-transferred lump-sum to the households. This assumption is not very realistic. Alternatively, we could have

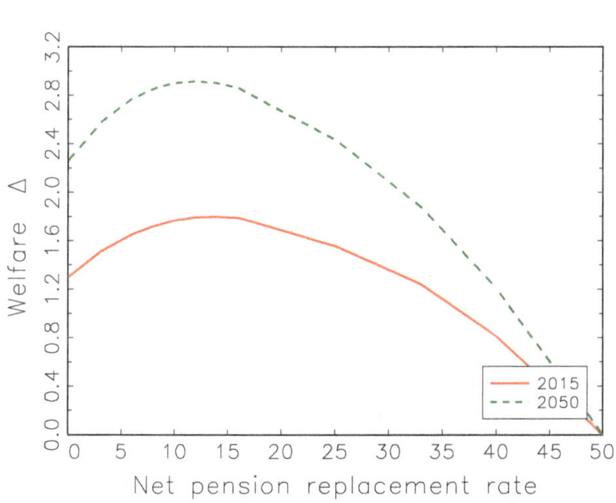

Fig. 6.12 Welfare effects of pension policy θ^p

Table 6.7 Decomposition of welfare effects

Household type	Δ
Low permanent productivity $\epsilon = 0.57$	
$\eta = 0.727$	−0.04%
$\eta = 1.273$	−1.11%
High permanent productivity $\epsilon = 1.43$	
$\eta = 0.727$	4.44%
$\eta = 1.273$	4.05%

Notes: The entries in the second column represent the consumption equivalent change that accrues to a 20-year-old household with the efficiency type $\{\epsilon, \eta\}$ resulting from a reduction of θ^p from 50% to the optimal rate $\theta^{p*} = 14\%$

assumed that the government reduces the wage income tax rate τ^w instead and thus that the distortion of the labor supply could have been further reduced. In this case, welfare gains would have been even higher, and you are asked to compute them in Problem 6.5.

Aggregate equilibrium values for the optimal pension policy $\theta^{p*} = 14\%$ are displayed in the fourth column of Table 6.6. Compared with the benchmark case, the economy with the optimal pension policy is characterized by higher capital stock $\tilde{K}$, aggregate labor $\tilde{L}$, and output $\tilde{Y}$. Comparing the second and the fourth columns, we observe that, for example, the capital stock increases by 23%, from $\tilde{K} = 1.110$ to $\tilde{K} = 1.369$. Accordingly, we should be careful to conclude from this comparative steady-state analysis that the government should decrease pensions from 50% to 14%. During the transition from the old steady state with $\theta^p = 50\%$ to the new steady state with $\theta^{p*} = 14\%$, households have to accumulate savings and forgo consumption, meaning that the generations during the transition between the two steady states suffer welfare losses. Recall also that we noted in the previous section that the transitional generations also suffer because they still have to pay the higher pensions for the retired workers, while they will only receive a smaller pension during their retirement. We will turn to the transition dynamics in the next section.

At this point, let us offer another word of caution. In Table 6.6, we report the average lifetime utility of newborns. However, some are fortunate and are endowed with high productivity at birth, while others have low productivity. If we reduce pension benefits from a replacement rate of 50% to the optimal rate of 14%, poor (or low-productivity) households will suffer disproportionally because they, in turn, benefit the most from PAYG pensions that are not defined contribution based. The low-income workers contribute relatively little in comparison with the rich households but receive the same lump-sum pension. In Table 6.7, we present the welfare gains or losses of the different productivity types $\epsilon\eta$. The workers with the high permanent productivity type, $\epsilon = 1.43$, benefit to a much larger extent from the reduction in social security benefits than do low-productivity workers and gain approximately 4% of their total consumption, while the workers with low productivity actually suffer welfare losses.

6.4.2.4 Aging and Optimal Pensions

The consequences of the demographic transition represent one of the greatest economic challenges facing modern economies. In the fourth column of Table 6.6, we present the equilibrium effects of an older population assuming that the population is stationary and characterized by the survival probabilities and birth rate projected by UN (2015) for the year 2050. The survival probabilities are presented by the broken green line in Fig. 6.8, and population growth falls to 0.41% in 2050, meaning that the implied stationary old-age dependency ratio OADR2 amounts to 36%.

As a consequence of aging, the share of workers in the population falls for a constant replacement rate $\theta^p = 50\%$, and thus, aggregate labor supply $\tilde{L}$ declines by 3.4%, from 0.248 to 0.239. The decline in aggregate (efficient) labor is smaller than the decline in the number of workers because (1) the age composition changes (such that the workers are more productive on average) and (2) the average working hours $\bar{l}$ increase by 3.0%, (from 0.298 to 0.307). If the pension policy remains unchanged and maintains a pension replacement rate of θ^p at 50%, the contribution rate τ^p has to increase from 8.53% to 10.92% between 2015 and 2050, and government transfers $\tilde{Tr}$ fall from 0.0229 to 0.0179 due to lower tax revenue. With respect to savings, we observe two opposing effects. On the one hand, we have a larger share of retired households who dissave. Furthermore, net income decreases because the pension contribution rate τ^p has to be augmented. On the other hand, the composition of workers changes, and the workforce ages. Since older workers have a higher savings rate than younger workers, aggregate savings increase because of this composition effect. Furthermore, all workers accumulate higher precautionary savings because they expect to live longer. The net effect is small but positive, and thus, the capital stock $\tilde{K}$ increases by a small amount, from 1.110 in 2015 to 1.145 in 2050. Furthermore, the inequality of the income distribution increases to a small extent, as the number of retired workers with a (relatively small) pension income rises.

How does a grayer population affect the optimal amount of pension payments? On the one hand, an increase in the old-age dependency ratio and a decline in the birth rate reduce the return from the pension system. On the other hand, retirees are becoming older on average, and thus, the (discounted) loss from old-age utility as a consequence of possible negative income shocks is decreased to a larger extent. The overall effect can only be computed numerically.

For the population in 2050 that is characterized by a higher old-age dependency ratio, the optimal pension policy consists of a net pension replacement rate that is somewhat smaller than the optimal one in 2015. In fact, the optimal pension rate θ^p falls from 14% in 2015 to 12% in 2050. The welfare effects of different pension policies θ^p are illustrated by the broken green line in Fig. 6.12. For the new optimum, $\theta^{p**} = 12\%$, consumption gains in 2050 are more than 50% larger than under the optimal policy in 2015 and amount to 2.91% of total consumption.

Therefore, we can conclude that pension reform policies are even more welfare-improving in economies with an older population than in younger economies.

In summary, we found two main results in this section, in which we focused on the steady state and neglected transitional dynamics. (1) Social security substantially decreases welfare for the present high levels of pensions that are observed in OECD countries and should be decreased to levels that are characterized by a net pension replacement rate θ^p below 20%. (2) The optimal level of pensions is somewhat smaller in an economy with a grayer population, and the quantitative welfare effects of an optimal pension policy are even more significant and in the amount of several percentage points of total consumption. This result is found to be rather robust in the literature. For example, Heer (2018) shows that the two results above continue to hold if we assume recursive preferences, a lower intertemporal elasticity of substitution, a lower Frisch labor supply elasticity,[50] or additional income uncertainty from unemployment.[51]

6.4.3 Transition Analysis

In the following, we also consider the transitional dynamics associated with a change in pension policy and during the demographic transition. To do so, we assume that the economy is in steady state in 2015 and that the population evolves according to the population projections for the US economy during the period 2015–2100. In particular, we assume that the survival probabilities and the population growth rates of the model population are equal to those forecasted by UN (2015). Starting in 2100, the demographic variables are constant and equal to those prevailing in the year 2100, with a population growth rate equal to 0.2%. Therefore, the dependency ratio increases from 27.0% in 2015 to 43.0% in the long run. The behavior of the dependency ratio is illustrated in Fig. 6.13. Notice that the dependency ratio does not stabilize until the year 2150, despite that we assume that

---

[50] İmrohoroğlu and Kitao (2009) also study the effect of the Frisch labor supply elasticity on aggregate labor and the labor-age profile. They distinguish between two different scenarios for the pension reform, consisting of the downsizing of the system by 50% or the total elimination of social security. İmrohoroğlu and Kitao show that the effect of pension reforms on aggregate labor is rather insensitive to the Frisch elasticity, while the profile of hours over the life-cycle is highly sensitive. They also find substantial welfare gains from the reduction in pensions even in the case of a low labor supply elasticity. According to their Table 6.2, the long-run welfare gain of half-privatization amounts to 4.3% of total consumption for a low Frisch elasticity equal to $\eta_{lw} = 0.5$. In contrast to our approach, however, they do not model permanent productivity differences between the workers, and thus, income heterogeneity is smaller in their model than in ours.

[51] We neglect one factor that might increase the welfare effects of social security, however. Fuster, İmrohoroğlu, and İmrohoroğlu (2003) find that in the case of two-sided altruism towards ancestors and descendants, the welfare effects of social security are enhanced. Altonij, Hayashi, and Kotlikoff (1997), however, present empirical evidence that rejects the implications of altruism for intergenerational risk-sharing behavior.

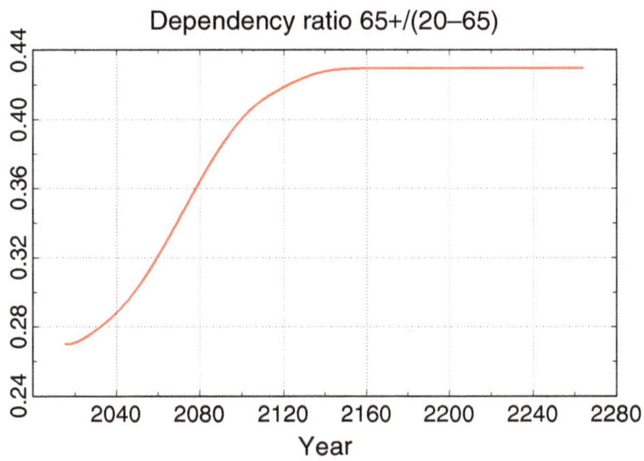

Fig. 6.13 Demographic transition in the OLG model

the survival probabilities and birth rates are constant after 2100. In our computation, we therefore assume that the transition to the new steady state is complete by the year 2250. In addition, the levels of government expenditures and the tax rates are held constant at the levels prevailing in the steady state in 2015 such that government transfers $\tilde{T}r$ adjust to balance the government budget (6.50).[52]

In 2015, the government announces an unexpected policy change that becomes effective in the same year. The new pension policy consists of a change in the replacement rate from 50% to $\tilde{\theta}^{p**} = 12\%$ that was found optimal for the stationary state population in 2050. We also present the results for the case of no policy change, $\tilde{\theta}^{p} = 50\%$, and one intermediate case, $\tilde{\theta}^{p} = 30\%$. In addition, we assume that the policy is implemented gradually and stretched over a period of $n_\theta = 45$ years such that the number of years accord with the length of the working life. Furthermore, the net pension replacement rate decreases linearly over the implementation period.[53] Figure 6.14 illustrates the pension policy θ^p for the case of a reduction to $\theta^{p**} = 12\%$.

Figure 6.15 plots the evolution of aggregate capital $\tilde{K}$, aggregate labor $\tilde{L}$, aggregate output $\tilde{Y}$, the pension contribution rate τ^p, and government transfers $\tilde{T}r$ (which are equal to household transfers $\tilde{t}r$) during the transition for the three policies with a long-run pension replacement rate θ^p equal to 50% (solid red line), 30% (broken green line), and 12% (broken and dotted blue line). The results are in line with those of similar transition experiments in the literature and our comparative steady-state analysis in the preceding section. During the demographic transition, the population is aging due to increasing life expectancy, and the old-age

---

[52]More precisely, we assume that the per capita government expenditures grow at the exogenous rate of technological growth.

[53]For example, Kitao (2014) also considers a linear adjustment over a period of 50 years.

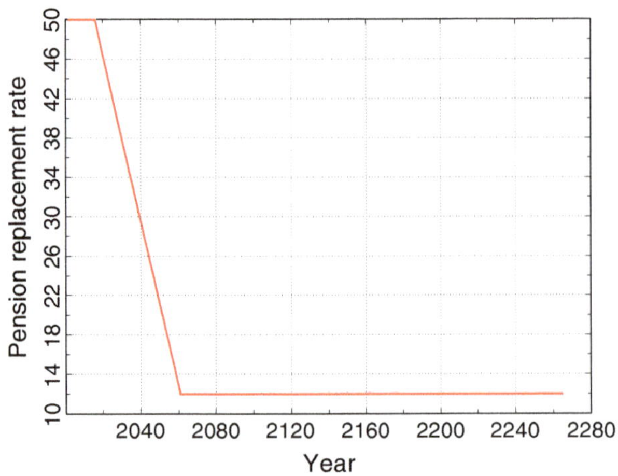

Fig. 6.14 Pension policy change and net pension replacement rate θ^p

dependency ratio of the 65+/(20–65) year-olds increases from 27% to 43% between 2015 and 2200. As a consequence, the labor force share and, hence, aggregate labor $\tilde{L}$ is shrinking.[54] There are multiple effects of the demographic transition on savings. On the one hand, a higher share of the population is retired and decumulates savings. On the other hand, households live longer, and therefore, workers accumulate more retirement savings. If pensions are also decreased, the latter effect compensates for the former effect, and aggregate savings and the capital stock increase over time. Since capital increases relative to labor over time, the real interest rate decreases, while the wage rate increases (not presented).

When the pension replacement rate θ^p is gradually reduced to its optimal steady-state rate $\theta^{p**} = 12\%$ for the year 2050, both pensions *pen* and the contribution rate τ^p (see the broken blue line in the middle-right panel of Fig. 6.15) fall until the year 2060. As a consequence, the net wage rate $(1 - \tau^w - \tau^p)w_t$ increases, and the individual augments his labor supply. Therefore, aggregate labor actually increases during the initial phase of the transition during the years 2015–2060 for the pension policies $\theta^p \in \{12\%, 30\%\}$. Thereafter, the effect of a shrinking labor force dominates, and $\tilde{L}$ declines to its new long-run equilibrium value. The dynamics of the capital stock over time are also hump-shaped for $\theta^p \in \{30\%, 50\%\}$, and aggregate savings peak later, around the year 2080, due to the sluggishness of the capital stock.[55] Therefore, aggregate output $\tilde{Y}$ also displays a hump-shaped profile over time. In the case of medium to high pensions, $\theta^p \in \{30\%, 50\%\}$, government

[54]Recall that aggregate labor L_t in period t is expressed relative to total population N_t.

[55]For example, young workers in the years 2050–2060 supply the highest number of working hours during the entire transition period, but their wealth peaks only at the end of their working life in later years.

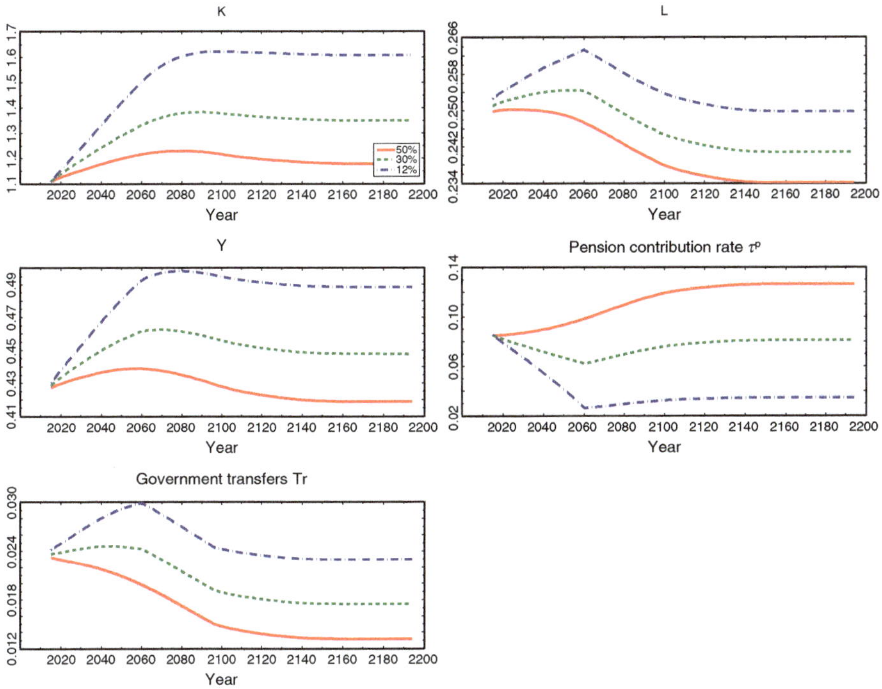

Fig. 6.15 Transition dynamics in the OLG model

transfers $\tilde{T}r$ to households have to decline in the medium and long run relative to the year 2015 because the share of government expenditures (government consumption) increases relative to GDP, while government revenues shrink.[56]

Figure 6.16 presents the welfare effects of a switch from the present policy, $\theta^p = 50\%$ to the two pension reform policies, $\theta^p \in \{30\%, 12\%\}$, for the individual generations that enter the labor force during the years 1946–2200. The first generation that is affected by this change in policy is the one that is still alive in the year 2015 and entered the labor force at age 20 in the year 1946. Since this generation is only affected in the last period of life, the effect on lifetime utility is negligible and close to zero. Later generations, however, suffer substantial welfare losses, which are the largest for those agents who enter the labor force around the year 1995 and are in the mid-period of their working life when the policy change is implemented in the year 2015. For those households, average lifetime utility declines by a consumption equivalent of 3.8% or 8.5% depending on the policy, $\theta^p \in \{30\%, 12\%\}$. These households receive a lower pension in old age but still

[56]Recall that we assumed that $\tilde{G}$ would remain at its 2015 level. In addition, government revenue from accidental bequests declines due to higher survival probabilities. The latter effect, however, is rather modest.

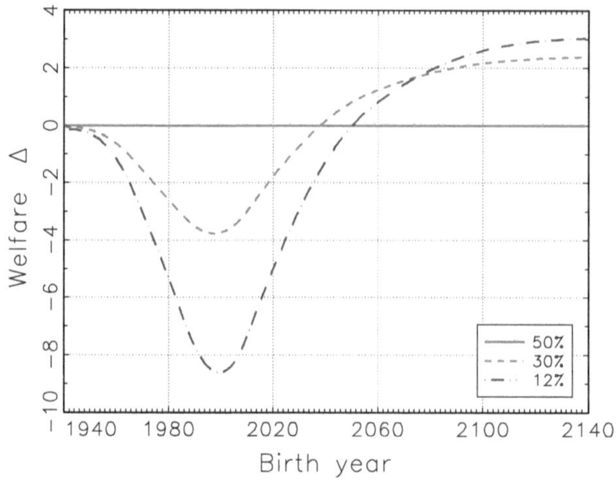

Fig. 6.16 Generational welfare and pension policy reform $\theta^p \in \{50\%, 30\%, 12\%\}$

have to provide for those retirees with the higher pension during the transition. Households born after the years 2040 and 2050 benefit from the new pension policy via a reduction in θ^p to 30% and 12%, respectively. Notice that the number of generations that will benefit from such a policy depends on the boldness of the reform. The stronger the reform, the longer it takes for its benefits to manifest as improved generational welfare. The implementation of a policy that maximizes steady-state lifetime utility, $\theta^p = 12\%$, therefore, implies significant costs for the current and even the initial future generations. Note further that if the government had asked the voters in 2015 if it should implement a pension policy change, it would not have obtained a majority, as no generation living would have benefitted from such a reform.[57,58]

6.5 Quantitative Studies of Pension Reform Proposals

The literature includes many quantitative studies on the effects of pension reforms akin to that in the previous section. The majority of these general equilibrium studies are based on OLG models with income uncertainty that are calibrated with respect

[57] Recall that we consider only the average lifetime utility of the individual generations. The welfare effects might vary considerably across the different productivity types (see also Table 6.7 for the steady-state analysis for the year 2015).

[58] A study that focuses on the political implementability of a transition from the status quo to a reduction in PAYGO pensions in the US is provided by Conesa and Krueger (1999). In accordance with our argument, they find that although the transition to a fully funded pension system would imply substantial welfare gains, a majority of voters would be worse off from this option and thus favor the status quo.

to the US economy. As one of the first such studies, the seminal work on US demographics and social security is Auerbach and Kotlikoff (1987), who study the effect of (1) a reduction in the pension replacement rate with respect to wages from 60% to 40% and (2) an increase in the retirement age from 65 to 67.[59] These policies result in considerable welfare effects of several percentage points.

While Auerbach and Kotlikoff only study two possible situations for the demographic transition (one with a sudden drop in the fertility rate and one with a boom and bust cycle), De Nardi, Imrohoroğlu, and Sargent (1999) use the 'medium' population projection of the Social Security Administration as an input into their model. In addition to the model in the previous section, they also introduce bequests to improve the modeling of the capital stock (relative to GDP) and its distribution. Bequests are introduced as a warm glow, meaning that instantaneous utility has an additive component that depends on the amount of bequests. To keep the model tractable and computable,[60] they assume a special function of utility from consumption and disutility of labor (both quadratic and additive). In addition, the insurance properties of the social security are not motivated by a temporary shock to individual labor productivity but rather a shock to the wealth endowment. As a consequence of these assumptions, individual policy functions (e.g., individual consumption) are a linear function of individual state variables (in particular, wealth), meaning that aggregation is straightforward and does not depend on the distribution of wealth (in contrast to our model above). De Nardi, Imrohoroğlu, and Sargent (1999) analyze different policies to finance additional expenditures on pensions due to the demographic transition, including raising different taxes (consumption, labor income), reducing benefits, or increasing the mandatory retirement age. In addition, the authors also account for the welfare of the cohorts during the transition. They find that the only policy of those considered in their paper that raises the welfare of all generations is one that switches to a purely defined contribution system. A model that is closely related to ours is provided by İmrohoroğlu, İmrohoroğlu, and Joines (1995). These authors find that the optimal social security replacement rate amounts to 30% and that the benefits are equal to 2.08% of GNP.[61] In contrast to our model, they assume that there are only two

[59]See pages 174–177 in Auerbach and Kotlikoff (1987).

[60]Recall (if your age allows for it) that computer technology in these years was less capable of handling such numerical problems with a high dimension of (individual) state variables.

[61]As noted by İmrohoroğlu, İmrohoroğlu, and Joines (1999), this high value for optimal pensions results from the fact that their model is characterized by dynamic inefficiency in the absence of social security. Higher pensions and, hence, lower savings actually increase total consumption at low replacement rates. In addition, İmrohoroğlu, İmrohoroğlu, and Joines (1999) argue that the US economy is dynamically efficient, as shown by Abel, Mankiw, Summers, and Zeckhauser (1989). In our model above, we only consider dynamically efficient economies in which the population growth rate is below the economic growth rate. İmrohoroğlu, İmrohoroğlu, and Joines (1999) also include land as a (constant) production factor in addition to capital and labor and, as a consequence, their economy is dynamically efficient. They find the optimal unfunded PAYG public pensions in the US to be zero in the stationary state.

types of households, employed and unemployed. Therefore, they do not model the heterogeneity of the wage rate distribution that is observed empirically except for the age-efficiency profile. Within cohorts, all workers have equal individual productivity. In addition to the within-cohort heterogeneity, we also endogenize the individual's labor supply decision and, therefore, the distortion to the labor supply, and thus, our optimal pension replacement rate is lower.[62] We already introduced technological growth into our benchmark, while İmrohoroğlu, İmrohoroğlu, and Joines (1995) only provide a sensitivity analysis for this case.[63]

In addition to our above problem of the optimal pension, Fehr, Kallweit, and Kindermann (2013) compute the optimal mix between flat and earnings-related pensions for the German pension system. They find that the flat-rate pension share should be equal to 30% of total pensions to optimize the trade-off between the increased labor supply distortion and the benefit from increased insurance provision against labor market risk. In addition to our model above, Fehr, Kallweit, and Kindermann (2013) endogenize the decision on the retirement age and also allow for disability risk, reflecting the fact that 20% of new entries into the German pension system are due to disability. In their analysis, however, the contribution rate τ^p is set to be constant; moreover, they abstract from population growth.

Many studies, e.g., Nishiyama and Smetters (2007) or Kitao (2014), compute the transition dynamics, as we did in the previous section. In addition, they go one step further in their analysis, following Auerbach and Kotlikoff (1987), and ask whether it is possible to compensate the losers who are alive in 2015 from a change in the social security system and still induce long-run welfare gains. In our setup in the previous section, we might wonder whether a transfer from the generations that are born after the year 2015 to those born before is possible such that these agents are indifferent between a pension reform that reduces the replacement rate from 50% to 12% and a constant pension policy. This transfer is also called *Hicksian compensation*. To finance these transfers, the government must accumulate debt (and sometimes assets under different policies) that has to be financed by the generations born after the year 2015.[64] For those generations that are born after 2015, the government debt is re-distributed among them such

[62]Another study with exogenous labor supply that focuses on the distortion of social security contributions affecting the accumulation of capital is Storesletten, Telmer, and Yaron (1999). The main channel emphasized in their model is the financing of pensions with a distortionary income tax that is levied on labor and capital income. Since labor supply is exogenous, the distortion only affects capital accumulation. The authors compare the current system (as of 1996) to alternative scenarios including the abolition of the social security system and a system that is partially PAYG and partially fully funded. They find the alternatives to imply significant welfare gains if general equilibrium effects are taken into account.

[63]İmrohoroğlu, İmrohoroğlu, and Joines (1995) find that the optimal level of social security "appears to be zero when ... we incorporate exogenous technological progress in the model".

[64]Notice that another asset variable in the form of government debt enters our model, and in general equilibrium, the sum of debt (equivalently, government assets) and capital is equal to aggregate savings.

that the welfare effect (as measured by the consumption equivalent change of the average newborn) is equalized across all these future generations. This change in welfare for the generations during the period 2015–2200 is called the Hicksian efficiency gain.[65] If future generations benefit from such a combination of pension and debt policy, the pension reform is welfare-enhancing. Using this approach, Nishiyama and Smetters (2007) analyze a 50% privatization of social security and find the welfare effects to be sensitive to the assumptions of a closed economy, missing annuities markets, and the progressivity of pensions. Similarly, Kitao (2014) compares four different financing policies to keep social security sustainable but does not derive the optimal pension. In particular, Kitao (2014) compares polices that (1) increase the payroll tax while keeping the benefit level constant, (2) keep the payroll tax constant, (3) increase the retirement age, and (4) introduce means-tested benefits. In accordance with our results, he finds that reducing the benefit is the most efficient policy in the long run.

In comparison to the demographics in the United States, many other industrialized countries are experiencing more rapid aging, especially in Japan or in many areas of Europe, e.g., in Italy and Germany. Modeling the US as an open economy, Krueger and Ludwig (2007) demonstrate in a multi-country OLG model that the stronger worldwide decline in the labor force relative to aggregate savings reduces the interest rate in the US economy even more markedly, while the increase in the wage rate is also reinforced.[66] The effect on factor prices is shown to be sensitive to the pension reform considered (cutting pensions or increasing the retirement age or contributions). They find that due to the compositional effects of an aging workforce, income inequality increases over the subsequent decades. The welfare effects of the demographic transition depend on the age of the household. Younger households gain because wage rates increase, while older households lose. Attanasio, Kitao, and Violante (2007) consider a two-region model that includes both a North (North America, Europe, Japan, Australia and New Zealand) and a South (Africa, Asia (excluding Japan), Latin America, and the Caribbean) region. While the former is at the end of the demographic transition, the latter is in the midst of it. They note that the welfare effects of pension policies diverge significantly between the closed- and the open-economy models.

The above studies all assume that productivity growth is constant and independent of the age distribution within the population. In contrast, Heer and Irmen (2014) endogenize growth. In their model, firms have a higher incentive to invest in labor-saving technological progress if labor becomes scarcer (relative to capital). Again, the quantitative effect of aging on the growth rate is sensitive to the particular

[65] You are asked to compute the Hicksian efficiency gain for the two-period OLG model from Sect. 6.3.2 in Problem 6.4.

[66] To keep the model tractable, Krueger and Ludwig (2007) assume that pensions depend only on the permanent efficiency type, not on the stochastic individual component. In addition, the authors study the transition dynamics under the assumption that contribution rates freeze in 2004. Beyond these assumptions, the model closely resembles that in the previous section.

pension reform considered. They find that the average annual growth rate, which amounted to 1.74% during the period 1990–2000, increases to 2.41% in 2100 when the replacement rate of pensions is held constant and financed by additional contributions. The effect is even larger if the contribution rate is frozen at its 2000 level.

In summary, the studies of the demographic transition above come to the conclusion that the quantitative effects of the demographic transition on income and its distribution are quantitatively significant. The government's policy responses to social security issues are crucial in this matter and make a substantial difference with respect to income, its distribution, welfare, and growth. The common factor of the studies reviewed above is that contribution rates should not be increased over the coming decades; rather pensions should be reduced and the retirement age be raised (or, at least, workers should be able to choose a longer working life).

6.6 Demography and the Fiscal Space

The demographic transition in modern industrialized (as well as developing) countries has significant effects on the sustainability of public finances. Dependency ratios, as presented in Sect. 6.2, are on the rise and will double in many countries between 2015 and 2050, implying twice as many retirees relative to workers. This development will seriously impact both sides of public finance, revenues and spending. On the one hand, a grayer population has a smaller number of workers and, hence, income tax payers. Tax revenue will decline. On the other hand, public expenditures on social security will increase. As we considered in the previous sections of this chapter, public PAYG pensions will rise relative to GDP (and wage income), and thus, pension contributions will have to rise if pension replacement rates are left unchanged. Similarly, health expenditures for the older population will rise relative to GDP.

To study the sustainability of public finance, we introduce the concept of *fiscal space*. Let us assume that (1) government expenditures also include both public pension and public debt payments and that (2) the government holds the debt level constant. Therefore, government expenditures are equal to tax revenue. In addition, we assume that the government only collects revenue from income taxes and social security contributions (on wage income). Figure 6.17 presents the tax revenue (including social security contributions) and government expenditures G_{2015} for different tax rates τ. Tax revenue is hump-shaped and equal to zero for a tax rate τ of 0% or 100%, as you learned in Sect. 5.5 on the Laffer curve. The level of government expenditures is assumed to be independent of the income tax rate τ and is illustrated by the line G_{2015}. To balance the budget (in 2015), the fiscal authority

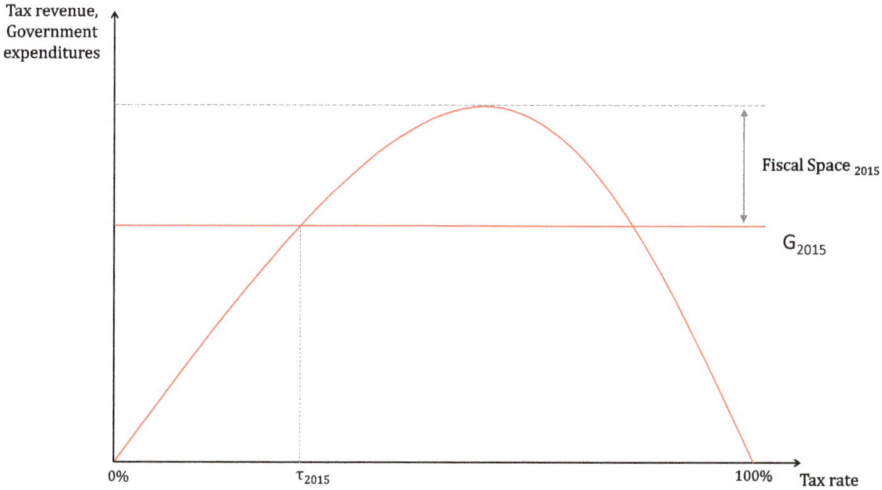

Fig. 6.17 The Laffer curve and the fiscal space

has to set the income tax rate equal to τ_{2015}.[67] In the depicted situation for the year 2015, the fiscal authority is still able to further increase tax revenue (and, hence, government spending). As a concept to measure fiscal sustainability, we define the fiscal space as the additional possible tax revenue relative to total tax revenue (or, alternatively, relative to GDP).

Aging will affect both curves, the Laffer and government expenditures curves. First, possible tax revenue will fall as the (relative) number of workers declines. Therefore, the Laffer curve moves inward between 2015 and 2050, as illustrated in Fig. 6.18. Second, government expenditures (on social security) will rise, and thus, the line for government expenditures in the year 2050, G_{2050}, lies above the line for present government expenditures in the year 2015, G_{2015}. As a consequence, the fiscal space shrinks during the demographic transition.

Heer, Polito, and Wickens (2017) study the sustainability of pension systems for the United States and 14-EU countries. To do so, they use an OLG model similar to that in Sect. 6.4.1 above.[68] They calibrate the model with respect to the individual characteristics of the 15 countries and use UN projections for the dependency ratios to estimate the year when the fiscal space will have shrunk to zero, meaning that public finances are no longer sustainable. The findings are rather mixed for the group of countries considered and depend on many factors including demographics, the

[67]There are two different tax rates that fulfill the condition of a balanced budget (the two points of interception of the Laffer curve and the line of government expenditures G_{2015}); naturally, the government chooses the lower tax rate on the upward-sloping side of the Laffer curve.

[68]However, they simplify the model by not considering income uncertainty.

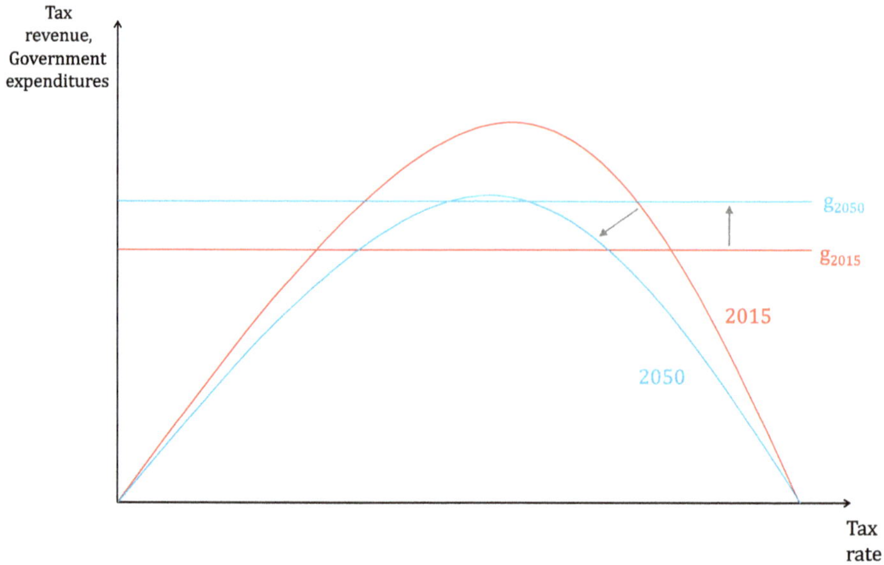

Fig. 6.18 Fiscal sustainability and aging

generosity of pension systems, and the level of total government expenditures. For the United States in 2010, for example, Heer, Polito, and Wickens (2017) estimate a fiscal space of 32% of additional tax revenue (holding the pension levels constant at their 2010 level), while the fiscal space only amounts to 6% on average in the EU-14 countries. Moreover, countries such as the United Kingdom, Ireland, and the United States are characterized by a very reassuring situation, and public finances in the United States, for example, do not become unsustainable over the time horizon 2015–2100. Many continental European countries, e.g., Germany, France, Italy, or Spain, reach the limit of their fiscal spaces over the next 20 years. Table 6.8 presents the estimates of Heer, Polito, and Wickens (2017) for two scenarios, an unchanged pension policy and an increase in the retirement age to 70. Evidently, reforming pension policy helps to improve fiscal sustainability in all countries, although the fiscal outlook remains rather bleak for Italy and Spain. The table also contains recent estimates from Braun and Joines (2015) for Japan. They find in a model that is very similar to ours[69] that even after the inclusion of planned increases in taxes and planned reductions in pensions, current fiscal policies are unsustainable. They project a sovereign debt crisis in Japan by 2039.[70]

[69]Their model is in some respects more elaborate than ours, in particular with respect to the projection of public medical expenditures; it also includes age-specific fertility rates. In addition, the authors assume that the interest rate on government debt is 1.145% lower than the rate of return on capital throughout the transition.

[70]We will discuss the role of debt in the next chapter.

Table 6.8 Years when the fiscal space is forecasted to be zero

Country	No reform	Retirement age 70
France	2025	2085
Germany	2035	—
United Kingdom	2090	—
Ireland	2075	—
Italy	2030	2040
Japan	2039	
The Netherlands	2035	—
Spain	2035	2050
US	—	—

Notes: Estimates are taken from Heer, Polito, and Wickens (2017) for Europe and the US and from Braun and Joines (2015) for Japan. A '—' indicates that the fiscal space remains strictly positive. For Japan, no estimates are available for the case of later retirement

Appendix 6.1: Accidental Bequests

To understand why accidental bequests in (6.56e) are given by the sum of next-period assets $\tilde{k}'$ and the interest $r_{t+1}\tilde{k}'$, we will consider a simplified two-period model. In particular, let $N_t(1)$ and $N_t(2)$ denote the sizes of the young and old generations in period t. The survival probability of the young is denoted by ϕ_1, and population grows at rate n, implying:

$$N_{t+1}(2) = \phi_1 N_t(1), \tag{6.61a}$$

$$N_{t+1}(1) = (1+n)N_t(1). \tag{6.61b}$$

Consequently, total population is given by

$$N_t = N_t(1) + N_t(2).$$

Total accidental bequests are confiscated by the government and redistributed lump-sum to the total population in the amount $N_{t+1}tr_{t+1}$ in period $t+1$:

$$Beq_{t+1} = N_{t+1}tr_{t+1}.$$

For the rest of the model, we stipulate that the standard equilibrium conditions of the two-period OLG model in Chap. 3 hold. Therefore, the aggregate capital stock K_{t+1} at the beginning of period $t+1$ is equal to total savings of the young generation at the end of period t, $N_t(1)s_t$; consumption of the young c_{t+1}^1 is equal to their wage income plus transfers minus savings; and consumption of the old c_{t+1}^2 is equal to savings plus interest and transfers. Finally, the factor prices are equal to their marginal products. These conditions are summarized by the following equations:

$$N_t(1)s_t = K_{t+1}, \tag{6.62a}$$

$$N_{t+1}(1)s_{t+1} = K_{t+2}, \tag{6.62b}$$

$$c_{t+1}^1 = w_{t+1}l_{t+1} + tr_{t+1} - s_{t+1}, \tag{6.62c}$$

$$c_{t+1}^2 = (1 + r_{t+1})s_t + tr_{t+1}, \tag{6.62d}$$

$$r_{t+1} = F_K\left(K_{t+1}, l_{t+1}N_{t+1}(1)\right) - \delta, \tag{6.62e}$$

$$w_{t+1} = F_L\left(K_{t+1}, l_{t+1}N_{t+1}(1)\right). \tag{6.62f}$$

As before, we assume that production is characterized by constant returns to scale, implying:

$$Y_t = F(K_t, l_t N_t(1)) = F_K\left(K_t, l_t N_t(1)\right) K_t + F_L\left(K_t, l_t N_t(1)\right) l_t N_t(1).$$

In the goods market equilibrium:

$$K_{t+1}(1-\delta) + Y_{t+1} = N_{t+1}(1)c_{t+1}^1 + N_{t+1}(2)c_{t+1}^2 + K_{t+2}. \tag{6.63}$$

Inserting the above equations into the equation for the goods market equilibrium, we derive

$$\begin{aligned}
&K_{t+1} + w_{t+1}l_{t+1}N_{t+1}(1) + r_{t+1}K_{t+1} \\
&= w_{t+1}l_{t+1}N_{t+1}(1) + N_{t+1}(1)tr_{t+1} - N_{t+1}(1)s_{t+1} \\
&+ N_{t+1}(2)\left[1 + r_{t+1}\right]s_t + N_{t+1}(2)tr_{t+1} + K_{t+2},
\end{aligned}$$

and therefore,

$$\begin{aligned}
K_{t+1} + r_{t+1}K_{t+1} &= N_{t+1}tr_{t+1} + N_{t+1}(2)\left[1 + r_{t+1}\right]s_t \\
&= N_{t+1}tr_{t+1} + \phi_1 N_t(1)\left[1 + r_{t+1}\right]s_t \\
&= N_{t+1}tr_{t+1} + \phi_1\left[1 + r_{t+1}\right]K_{t+1}.
\end{aligned}$$

Consequently,

$$Beq_{t+1} = N_{t+1}tr_{t+1} = (1 - \phi_1)\left[1 + r_{t+1}\right]K_{t+1}. \tag{6.64}$$

Appendix 6.2: Computation of the Large-Scale OLG Model in Sect. 6.4

Solving for the stationary equilibrium is a computational challenge. Recall how we solved the two-period OLG model with a PAYG pension system and defined contribution benefits in Sect. 6.3.2. We were able to directly solve the first-order conditions of the households together with the equilibrium equations. In particular, the optimality condition with respect to the labor supply was given by (6.33c). In the large-scale OLG model in Sect. 6.4, the computational problem is much more complicated. We cannot solve for the optimal labor supply for all workers alive because of their high numbers. For this reason, consider the first period of life, $s = 1$. In this cohort, we have four different types of workers with the productivity types $\eta_i \epsilon_j$ with $i = 1, 2$ and $j = 1, 2$. They all accumulate different amounts of savings. At age 2, they may change their productivity type such that the number of heterogeneous workers increases to $4 \times 2 = 8$. Since the workers accumulate different amounts of savings depending on their employment history, the workers in a cohort are different (with respect to their savings) even for the same productivity type $\epsilon_i \eta_j$. Prior to retirement at age $s = 45$, we observe $4 \times 2^{44} = 7.0 \cdot 10^{13}$ different workers. We cannot solve this problem with the help of direct computation for such a large number of agents. Instead, we use value function iteration.

Value Function Iteration

To describe the optimization problem, we use a recursive representation of the consumer's problem, following Stokey, Lucas, and Prescott (1989). This specification is very amenable to the solution methods described below. Let $V_t(k_t^s, s, \epsilon, \eta)$ be the value of the objective function of the s-year-old agent with wealth k_t^s, age s, permanent efficiency type ϵ, and individual productivity η in period t. The value function $V_t(.)$ is equal to the optimized discounted expected lifetime utility. Thus, for the individual during the last period of his life, $s = 70$, the stationary value function is simply given by:

$$V_t(k_t^{70}, 70, \epsilon, \eta) = \max_{c_t^{70}, k_{t+1}^{71}} u(c_t^{70}, 1)$$

subject to $k_{t+1}^{71} \geq 0$ and the budget constraint (6.55) noticing that $c_t^s = A_t \tilde{c}_t^s$ and $k_t^s = A_t \tilde{k}_t^s$. Obviously, the optimal policy is given by $\tilde{k}_{t+1}^{71} = 0$ and

$$\tilde{c}_t^{70} = \tilde{pen}_t + \left[1 + (1 - \tau^K) r_t\right] \tilde{k}_t^{70} + \tilde{tr}_t.$$

The household completely consumes its income (from pensions and interest) and wealth during the last period of life.

In the second-to-last period of life, $s = 69$, the value function is given by the following equation:

$$V_t(k_t^{69}, 69, \epsilon, \eta) = \max_{c_t^{69}, k_{t+1}^{70}, c_{t+1}^{70}, k_{t+2}^{71}} \left\{ u(c_t^{69}, 1) + \beta \mathbb{E}_t u(c_{t+1}^{70}, 1) \right\}$$

subject to the budget constraint (6.55) in periods t and $t+1$. It will be convenient to transform the above equation into one with stationary values. For this reason, divide the equation by $A_t^{\iota(1-\sigma)}$, which results in

$$\begin{aligned} &\tilde{V}_t(\tilde{k}_t^{69}, 69, \epsilon, \eta) \\ &= \max_{\tilde{c}_t^{69}, \tilde{k}_{t+1}^{70}, \tilde{c}_{t+1}^{70}, \tilde{k}_{t+2}^{71}} \left\{ u(\tilde{c}_t(69), 1) + (1+\gamma)^{\iota(1-\sigma)} \beta \mathbb{E}_t u(\tilde{c}_{t+1}^{70}, 1) \right\}, \end{aligned}$$

with $\tilde{V}_t \equiv V_t / A_t^{\iota(1-\sigma)}$. In addition, we have used the fact that

$$\frac{u(c_{t+1}^{70}, 1)}{A_t^{\iota(1-\sigma)}} = \frac{(c_{t+1}^{70})^{\iota(1-\sigma)}}{1-\sigma} \frac{1}{A_t^{\iota(1-\sigma)}} = \frac{(\tilde{c}_{t+1}^{70})^{\iota(1-\sigma)}}{1-\sigma} (1+\gamma)^{\iota(1-\sigma)}.$$

Equivalently, the above dynamic equation can be restated as a recursive equation as follows:

$$\begin{aligned} \tilde{V}_t(\tilde{k}_t^{69}, 69, \epsilon, \eta) = \max_{\tilde{c}_t^{69}, \tilde{k}_{t+1}^{70}} &\left\{ u(\tilde{c}_t^{69}, 1) \right. \\ &\left. + (1+\gamma)^{\iota(1-\sigma)} \beta \mathbb{E}_t \tilde{V}_{t+1}(\tilde{k}_{t+1}^{70}, 70, \epsilon, \eta) \right\}. \end{aligned}$$

For the household aged s, we can use this recursive formulation more generally as follows:

$$\tilde{V}_t(\tilde{k}_t^s, s, \epsilon, \eta) = \begin{cases} \max\limits_{\tilde{k}_{t+1}^{s+1}, \tilde{c}_t^s, l_t^s} \left[u\left(\tilde{c}_t^s, l_t^s\right) \right. \\ \left. + (1+\gamma)^{\iota(1-\sigma)} \beta \mathbb{E}_t \tilde{V}_{t+1}(\tilde{k}_{t+1}^{s+1}, s+1, \epsilon, \eta') \right], \\ \quad s = 1, \ldots, R-1 \\ \\ \max\limits_{\tilde{k}_{t+1}^{s+1}, \tilde{c}_t^s} \left[u\left(\tilde{c}_t^s, 1\right) + \right. \\ \left. (1+\gamma)^{\iota(1-\sigma)} \beta \mathbb{E}_t \tilde{V}_{t+1}(\tilde{k}_{t+1}^{s+1}, s+1, \epsilon, \eta) \right], \\ \quad s = R, \ldots, J, \end{cases} \tag{6.65}$$

subject to (6.55) and (6.46). Equation (6.65) is also known as the *Bellman equation*. We will exploit its recursive nature to compute the optimal policy functions of the households (for given factor prices $\{w_t, r_t\}$, government transfers $\tilde{tr}_t$, and pension policies $\tilde{pen}_t$).

We solve for the optimization problem of the household starting in the last period of its life, working our way back to the first period of the household's life. As we do not know the exact value of $\tilde{k}_t^{70}$, we compute the value function $\tilde{V}(\tilde{k}, 70, \epsilon, \eta)$ over a range of $\tilde{k} \in [\tilde{k}_{min}, \tilde{k}_{max}]$. Since we assume a liquidity constraint $\tilde{k} \geq 0$, we choose $\tilde{k}_{min} = 0$. Finding a good value for the upper limit of the interval $\tilde{k}_{max}$ is more difficult. Since we do not yet know the equilibrium values of w, r, $\tilde{tr}$, or $\tilde{pen}$, we do not know the maximum income of households that is required to obtain a possible guess for the upper limit of savings. We, instead, advocate for a rather pragmatic procedure. We will calibrate our model such that the real interest rate r is approximately equal to 4% as observed in the US economy. In addition, we know the mass of working agents in our model, which is approximately 2/3. Assuming that average productivity is equal to one and agents work approximately 30% of their available time (we will calibrate the model accordingly below), we obtain a rough estimate of $\tilde{L} \approx 0.2$. From the first-order condition of the firm, we know that $r = \alpha \tilde{k}^{\alpha-1} \tilde{L}^{1-\alpha} - \delta$. Consequently, we can compute an approximate value of $\tilde{k} = 6.19$. Since savings initially increase over the working life, wealth $\tilde{k}_t^s$ will also be hump-shaped over the life-cycle and may well exceed average wealth. We, therefore, choose an initial value of $\tilde{k}_{max} = 20$ and find it to be non-binding in our computations.

Since we cannot compute the value function of the 70-year-old at each point of the interval $[\tilde{k}_{min}, \tilde{k}_{max}]$ (the number of all points is infinite), we only perform this calculation at certain grid points. We choose equispaced grids with $n_k = 100$ points. In the computation of the value function of the 69-year-old, we will need the value of the value function at age $s = 70$ between grid points. To obtain this, we will interpolate linearly between grid points if necessary.

Computing the value function $\tilde{V}(\tilde{k}, s, \epsilon, \eta)$ at a grid point $(\tilde{k}^i, 70, \epsilon, \eta)$, $i = 1, \ldots, n_k$, is straightforward and follows from the budget constraint and the definition of the utility function:

$$\tilde{V}_t(\tilde{k}^i, 70, \epsilon, \eta) = \frac{\left(\tilde{pen}_t + \left[1 + (1 - \tau^K) r_t\right] \tilde{k}^i + \tilde{tr}_t\right)^{\iota(1-\sigma)}}{1 - \sigma}.$$

Notice that the value of the value function is the same for all productivity types (ϵ, η) since they no longer affect income as long as households have accumulated the same wealth $\tilde{k}^i$.

For the retired households with age $s = 69, 68, \ldots, 46$, we have to solve an optimization problem in one variable. To do so, let us consider the *Bellman equation* at age $s = 69$ at a grid point $(\tilde{k}^i, 69, \epsilon, \eta)$[71]:

$$\tilde{V}_t(\tilde{k}^i, 69, \epsilon, \eta) = \max_{\tilde{c}_t^{69}, \tilde{k}_{t+1}^{70}} \left\{ u(\tilde{c}_t^{69}, 1) + (1+\gamma)^{\iota(1-\sigma)} \beta \mathbb{E}_t \tilde{V}_t(\tilde{k}_{t+1}^{70}, 70, \epsilon, \eta) \right\}.$$

If we substitute for $\tilde{c}_t^{69}$ from the budget constraint (6.55), the maximand within the brackets is a function of $\tilde{k}_{t+1}^{70}$. There are various numerical procedures to solve such a maximization problem. In the Gauss program *Ch6_optimal_pension.g*, we use the so-called *golden section search.*[72] The basic idea of this method is to bracket the maximum $\tilde{k}^- \leq \tilde{k}_{t+1}^{70} \leq \tilde{k}^+$ and let this interval shrink against zero. The limit points are easy to choose, e.g., $\tilde{k}^- = 0$ for the lower limit and $\tilde{k}^+$ as the value for which $\tilde{c}_t = 0$. In the next step, two points within the interval are selected, and the one with the lower value for the right-hand side of the above equation becomes the new limit point of the interval. The golden section search method optimizes the choice of these new points. This procedure is iterated until the interval length is sufficiently small, e.g., 10^{-6}, and we then stop.

For the working agent, the maximization problem is more complicated because he also chooses his optimal labor supply l_t^s. There are various numerical methods that are able to compute this two-dimensional optimization problem. We have chosen to transform the problem into two nested one-dimensional optimization problems and apply golden section search in the outer loop over the next-period capital stock $\tilde{k}_{t+1}$ and direct computation from the first-order condition with respect to labor l_t with the help of the Gauss-Newton algorithm in the inner loop.

Once we have solved the individual optimization problem, we can aggregate individual savings and labor supply to derive aggregate quantities and update our initial guesses of $\tilde{K}$, $\tilde{L}$, and the factor prices w and r. The budgets of the government and the pension system imply the values for $\tilde{tr}$ and τ^p. We update the old values by taking a weighted average of the two and iterating until convergence. The complete algorithm is described in Algorithm 6.1.

Algorithm 6.1 (Computation of the Stationary Equilibrium of the OLG Model in Sect. 6.4.1)

Purpose: *Computation of the stationary equilibrium.*

Steps:

Step 1: *Make initial guesses of the steady-state values of the aggregate capital stock $\tilde{K}$, efficient labor $\tilde{L}$, and aggregate accidental bequests $\tilde{Beq}$.*

---

[71] In fact, we could drop the expectational operator $\mathbb{E}_t$ in the Bellman equation for the retired and replace it by the survival probability $\phi_{s,t}$ because they do not face income uncertainty (in contrast to workers).

[72] A detailed description can be found in Chapter 11.6.1 in Heer and Maußner (2009).

Step 2: *Compute the values w, r, $\tilde{tr}$, τ^p, and pen_t, which solve the firm's Euler equations and the budgets of the government and social security.*

Step 3: *Compute the optimal policy functions for consumption, savings, and labor supply for the newborn generation by value function iteration.*

Step 4: *Compute the aggregate capital stock $\tilde{K}$, efficient labor $\tilde{L}$, and aggregate accidental bequests $\tilde{B}eq$.*

Step 5: *Update $\tilde{K}$, $\tilde{L}$, and $\tilde{B}eq$, and return to step 2 until convergence.*

We compute the transition dynamics for the US economy as described in Algorithm 9.2.1 in Heer and Maußner (2009). We first choose a number of transition periods under the assumption that the transition is complete by 2250.[73] Next, we compute the initial and final steady states and project a trajectory for the endogenous values of $\{\tilde{K}_t, \tilde{L}_t, \bar{l}_t, \tau_t^w, \tau_t^p, \tilde{tr}_t\}_{t=2015}^{2250}$. As our initial guess, we postulate a linear adjustment path for all endogenous variables. We assume that the economy is in steady state in and prior to 2015. For given path of $\{\tilde{K}_t, \tilde{L}_t, \bar{l}_t, \tau_t^w, \tau_t^p, \tilde{tr}_t\}_{t=2015}^{2250}$, we compute the individual policy functions in each year and aggregate individual labor supply and consumption. With the help of the consistency conditions and the fiscal budget constraints, we are able to provide a new guess for the path of $\{\tilde{K}_t, \tilde{L}_t, \bar{l}_t, \tau_t^w, \tau_t^p, \tilde{tr}_t\}_{t=2015}^{2250}$. Again, we use a simple dampening iterative scheme, as described in Section 3.9 of Judd (1998), to update the sequence $\{\tilde{K}_t, \tilde{L}_t, \bar{l}_t, \tau_t^w, \tau_t^p, \tilde{tr}_t\}_{t=2015}^{2250}$ until the sequence converges (with an accuracy equal to 10^{-6}).

The run time of the computer program *Ch6_optimal_pension.g* depends sensitively on the amount of grid points n_k and is considerable. Using Windows 7 on a computer with a 64-BIT system, 32 MB RAM, and an Intel(R) Xeon(R) 2.90 GHz processor, the stationary equilibrium is computed within 4 min, while the computation of the transition takes approximately 14 h.

Appendix 6.3: Data Sources

The data on population are taken from the UN, while the pensions-related data are retrieved from the OECD.

- **Old-age dependency ratio** The data presented in Fig. 6.1 is published by the UN in its *World Population Prospects: The 2017 Revision*, 'File POP/13-D: Old-age dependency ratio 65+/(25–64) by region, subregion and country, 1950–2100 (ratio of population 65+ per 100 population 25–64)'. The UN provides projections of the OADR for a 'low', 'medium', and 'high' fertility variant. If not mentioned otherwise, we use the 'medium' fertility variant.

[73]This value for the final year is found by trial and error. We choose 2250 because the transition of the endogenous values is complete by then. In Fig. 6.15, we drop the presentation of the final periods to better illustrate the transition.

- **Pension spending** The data displayed in Fig. 6.3 are taken from the OECD (2015), Pensions Statistics: Pensions at a Glance (Accessed on February 15, 2018).
http://data.oecd.org/socialexp/pension-spending.htm.
- **Pension replacement rates** The data in Fig. 6.4 presents the gross pension replacement rates of men as a percentage of pre-retirement earnings and can be retrieved from OECD (2017), Gross pension replacement rates (indicator). doi: 10.1787/3d1afeb1-en (Accessed on February 15, 2018).
https://data.oecd.org/pension/gross-pension-replacement-rates.htm.

Problems

6.1. Recompute the solution to Numerical Examples in Sect. 6.3 with the following changes:

1. Check the robustness of the results with respect to the intertemporal elasticity of substitution $1/\sigma$, with $\sigma \in \{2, 4\}$.
2. Assume that capital depreciates completely so that the real interest rate is represented by

$$r_t = \alpha k_t^{\alpha-1} l_t^{1-\alpha} - 1.0.$$

6.2. Contribution-Based Pensions in the Three-Period OLG Model Assume that an agent lives three periods. Each period length is equal to 20 years. In the first two periods, the agent is working, and in the third period, he receives a pension. Each generation has mass $1/3$. We will only consider the steady state.

Lifetime utility is given by

$$U = \sum_{s=1}^{3} \beta^{s-1} u(c^s, 1 - l^s). \tag{6.66}$$

Instantaneous utility is presented by

$$u(c, 1 - l) = u(c, 1 - l) = \frac{(c(1 - l)^\iota)^{1-\sigma}}{1 - \sigma},$$

with $\iota = 2.0$ and $\sigma = 2.0$. Assume that $\beta = 0.90$. Time is allocated to either work or leisure.

During the first two periods, households work; in the third period, they retire ($l^3 \equiv 0$). Agents are born without assets, $k^1 = 0$. In addition, the workers pay contributions to the pension system equal to $\tau = 10\%$ of their gross labor income. Therefore, the budget constraint at age $s = 1, 2$ is given by

$$(1 - \tau)wl^s + (1 + r)k^s = c^s + k^{s+1}.$$

During retirement, agents receive pensions that depend on past earnings

$$d = \sum_{s=1}^{2} \tau wl^s.$$

In particular, the pension system does not pay any interest on accumulated contributions. The pension depends on accumulated contributions as follows:

$$pen(d) = pen_{min} + \rho_{pen} d.$$

Therefore, the budget constraint of the retired worker is given by:

$$pen(d) + (1 + r)k^3 = c^3.$$

In addition, we assume that the government runs a balanced budget:

$$\tau w \frac{l^1 + l^2}{3} = \frac{1}{3}\left(pen_{min} + \rho_{pen} d\right).$$

Assume that $\rho_{pen} = 0.5$, implying

$$pen_{min} = 0.5\tau w(l^1 + l^2).$$

Production is modeled as in Sect. 6.3:

$$Y = K^\alpha L^{1-\alpha},$$

with

$$L = \frac{l^1 + l^2}{3}, \quad K = \frac{k^2 + k^3}{3},$$

and $\alpha = 0.36$.
Factors are rewarded by their marginal products:

$$w_t = (1 - \alpha)\left(\frac{K_t}{L_t}\right)^\alpha,$$

$$r_t = \alpha\left(\frac{K_t}{L_t}\right)^{\alpha - 1} - \delta.$$

The depreciation rate is set equal to $\delta = 0.5$.

1. Solve the problem with the help of direct computation (solving a system of non-linear equations). Show that the first-order conditions are given by

$$\lambda^1 = \left(c^1\right)^{-\sigma}\left(1-l^1\right)^{\iota(1-\sigma)},$$
$$\lambda^2 = \left(c^2\right)^{-\sigma}\left(1-l^2\right)^{\iota(1-\sigma)},$$
$$\lambda^3 = \left(c^3\right)^{-\sigma},$$
$$\iota\left(c^1\right)^{1-\sigma}\left(1-l^1\right)^{\iota(1-\sigma)-1} = \lambda^1(1-\tau)w + \beta^2\lambda^3\rho_{pen}\tau w,$$
$$\iota\left(c^2\right)^{1-\sigma}\left(1-l^2\right)^{\iota(1-\sigma)-1} = \lambda^2(1-\tau)w + \beta\lambda^3\rho_{pen}\tau w,$$
$$\lambda^1 = \beta\lambda^2(1+r),$$
$$\lambda^2 = \beta\lambda^3(1+r).$$

   For given aggregate variables w and r, the 7 first-order conditions together with the 3 budget constraints are a system of non-linear equations in 10 unknowns c^s and λ^s, $s = 1,2,3$, l^1, l^2, k^2, k^3. Use numerical methods to solve the system. Start with educated guesses for K and L, compute the individual policy functions, and update K and L accordingly. Compute the implied gross pension replacement rate with respect to the earnings in the second period. How does the abolition of social security affect output, labor, and welfare?
2. Assume that the government switches from a defined contribution to a defined benefit system and that it applies the same gross pension replacement rate with respect to the earnings in the second period as above (for the case with $\tau = 10\%$). What are the effects on labor supply, savings, output, and the social security contribution rate τ?

6.3. Quasi-Hyperbolic Discounting Recompute the three-period OLG model from Problem 6.2. However, instead assume (1) that pensions are not contribution-based but provided lump-sum ($\rho_{pen} = 0$) and (2) that the household behaves inconsistently and in a naive way. Therefore, let the household at age 1 assume that its lifetime utility is represented by

$$U = u(c^1, 1-l^1) + \mu\beta u(c^2, 1-l^2) + \mu\beta^2 u(c^3, 1-l^3), \quad \mu < 1, \tag{6.67}$$

rather than by Eq. (6.66), where μ denotes the hyperbolic discounting parameter. The household maximizes its utility for $\mu = 0.85$ in period 1, where the first-order conditions are given by:

$$\lambda^1 = \left(c^1\right)^{-\sigma} \left(1 - l^1\right)^{\iota(1-\sigma)},$$

$$\lambda^2 = \left(c^2\right)^{-\sigma} \left(1 - l^2\right)^{\iota(1-\sigma)},$$

$$\lambda^3 = \left(c^3\right)^{-\sigma},$$

$$\iota \left(c^1\right)^{1-\sigma} \left(1 - l^1\right)^{\iota(1-\sigma)-1} = \lambda^1(1-\tau)w,$$

$$\iota \left(c^2\right)^{1-\sigma} \left(1 - l^2\right)^{\iota(1-\sigma)-1} = \lambda^2(1-\tau)w,$$

$$\lambda^1 = \mu\beta\lambda^2(1+r),$$

$$\lambda^2 = \beta\lambda^3(1+r).$$

Solve this system of equations for 10 unknowns c^s and λ^s, $s = 1, 2, 3, l^1, l^2, k^2, k^3$ and denote the solutions by $\tilde{c}^s$, $\tilde{\lambda}^s$, $\tilde{k}^s$, and $\tilde{l}^s$.

In period 2, however, the household behaves in an inconsistent way and, for a given k^2, re-maximizes[74]

$$U = u(c^2, 1 - l^2) + \mu\beta u(c^3, 1 - l^3).$$

The first-order conditions with respect to c^2, c^3, l^2, and k^3 are given by:

$$\lambda^2 = \left(c^2\right)^{-\sigma} \left(1 - l^2\right)^{\iota(1-\sigma)},$$

$$\lambda^3 = \left(c^2\right)^{-\sigma},$$

$$\iota \left(c^2\right)^{1-\sigma} \left(1 - l^2\right)^{\iota(1-\sigma)-1} = \lambda^2(1-\tau)w,$$

$$\lambda^2 = \mu\beta\lambda^3(1+r).$$

[74]Our households are naive in the sense that they ignore their future behavior in the optimization decision in period 1; in other words, in period 1, they do not realize that they will behave the same way in period 2 (applying quasi-hyperbolic discounting to the discounted utility of the remaining lifetime).

Denote the solutions to these equilibrium conditions as $\hat{c}^s$ and $\hat{\lambda}^s$, $s = 2, 3$, $\hat{l}^2$, and $\hat{k}^3$. The household suffers from inconsistent behavior and chooses smaller savings for $\mu < 1$ in period 2 than it would have chosen in period 1.[75]

In general equilibrium, aggregate capital and labor are given by

$$K = \frac{\tilde{k}^2 + \hat{k}^3}{3},$$

$$L = \frac{\tilde{l}^1 + \hat{l}^2}{3}.$$

In a large-scale OLG model with individual income uncertainty, İmrohoroğlu, İmrohoroğlu, and Joines (2003) show that quasi-hyperbolic discounting at the rate of 15% lowers the capital stock by approximately 20% at any social security contribution rate.[76] Can you verify this result in the above example for pension replacement rates $\theta_{pen} \in \{0, 50\%\}$?

6.4. Hicksian Compensation Consider Fig. 6.5. Recompute the dynamics of the model under the assumption that the two cohorts alive at the time of the policy changes in period 1 are compensated such that the consumption equivalent change in comparison to the cohorts in the economy without the policy change is zero. Further assume that future generations receive transfers or pay lump-sum taxes such that (1) they all have equal lifetime utility and (2) the net present value of these transfers is equal to the payments of transfers to the cohorts in period 1. How large is the *Hicksian efficiency* gain?

6.5. Compute the optimal pension in the model of Sect. 6.4.1 under the assumption that the government holds both government consumption $\tilde{G}$ and transfers $\tilde{Tr}$ constant at their benchmark equilibrium values. Adjust the wage income tax rate τ^w such that the government budget (6.50) is balanced.

6.6. A Simple Auerbach-Kotlikoff Model[77] Consider a 60-period OLG model in the tradition of Auerbach and Kotlikoff (1987). Three sectors can be depicted: households, production, and the government.

Households Every year, a generation of equal measure is born. The total measure of all generations is normalized to one. Their first period of life is period 1.

[75] A seminal paper that introduces you to quasi-hyperbolic discounting and commitment technologies is Laibson (1997).

[76] In addition, İmrohoroğlu, İmrohoroğlu, and Joines (2003) find that social security is not effective in correcting for under-saving that results from time-inconsistent preferences.

[77] The following description is taken from Chapter 9.1 in Heer and Maußner (2009).

Households live $J = 60$ years. Consequently, the measure of each generation is 1/60. During their first $T = 40$ years, agents supply labor l_t^s at age s in period t enjoying leisure $1 - l_t^s$. After 40 years, retirement is mandatory ($l_t^s = 0$ for $s > 40$) for the remaining $T^R = 20$ years. Agents maximize lifetime utility at age 1 in period t:

$$\sum_{s=1}^{J} \beta^{s-1} u(c_{t+s-1}^s, 1 - l_{t+s-1}^s),$$

where β denotes the discount factor. Instantaneous utility is a function of both consumption and leisure:

$$u(c, 1-l) = \frac{(c(1-l)^{\iota})^{1-\sigma} - 1}{1-\sigma}. \tag{6.68}$$

Agents are born without wealth, $k_t^1 = 0$, and do not leave bequests, $k_t^{61} = 0$. Agents receive income from capital k_t^s and labor l_t^s. The real budget constraint of the working agent is given by

$$k_{t+1}^{s+1} = (1 + r_t)k_t^s + (1 - \tau_t)w_t l_t^s - c_t^s, \quad s = 1, \ldots, T,$$

where r_t and w_t denote the interest rate and the wage rate in period t, respectively. Wage income in period t is taxed at rate τ_t. We can also interpret $\tau_t w_t l_t^s$ as the worker's social security contributions.

The first-order conditions of the working household are given by:

$$(1 - \tau_t)w_t = \frac{u_{1-l}(c_t^s, 1 - l_t^s)}{u_c(c_t^s, 1 - l_t^s)} = \iota \frac{c_t^s}{1 - l_t^s}, \tag{6.69}$$

$$\begin{aligned} \frac{1}{\beta} &= \frac{u_c(c_{t+1}^{s+1}, 1 - l_{t+1}^{s+1})}{u_c(c_t^s, 1 - l_t^s)} [1 + r_{t+1}] \\ &= \frac{\left(c_{t+1}^{s+1}\right)^{-\sigma} \left(1 - l_{t+1}^{s+1}\right)^{\iota(1-\sigma)}}{\left(c_t^s\right)^{-\sigma} \left(1 - l_t^s\right)^{\iota(1-\sigma)}} [1 + r_{t+1}]. \end{aligned} \tag{6.70}$$

During retirement, agents receive public pensions pen irrespective of their employment history, and the budget constraint of the retired worker is given by

$$k_{t+1}^{s+1} = (1 + r_t)k_t^s + pen - c_t^s, \quad s = T + 1, \ldots, T + T^R.$$

The first-order condition of the retired worker is given by (6.70) with $l_t^s = 0$.

Production Firms are of measure one and produce output Y_t in period t with labor L_t and capital K_t. Labor L_t is paid wage w_t. Capital K_t is hired at rate r_t and depreciates at rate δ. Production Y_t is characterized by constant returns to scale and assumed to be Cobb-Douglas:

$$Y_t = K_t^{\alpha} L_t^{1-\alpha}. \tag{6.71}$$

In factor market equilibrium, factors are rewarded with their marginal products:

$$w_t = (1-\alpha) K_t^{\alpha} L_t^{-\alpha}, \tag{6.72}$$

$$r_t = \alpha K_t^{\alpha-1} L_t^{1-\alpha} - \delta. \tag{6.73}$$

Government The government uses the revenues from taxing labor to finance its expenditures on social security:

$$\tau_t w_t L_t = \frac{T^R}{T+T^R} pen. \tag{6.74}$$

Following a change in the provision of public pensions pen or in gross labor income $w_t L_t$, the labor income tax rate τ_t adjusts to keep the government budget balanced.

Equilibrium An equilibrium for a given government policy pen and initial distribution of capital $\{k_0^s\}_{s=1}^J$ is a collection of individual policy rules $c(s, k_t^s, K_t, L_t)$, $l(s, k_t^s, K_t, L_t)$, and $k'(s, k_t^s, K_t, L_t)$, and relative prices of labor and capital $\{w_t, r_t\}$, such that:

(i) Individual and aggregate behavior are consistent:

$$L_t = \sum_{s=1}^{40} \frac{l_t^s}{60},$$

$$K_t = \sum_{s=1}^{60} \frac{k_t^s}{60}.$$

The aggregate labor supply L_t is equal to the sum of the labor supplies of each cohort, weighted by their mass $1/J = 1/60$. Similarly, the aggregate capital supply is equal to the sum of the capital supplies of all cohorts.

(ii) Relative prices $\{w_t, r_t\}$ solve the firm's optimization problem.

(iii) Given relative prices $\{w_t, r_t\}$ and the government's policy pen, the individual policy rules $c(.)$, $l(.)$, and $k'(.)$ solve the consumer's optimization problem.

(iv) The goods market clears:

$$K_t^{\alpha} L_t^{1-\alpha} = \sum_{s=1}^{60} \frac{c_t^s}{60} + K_{t+1} - (1-\delta) K_t.$$

(v) The government budget is balanced.

Calibration The benchmark case is characterized by the following calibration: $\sigma = 2$, $\beta = 0.99$, $\alpha = 0.3$, $\delta = 0.1$, replacement rate $\theta^p = \frac{pen}{(1-\tau)w\bar{l}} = 0.3$ (where $\bar{l}$ denotes the average labor supply in the economy). ι is chosen to imply a steady-state labor supply of the working agents approximately equal to $\bar{l} = 35\%$ of available time and amounts to $\iota = 2.0$.

1. Compute the steady state using numerical methods. To do so, you have to consider a system of 99 non-linear equations in the variables k^s, $s = 2, \ldots, 60$ and l^s, $s = 1, \ldots, 40$. Show that the shape of the consumption-age profile resembles that presented in Fig. 6.11 and displays a downward jump at the beginning of retirement. Show that the steady-state level of output is equal to $Y = 0.3842$.
2. Introduce a consumption tax of 10% that is retransferred lump-sum to the household. How does the consumption tax affect steady-state values?
3. Introduce a government sector that consumes 20% of steady-state output Y, $G = 0.0768$. Compare the following three tax instruments to finance G: (1) a tax on consumption τ^c, (2) a tax on capital income τ^K (assuming that depreciation is tax-deductible), and (3) a wage income tax rate τ^w. Assume further that government consumption provides utility to the household that is additively separated from the utility in consumption and leisure such that the first-order conditions of the household are not affected by government consumption G. What are the steady-state effects of these three tax instruments on capital, labor, and output? What is the tax instrument that results in the lowest welfare losses? To answer these questions, compute the consumption equivalent change of each policy measure with respect to the case without government consumption.
4. Introduce consumption habits

$$u(c^s, l^s; c^{s-1}) = \frac{\left[\left(c^s - \kappa c^{s-1}\right)(1 - l^s)^{\iota}\right]^{1-\sigma}}{1 - \sigma}, \quad \kappa \in [0, 1)$$

with $\kappa = 0.7$. In addition, assume that the utility of the newborn generation is given by $u(c^1, l^1) = \frac{\left[c^1(1-l^1)^{\iota}\right]^{1-\sigma}}{1-\sigma}$. Notice that you cannot solve the problem recursively but have to solve the system of 99 variables simultaneously. Use your solution from the case without habits as an initial value for $\{l^1, l^2, \ldots, l^{40}, k^2, k^3, \ldots, k^{60}\}$. Do consumption habits help to make the consumption-age profile smoother? In particular, does the downward jump in consumption at the beginning of retirement disappear?
5. In the literature, two types of PAYG pension systems are often distinguished, the *Bismarck* versus *the Beveridge* system. In the former, only employed workers contribute to the pension system, and the contribution is levied on the wage income. In the latter, all households contribute prior to their retirement age, and contributions are based on total income, i.e., including capital income. Moreover, while in the Bismarck system, pensions are closely linked to contributions,

the Beveridge pension system provides a guaranteed minimum income during retirement and redistributes strongly.

In the above model, introduce a pension system that provides a lump-sum payment to the retired worker that is financed by a tax on both labor and capital income. Compare this with the average lifetime utility under a PAYG pension system that collects contributions that are levied only on labor income.

References

Abel, A. B., Mankiw, N. G., Summers, L. H., & Zeckhauser, R. J. (1989). Assessing dynamic efficiency: Theory and evidence. *Review of Economic Studies, 56*, 1–20.

Alkema, L., Raftery, A. E., Gerland, P., Clark, S. J., Pelletier, F., Buettner, T., & Heilig, G. K. (2011). Probalistic projections of the total fertility rate for all countries. *Demography, 48*, 815–839.

Altonij, J. G., Hayashi, F., & Kotlikoff, L. (1997). Parental altruism and intervivos transfers: Theory and evidence. *Journal of Political Economy, 105*, 1121–1166.

Attanasio, O., Kitao, S., & Violante, G. L. (2007). Global demographic trends and social security reform. *Journal of Monetary Economics, 54*, 144–198.

Auerbach, A. J., & Kotlikoff, L. (1987). *Dynamic fiscal policy*. New York: Cambridge University Press.

Barr, N., & Diamond, P. (2008). *Reforming pensions—Principles and policy choices*. Oxford: Oxford University Press.

Blanchard, O. J., & Fischer, S. (1989). *Lectures on macroeconomics*. Cambridge: MIT Press.

Börsch-Sopan, A. H., & Winter, J. (2001). *Population aging, savings behavior and capital markets* (NBER working paper No. 8561).

Braun, R. A., & Joines, D. H. (2015). The implications of a graying Japan for government policy. *Journal of Economic Dynamics and Control, 57*, 1–23.

Budría Rodriguez, S., Díaz-Giménez, J., & Quadrini, V. (2002). Updated facts on the U.S. distributions of earnings, income, and wealth. *Federal Reserve Bank of Minneapolis Quarterly Review, 26*, 2–35.

Caliendo, F. N., Guo, N. L., & Hosseini, R. (2014). Social security is NOT a substitute for annuity markets. *Review of Economic Dynamics, 17*, 739–755.

Conesa, J. C., & Krueger, D. (1999). Social security reform with heterogeneous agents. *Review of Economic Dynamics, 2*, 757–795.

De Nardi, M. (2015). *Quantitative models of wealth inequality: A survey* (NBER Working Paper, 21106).

De Nardi, M., & Yang, F. (2016). Wealth inequality, family background, and estate taxation. *Journal of Monetary Economics, 77*, 130–145.

De Nardi, M., Imrohoroğlu, S., & Sargent, T. J. (1999). Projected US demographics and social security. *Review of Economic Dynamics, 2*, 575–615.

Fehr, H. (1999). *Welfare effects of dynamic tax reforms*. Tübingen: Mohr Siebeck.

Fehr, H., Kallweit, M., & Kindermann, F. (2013). Should pensions be progressive? *European Economic Review, 63*, 94–116.

Fernández-Villaverde, J., & Krueger, D. (2007). Consumption over the life cycle: Facts from consumer expenditure survey data. *Review of Economics and Statistics, 89*(3), 552–565.

Fuster, L., İmrohoroğlu, A., & İmrohoroğlu, S. (2003). A welfare analysis of social security in a dynastic framework. *International Economic Review, 44*, 1247–1274.

Gerland, P., Raftery, A. E., Sevčíková, H., Li, N., Gu, D., Spoorenberg, T., Alkema, L., Fosdick, B. K., Chunn, J., Lalic, N., Bay, G., Buettner, T., Heilig, G. K., & Wilmoth, J. (2014). World population stabilization unlikely this century. *Science, 346*, 234–237.

Guvenen, F., Karahan, F., Ozkan, S., & Sang, J. (2015). *What do data on millions of U.S. workers reveal about life-cycle earnings risk?* (NBER Working Paper Series, 20913).
Hansen, G. (1993). The cyclical and secular behavior of the labor input: Comparing efficiency units and hours worked. *Journal of Applied Econometrics, 8*, 71–80.
Hansen, G., & İmrohoroğlu, S. (2008). Consumption over the life cycle: The role of annuities. *Review of Economic Dynamics, 11*(3), 566–583.
Heer, B. (2018). Optimal pensions in aging economies. *B.E. Journal of Macroeconomics, 18*, 1–19.
Heer, B., & Irmen, A. (2014). Population, pensions and endogenous economic growth. *Journal of Economic Dynamics and Control, 46*, 50–72.
Heer, B., & Maußner, A. (2009). *Dynamic general equilibrium modeling: Computational methods and applications* (2nd ed.). Heidelberg: Springer.
Heer, B., Polito, V., & Wickens, M. R. (2017). *Population aging, social security and fiscal limits* (CEPR Discussion Paper Series, DP11978).
Hubbard, R. G., & Judd, K. L. (1987). Social security and individual welfare: Precautionary saving, borrowing constraints, and the payroll tax. *American Economic Review, 77*, 630–646.
Huggett, M. (1996). Wealth distribution in life-cycle economies. *Journal of Monetary Economics, 38*, 373–396.
Hurd, M. D. (1989). Mortality risk and bequests. *Econometrica, 57*, 799–813.
İmrohoroğlu, S., & Kitao, S. (2009). Labor supply elasticity and social security reform. *Journal of Public Economics, 93*, 867–878.
İmrohoroğlu, A., İmrohoroğlu, S., & Joines, D. H. (1995). A life cycle analysis of social security. *Economic Theory, 6*, 83–114.
İmrohoroğlu, A., İmrohoroğlu, S., & Joines, D. H. (1999). Social security in an overlapping generations economy with land. *Review of Economic Dynamics, 2*, 638–665.
İmrohoroğlu, A., İmrohoroğlu, S., & Joines, D. H. (2003). Time-inconsistent preferences and social security. *Quarterly Journal of Economics, 118*, 745–784.
Judd, K. L. (1998). *Numerical methods in economics*. Cambridge: MIT Press.
Kitao, S. (2014). Sustainable social security: Four options. *Review of Economic Dynamics, 17*, 756–779.
Krueger, D., & Ludwig, A. (2007). On the consequences of demographic change for rates of returns to capital, and the distribution of wealth and welfare. *Journal of Monetary Economics, 54*, 49–87.
Krueger, D., & Perri, F. (2006). Does income inequality lead to consumption inequality? Evidence and theory. *Review of Economic Studies, 73*, 163–193.
Laibson, D. (1997). Golden eggs and hyperbolic discounting. *Quarterly Journal of Economics, 112*, 443–477.
Mendoza, E. G., Razin, A., & Tesar, L. L. (1994). Effective tax rates in macroeconomics: Cross country estimates of tax rates on factor incomes and consumption. *Journal of Monetary Economics, 34*, 297–323.
Nishiyama, S., & Smetters, K. (2007). Does social security privitization produce efficiency gains? *Quarterly Journal of Economics, 122*, 1677–1719.
OECD. (2015). *Pensions at a Glance 2015: OECD and G20 indicators*. Paris: OECD Publishing.
OECD. (2017). *Pensions at a Glance 2017: OECD and G20 indicators*. Paris: OECD Publishing.
Quadrini, V. (2000). Entrepreneurship, saving, and social mobility. *Review of Economic Dynamics, 3*, 1–40.
Raftery, A. E., Li, N., Sevčíková, H., Gerland, P., & Heilig, G. K. (2012). Bayesian probabilitic population projections for all countries. *Proceedings of the National Academy of Sciences, 109*, 13915–13921.
Raftery, A. E., Chunn, J., Gerland, P., & Sevčíková, H. (2013). Bayesian probabilitic projections of life expectancy for all countries. *Demography, 50*, 777–801.
Stokey, N., Lucas, R. E., & Prescott, E. C. (1989). *Recursive methods in economic dynamics*. Cambridge: Harvard University Press.

Storesletten, K., Telmer, C. I., & Yaron, A. (1999). The risk-sharing implications of alternative social security arrangements. *Carnegie-Rochester Series on Public Policy, 50*, 213–259.

Storesletten, K., Telmer, C. I., & Yaron, A. (2004). Consumption and risk sharing over the business cycle. *Journal of Monetary Economics, 51*, 609–633.

UN (2015). *World population prospects: The 2015 revision, Methodology of the United Nations population estimates and projections* (ESA/P/WP. 242).

7 Public Debt

7.1 Introduction

During the 1980s and 1990s, we observed many public debt crises in Latin America and Asia. Prior to, during, and after the financial crisis of 2007–2008, we also observed many countries in the European Monetary Union (EMU) with severe public debt problems. Government debt has increased to unprecedented levels in the post-World War II era in the so-called "GIIPS" countries (Greece, Ireland, Italy, Portugal, and Spain). The Greek, Spanish, and Italian governments have had to pay premiums of 16, 3, and 3 percentage points on their public debt relative to Germany. Consequently, the president of the European Central Bank (ECB), Mario Draghi, initiated a program whereby the ECB extended 3-year loans in the amount of 1 trillion (!) euros to the Eurozone banking sector. Interest rates subsequently converged, but debt levels remain high and still amounted to 179%, 100%, and 132% of GDP in Greece, Spain, and Italy in 2015, almost 10 years after the onset of the crisis. As a second major second case of severe public debt, Japan had accumulated a gross public debt equal to approximately 240% of GDP by 2015.

Two natural questions arise in face of the present Eurozone debt crisis (and the high debt in Japan): (1) Are these levels of high public debt sustainable, and are the governments able to repay and service their debt? (2) What are the empirical and quantitative effects of high debt on output and investment? To answer these two questions, in this chapter, we will consider the US economy, which is currently characterized by a sustainable level of debt. We simulate the demographic transition until 2100 and find that the US government can still increase its debt to a level above 200% of GDP. The costs of this fiscal policy in terms of output, investment, and consumption are dramatic. For example, we find that if the increased pension payments over the next 50 years are financed by debt rather than taxes, output falls by more than an additional 10 percentage points until the end of the demographic transition.

B. Heer, *Public Economics*, Springer Texts in Business and Economics,
https://doi.org/10.1007/978-3-030-00989-2_7

This chapter is structured as follows. In Sect. 7.2, we present empirical facts on government debt. We distinguish among net, gross, and implicit debt. Accounting for off-balance-sheet obligations of the government in the form of future pensions and medical expenditures, we observe (implicit) debt levels in excess of 700% of GDP for some European countries, including Ireland and Spain. We also present the literature on and the controversy surrounding the empirical result established by Reinhart and Rogoff (2009) that growth is reduced once a certain threshold value of government debt is passed (estimated to be 90% of GDP in explicit gross debt). In Sect. 7.3, we present simple debt arithmetic and discuss the idea of the Eurozone's Stability and Growth Pact.[1] In Sect. 7.4, we present the concept of Ricardian equivalence and show that government financing decisions affect the real economy in the standard overlapping generations (OLG) model in which households have a finite lifetime, and parents are not altruistic. We also derive quantitative effects and show that public debt significantly crowds out investment. Section 7.5 presents a conglomeration of our results from Chaps. 6 and 7. We study a large-scale OLG model of the demographic transition in which the aging of the population increases the pension burden over the next century. We will consider different pension policies and financing decisions of the US government and study the transition dynamics with respect to output, investment, and (generational) welfare. In the concluding epilogue in Sect. 7.6, we review the studies on debt default in quantitative models. In contrast to the material presented in the previous sections, the (overwhelming part of the) literature on both sovereign and domestic debt default assumes that fiscal policy is sustainable, but a sovereign nevertheless might find it optimal to default.

7.2 Empirical Facts: Government Debt

At present, governments in many industrialized countries have accumulated record levels of public debt relative to rest of the period since World War II. Table 7.1 presents the debt levels relative to GDP in 2015 for a cross-section of countries.[2] Among the industrialized countries, Japan, Greece, and Italy have the highest indebtedness and were characterized by gross debt levels of 238%, 179%, and 132% of GDP in 2015, respectively.

Figure 7.1 displays the accumulated public debt in the United States during the period 1966–2015.[3] In 1981, the debt-GDP ratio reached an all-time low for the post-World War II period, amounting to 30.6%. In 1946, the US government had accumulated wartime debt equal to 119% of GDP, which it efficiently reduced

[1]The former President of the European Commission, Romano Prodi, referred to this pact as the "stupidity pact".

[2]The data for the gross and net debt-GDP ratios are taken from the IMF World Economic Outlook database. A more detailed description is presented in Appendix 7.3.

[3]The empirical figures in this section are computed with the help of the Gauss program `Ch7_data.g` which is available as a download from the author's homepage.

Table 7.1 Debt-GDP ratios in 2015

	Debt-GDP ratio in 2015	
Country	Gross debt	Net debt
Argentina	56.0%	
Australia	37.9%	17.9%
Canada	91.6%	25.2%
China	41.1%	
France	95.6%	86.9%
Germany	70.9%	50.5%
Greece	179.4%	
Italy	132.1%	119.8%
Japan	238.1%	118.4%
Mexico	53.7%	47.2%
Spain	99.8%	86.0%
Turkey	27.5%	23.0%
UK	89.0%	80.3%
US	105.2%	80.2%

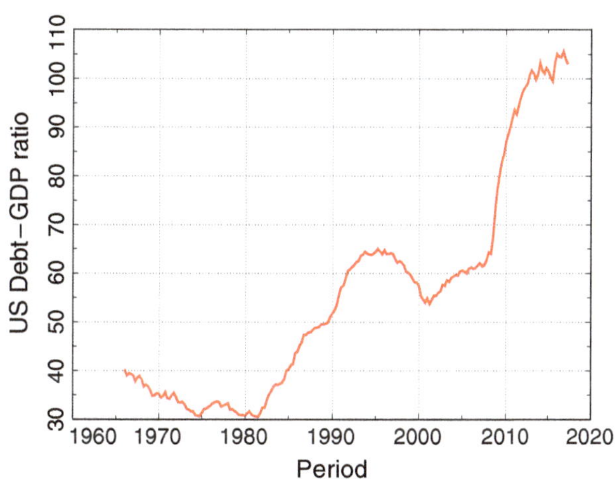

Fig. 7.1 US debt-GDP ratio during the period 1966–2015

during the subsequent three decades. Notice that even during the severe recessions, namely, in the form of the oil crises during the 1970s, total federal debt fell. At the beginning of the 1980s, the Reagan administration implemented a major income tax reform, which resulted in both an economic boom and a shortfall in government revenue. In the following decade 1980–1990, federal debt increased from 30% to approximately 60% of GDP. The next dramatic change in public debt resulted from the increase in fiscal spending during the financial crisis of 2007–2008. From 2007 to 2011, federal debt increased from 62% to 94% and reached 105% of GDP in 2015. At the end of 2017, the Trump income tax reform was approved by Congress.

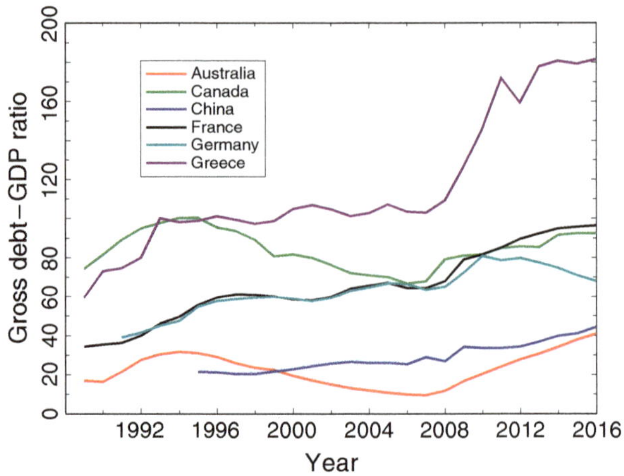

Fig. 7.2 Gross debt-GDP ratios, 1/2

According to estimates from the Committee for a Responsible Federal Budget, the plan will add approximately $2.2 trillion to deficits over the next decade, which amounts to approximately 12% of GDP.[4]

Many European and other industrialized countries have also experienced similar significant increases in public debt during and in the aftermath of the financial crisis of 2007–2008. As presented in Figs. 7.2 and 7.3, debt increased by more than 20 percentage points during the period 2005–2015 in Eurozone countries France, Greece, Italy, and Spain, as well as in the Anglo-American countries (the UK, the US, Australia, and Canada). In China and, in particular, Germany, however, the change in debt was much more moderate. In these two countries, federal debt increased from 26% to 41% and from 67% to 71% of GDP between 2005 and 2015, respectively.

The debt dynamics are reflected in the emergence of large fiscal budget deficits during the recent financial crisis of 2007–2008. Figures 7.4 and 7.5 display the fiscal deficits as a share of GDP during the period 1988–2015 for a cross-section of countries. The Anglo-American countries the UK and the US display government budget deficits of approximately 10% and 12% in 2009, which slowly recovered to approximately 4–5% by 2016. Australia, which had actually run a government surplus prior to the crisis, has been characterized by relatively mild deficits of 3–5% since 2007–2008. The Eurozone countries Italy, Spain, and Greece, which were hit the hardest by the financial crisis[5] are running deficits equal to 5%, 10%, and 15%,

[4]Annual nominal GDP amounted to $18.6 trillion in 2016.

[5]Comparing the per capita incomes (as measured by the per capita GDP in constant prices) for these countries between the year 2000, prior to the introduction of the euro, and the year 2015 (7 years after the onset of the financial crisis), we find that it fell in Greece (−3.5%) and Italy (−7.1%),

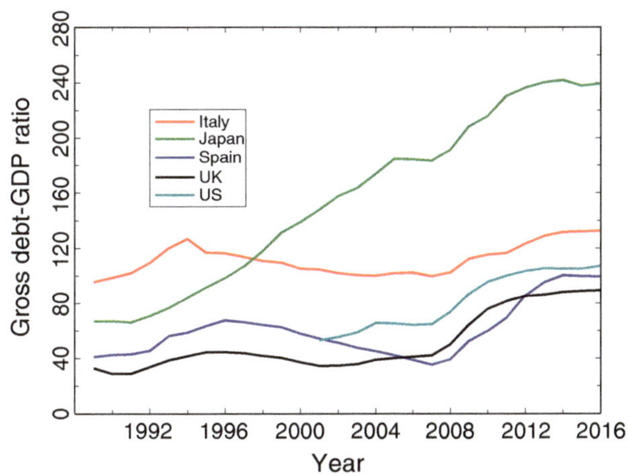

Fig. 7.3 Gross debt-GDP ratios, 2/2

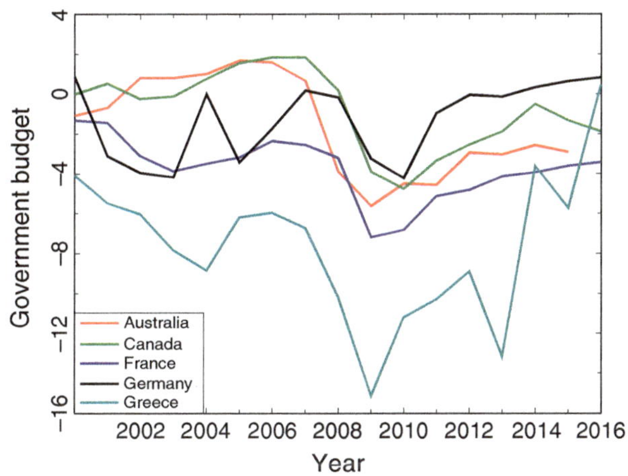

Fig. 7.4 Government budget (relative to GDP), 1/2

clearly violating the Stability and Growth Pact of the European Union that provides for a deficit ceiling of 3%. Moreover, the two heavyweights in the Eurozone, France and Germany, also did not meet the deficit ceiling of 3% in 2008 or 2009.

In the face of these unprecedented post-World War II debt levels, it is natural to wonder whether there is a debt level threshold beyond which government debt becomes unsustainable. Government default is not only a merely academic question

---

while it only slightly increased in Spain (+7.5%). By comparison, in Germany, per capita income increased by 18.3% during the period 2000–2015.

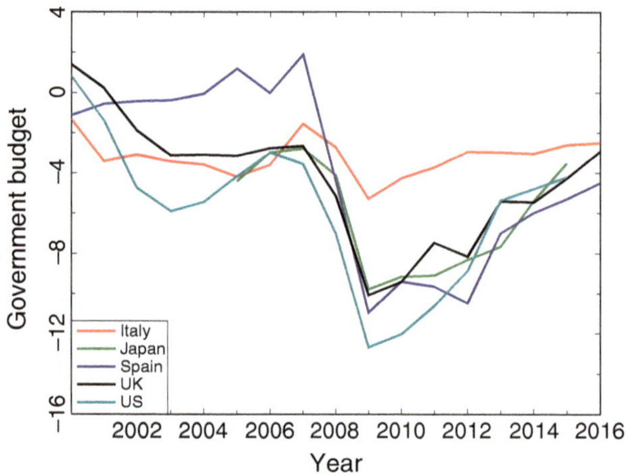

Fig. 7.5 Government budget (relative to GDP), 2/2

but has been observed throughout history. Reinhart and Rogoff (2009) provide a narrative of these episodes in their book entitled "This Time is Different". In Table 7.2, we replicate some of their data on default periods. For example, Argentina defaulted on its debt in 1890 and 2001. By 1890, the Argentine government had accumulated massive levels of government debt, on which it defaulted. The crisis was reinforced by a lack of coordination in monetary and fiscal policy and eventually also lead to a financial crisis. Another government default occurred at the end of the twentieth century, when the Argentine peso was pegged to the US dollar. Starting in 1998, Argentina experienced a severe recession and had to seek assistance from the IMF. The appreciation of the dollar and inadequate adjustments through fiscal and economic reforms resulted in a massive sell-off of Argentine government bonds. By the end of 2001, the Argentine government defaulted on its international debt obligations, amounting to $95 billion. At the time, this was the largest default in history.

When they defaulted, the governments considered by Reinhart and Rogoff had accumulated large amounts of public debt. In the two rightmost columns of Table 7.2, government debt is presented relative to government revenue. The reason to report debt relative to revenue rather than GDP is twofold. First, data on GDP prior to World War II are of limited accuracy or not readily available. Second, government revenues are a very good measure of the government's ability to service its debt, both at present and in the future. We observe a wide range of threshold debt levels at which government default sets in, amounting to a multiple of 2.4–15.8 of government revenue. Comparing these numbers to the present debt-revenue ratios (as presented in the right column of Table 7.2), some countries, such as the US and Italy, are already in the lower range of the interval, while Japan, with a debt-

Table 7.2 Debt default periods

Country	Default year	Debt-revenue ratio default year	Debt-revenue ratio in 2015
Argentina	1890	12.46	1.57
	2001	2.62	
China	1939	8.96	1.44
Germany	1932	2.43	1.59
Mexico	1827	4.20	2.33
	1982	5.06	
Spain	1877	15.83	2.58
Turkey	1978	2.69	0.86
Australia			1.09
Canada			2.33
US			3.33
UK			2.49
France			1.80
Italy			2.76
Japan			7.18

Notes: The data for the entries in columns 2 and 3 are taken from Table 8.1 in Reinhart and Rogoff (2009). The data in the right column are taken from the IMF World Economic Outlook database at https://www.imf.org/external/pubs/ft/weo/2017/02/weodata/index.aspx

revenue ratio of 7.2, is in the middle of the observed default ratios.[6] In addition, we recognize that the level of debt seems to be more tolerable in the case of industrialized countries than in the case of emerging countries. For example, at the time of default, the (gross) debt-GDP ratio only amounted to 49% in Argentina in 2001.

Regarding the effects of high debt levels on economic performance, one of the most influential and controversial research articles in economics in this century is a study by Carmen Reinhart and Kenneth Rogoff, "Growth in Times of Debt," that was published in the *American Economic Review* in 2010. Considering a cross-section of 20 industrialized countries in the post-war period, they find that a country's annual growth declined by 2% if gross external debt reached 60% of GDP and was "roughly cut in half" for levels of external debt in excess of 90%. As this result was published shortly after the financial crisis of 2007–2008, it has often been used as an argument in the political debate to justify or demand pro-austerity policies.

[6]When you compare the present values of the debt-revenue ratio with those from the default years, bear in mind that prior to World War II, government revenue constituted a much smaller share of GDP than at present. In the US, for example, government revenue only amounted to 11–12% of GDP during the 1920s, while it was equal to 23.5% in 2015. Consequently, the debt-GDP ratios of the defaulting countries in the default years and in 2015 are even closer to one another.

The study of Reinhart and Rogoff (2010) has been challenged on many grounds. (1) Reinhart and Rogoff made their data available to Thomas Herndon, Michael Ash and Robert Pollin, who found coding errors and selective data omissions[7] that resulted in sample bias.[8] Correcting for these data-handling errors, they found the 90% debt threshold level to imply no significant changes in a country's growth performance. Subsequently, Reinhart and Rogoff (2012) responded to these critiques and re-estimated their results. They find that periods of public debt overhang (above the 90% debt threshold) to be characterized by significantly lower economic growth, and they maintain their initial position despite that the threshold's impact on economic growth is lower than they originally found (but still amounts to more than 1% lower growth).[9] (2) The study might suffer from reverse causality; in other words, periods of weak growth result in high debt accumulation. (3) Other control variables such as the exchange rate regime or the denomination of public debt in either domestic or foreign currency were not considered. In summary, it is fair to say that the threshold controversy remains unsettled, and at present, no consensus has been reached among economists.

Nevertheless, many Eurozone countries have reached the 90% public debt threshold, and in particular, countries such as Italy and Greece have far surpassed this point. As the financial crisis continued to depress the economic and fiscal situations in these countries, investors began to demand higher interest rates on the government bonds of the GIIPS countries. Figure 7.6 displays the yield on the 10-year government bonds for the countries in the EMU. Evidently, the interest rate spreads increased after the financial crisis of 2007–2008 and reached a maximum during the years 2011–2012. At the beginning of 2012, Greece, Portugal, Ireland, Italy, and Spain had to pay premiums of 29, 10, 4, 3, and 3 percentage points vis-à-vis Germany. Since all government bonds are denominated in the same currency, the euro, this premium reflects investors' concern that either the Eurozone will break up and government bonds will be repaid in the former currencies and/or that governments default on the repayment of their debt. Of course, the accumulation of higher public debt and its sustainability were a driving force of this interest rate behavior.[10] Nevertheless, there does not seem to be an automatic relationship between higher debt and default probabilities. After 2012, interest rates on government bonds in the Eurozone converged, and by 2014, interest rate

[7]For example, data for Australia, New Zealand, and Canada in the late 1940s were excluded.

[8]The critique of Herndon, Ash, and Pollin (2014) is only directed at the 1949–2009 dataset for the 20 industrialized countries. Reinhart and Rogoff (2010) also consider two other datasets, including one for an extended period from 1791 to 2009.

[9]In addition, they find that the lower growth performance prevails even for those countries that do not experience higher real interest rates because of their indebtedness. Accordingly, the growth-reducing effects do not stem exclusively from higher real interest rates.

[10]In this regard, Polito and Wickens (2015) present a measure of sovereign credit ratings that is derived solely from the fiscal position of a country and its ability to repay future debt obligations. It is capable of identifying the European debt crisis and the deterioration of credit rating quality among GIIPS countries even 2 years prior to the release of official credit ratings.

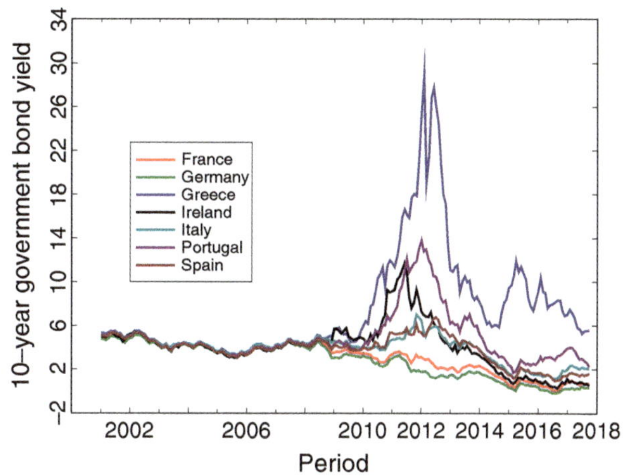

Fig. 7.6 Interest rates in the EMU

differentials amounted to 2 percentage points or less.[11] Accordingly, although debt levels (relative to GDP) were elevated (and rising) in countries with government debt in excess of 90% of GDP, default probabilities were shrinking.

Before we conclude this section, two comments on the measurement of debt and implications for its sustainability are worthwhile. First, we presented data on the gross public debt to assess the wealth position of a government. However, some part of government lending is between the different local, state, and federal layers of the government. A measure that corrects for such intra-government debt is given by the net debt-GDP ratio that is presented in the right column of Table 7.1.[12] The relationship between the gross debt and net debt-GDP ratios in the table is close but not perfect, with a correlation coefficient equal to 0.80. For example, the UK and the US have the same net debt-GDP ratio, but their gross debt ratios amount to 89% and 105%, respectively. Furthermore, in the case of Japan, the net debt-GDP ratio is only half the amount of the gross debt GDP-ratio.[13]

[11] In February 2012, the ECB president, Mario Draghi, started the tender "Big Bertha" whereby the ECB offered one trillion euros in 3-year loans to the banking sector. By mid-year, Mario Draghi went one step further and announced new measures, promising that they would be sufficient. In particular, Mario Draghi said, "within our mandate, the ECB is ready to do whatever it takes to preserve the euro. And believe me, it will be enough."

[12] For the countries with the empty entries in Table 7.1 – Argentina, China and Greece – no data on net debt are available in the IMF statistics.

[13] Another debt-ratio measure of the sustainability of public finances that is the most relevant for the evaluation of emerging countries is the amount of external debt or debt denominated in foreign currency. A detailed description of external debt and historical periods of high external debt is provided by Reinhart and Rogoff (2009).

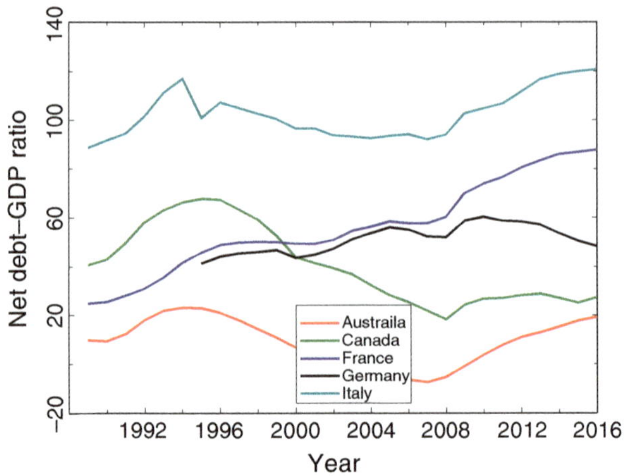

Fig. 7.7 Net debt-GDP ratios, 1/2

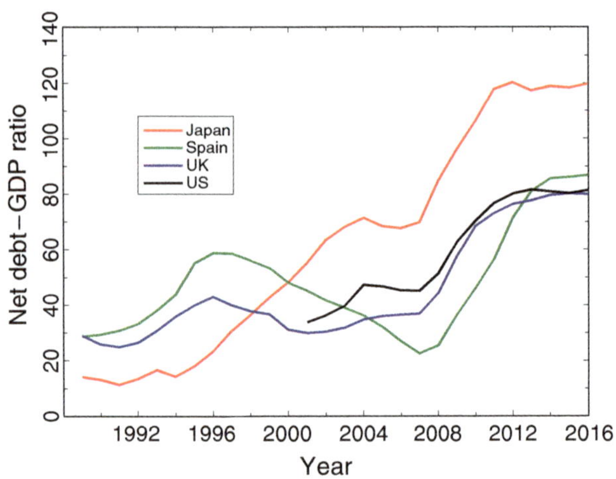

Fig. 7.8 Net debt-GDP ratios, 2/2

The dynamics of the net debt-GDP ratios during the period 1989–2016 are illustrated in Figs. 7.7 and 7.8. The characteristics of the time profiles are in close accordance with those presented for the gross debt-GDP ratios. In particular, there is a steep increase in the net debt-GDP ratios at the onset of the financial crisis of 2007–2008 – the value rose by more than 20 percentage points during the period 2007–2012 in the EMU countries France, Italy, and Spain and in the US, the UK, and Japan. Throughout the period, the most dramatic increase in net indebtedness is observed in the case of Japan, where the net debt-GDP ratio increased from 12% in 1992 to 120% in 2012.[14]

Second, a large part of public debt is only reported off the balance sheet and not contained in the official government debt data, for example unfunded public pensions. As you learned in Chap. 6, the nature of an unfunded public pay-as-you-go pension is basically the same as that of a debt contract. The government collects payments from its workers today that it repays in the future. A similar reasoning applies to some parts of public health services. Evaluating these future claims on the government is difficult for at least three reasons. (1) Future payments depend on the (projected) survival probabilities of workers and retirees. (2) Future payments need to be expressed in present value, meaning that the assumed discount rate has a significant impact. (3) Assumptions of future indexation and future effective retirement age enter the computation of the liabilities in a non-trivial way, i.e., the extent to which pensions and contributions are adjusted for inflation and productivity growth over the upcoming decades and when workers will (effectively) retire.

These off-balance-sheet government liabilities are captured in the concept of "implicit debt". Table 7.3 presents European Commission estimates of the implicit debt of European countries.[15] Notice that in European countries, the net present value of future liabilities varies considerably, and implicit debt is often a multiple of gross debt. For example, implicit debt amounted to 709% and 665% of GDP in Spain and Ireland in 2016, while gross debt only amounted to 79% and 100%, respectively. Italy is actually one of the countries with the lowest total debt in the EMU if implicit debt is accounted for, while the fiscal sustainability of Ireland, Luxembourg, and Spain is much more problematic in light of their future pension obligations.[16]

In conclusion, some measures of fiscal sustainability such as the debt-GDP ratio and the current budget deficit as a share of GDP bear little relationship with the sustainability of fiscal policy. Some countries such as Italy appear to be on relatively sustainable paths in terms of fiscal and pension policy despite challenging short-

[14]At the end of 1991 and early 1992, an asset price bubble burst in Japan and was followed by a long period of stagnation. The period 1990–2010 is sometimes referred to as Japan's "Lost 20 years".

[15]The introduction of the European System of National and Regional Accounts (ESA 2010) obliged the EU countries to publish their implicit public debt.

[16]Notice that estimates for implicit debt are subject to much stronger volatility than the official numbers on gross or net debt. Numbers on implicit debt are very sensitive with regard to reforms of the pension or health system and changes in the growth dynamics of a country. For example, Hagist, Moog, Raffelhüschen, and Vatter (2009) use the method of generational accounting to provide estimates of the implicit debt-GDP ratio in 2004 at the amount of 254% (France), 252% (Germany), 35% (Spain), 510% (UK), and 350% (US). Compared to the values in Table 7.3, their numbers are higher and the ordering of the implicit debt numbers is the same except for Spain, for which the authors report a drastically lower implicit debt-GDP ratio.

Table 7.3 Explicit and implicit debt levels in 2016

	Debt-GDP ratio 2016		
Country	Gross debt	Implicit debt	Total debt
France	96%	170%	286%
Germany	71%	90%	161%
Greece	177%	154%	331%
Ireland	79%	709%	788%
Italy	132%	−25%	107%
Luxembourg	22%	803%	825%
Spain	100%	665%	765%
UK	89%	301%	390%

run statistics, while for others, e.g., Luxembourg and Ireland, favorable short-term measures mask very substantial long-term problems.[17]

7.3 Debt Arithmetic

Let $\tilde{B}_t$ denote nominal government debt at the end of period $t-1$ (equivalently, the beginning of period t).[18] For simplicity, we assume that the government only issues zero-coupon bonds with a maturity of one period. The face value of the bonds is unity, meaning that $\tilde{B}_t$ also measures the number of bonds. At the beginning of period t, the government issues new government bonds $\tilde{B}_{t+1}$ at price P_t^B and thus borrows $P_t^B \tilde{B}_{t+1}$. The nominal interest rate i_t^B on the government bond in period t is therefore

$$1 + i_t^B = \frac{1}{P_t^B}. \tag{7.1}$$

[17] We should be extremely careful to assess the accuracy of the implicit debt numbers since all the data are self-reported by the individual EU countries. In January 2010, for example, the European Commission condemned Greece for falsifying its data on public finances and deliberate misreporting after the official 2009 deficit numbers had been revised from 3.7% to 12.5% of GDP in fall 2009 by the newly elected Greek government (which was led by the center-left PASOK).

[18] Sometimes, $\tilde{B}_{t-1}$ or B_{t-1} is used as the notation for the beginning-of-period-t government debt level because it is equal to the government debt at the end of period $t-1$. We use the notation $\tilde{B}_t$ such that the time index is in accordance with that of the capital stock K_t at the beginning of period t in the Ramsey and OLG models from the previous chapters.

Nominal government debt increases by the government budget deficit[19]:

$$P_t^B \tilde{B}_{t+1} - \tilde{B}_t = P_t G_t - P_t T_t. \tag{7.2}$$

The difference between government expenditures (excluding interest payments) and taxes, $P_t G_t - P_t T_t$, is called the primary budget deficit.

Let B_t denote aggregate real government bonds, $B_t \equiv \tilde{B}_t / P_t$.[20] To derive the government budget constraint in real terms, we divide (7.2) by the price level P_t in period t

$$P_t^B \frac{P_{t+1}}{P_t} \frac{\tilde{B}_{t+1}}{P_{t+1}} = \frac{\tilde{B}_t}{P_t} + G_t - T_t.$$

Recalling the definition of the inflation rate, $\pi_t = (P_t - P_{t-1})/P_{t-1}$, and using the nominal interest rate i_t^B from (7.1), we can derive the real government budget constraint:

$$\frac{1}{1+r_t^B} B_{t+1} = B_t + G_t - T_t, \tag{7.3}$$

where the real interest rate on government bonds r_t^B follows with the help of the Fisher equation

$$1 + i_t^B = (1 + r_t^B)(1 + \pi_{t+1}). \tag{7.4}$$

By induction, we can solve this equation forward to derive the intertemporal government budget constraint:

$$B_0 + \sum_{t=0}^{\infty} \frac{1}{\prod_{s=0}^{t}(1 + r_{s-1}^B)} G_t = \sum_{t=0}^{\infty} \frac{1}{\prod_{s=0}^{t}(1 + r_{s-1}^B)} T_t,$$

[19]In Appendix 7.1, we also consider the case in which the government finances its deficit with the help of money, so-called seignorage. Since seignorage, however, only constitutes a small share of government finance in industrialized countries, we neglect it in the remainder of the main text. For example, King and Plosser (1985) estimate that seignorage amounted to 0.3% and over 2% of GDP in the US and Italy during the period 1952–1982, respectively.

[20]In the literature, real government debt is often denoted by the lower-case variable b_t. Our notational convention in this book is that upper-case variables denote aggregate variables, while lower-case variables denote individual variables. For example, we will use the notation b_t^s and b_t for the real bonds of the s-year-old individual and the per capita bonds in period t in subsequent sections.

or

$$B_0 = \sum_{t=0}^{\infty} \frac{1}{\prod_{s=0}^{t}(1 + r_{s-1}^B)} [T_t - G_t], \tag{7.5}$$

with $r_{-1}^B \equiv 0$. Accordingly, the budget constraint (7.5) states that the government must run (in present value) primary budget surpluses large enough to offset its initial debt B_0. In the derivation of this intertemporal budget constraint, we have ruled out Ponzi schemes by imposing the transversality condition[21]

$$\lim_{t\to\infty} \frac{B_t}{\prod_{s=0}^{t}(1 + r_{s-1}^B)} = 0. \tag{7.6}$$

Notice that a sustainable fiscal policy for given real debt B_0 requires the net present value of primary surpluses to satisfy (7.5). Therefore, the budget surplus, $P_tT_t - i_t^B B_t - P_tG_t$, is not a good measure to gauge fiscal sustainability in times of volatile inflation. In times of high inflation, the nominal interest rate, $i_t^B = r_t^B + \pi_{t+1}$, increases, and thus, the budget deficit also increases, ceteris paribus. However, higher inflation also reduces the amount of nominal debt equally and thus does not affect (7.5) if the real interest rate r_t^B remains constant.

Often, it will be convenient to relate the government debt level to GDP, particularly if we study a growing economy or compare different countries. Dividing (7.2) by nominal GDP P_tY_t results in

$$\frac{1}{1 + i_t^B} \frac{Y_{t+1}}{Y_t} \frac{P_{t+1}}{P_t} \frac{\tilde{B}_{t+1}}{P_{t+1}Y_{t+1}} = \frac{\tilde{B}_t}{P_tY_t} + \frac{G_t - T_t}{Y_t}. \tag{7.7}$$

Let the economic growth rate be defined as the growth rate of real output, $\gamma_t = (Y_t - Y_{t-1})/Y_{t-1}$. We can simplify this expression to

$$\frac{1 + \gamma_{t+1}}{1 + r_t^B} \frac{B_{t+1}}{Y_{t+1}} = \frac{B_t}{Y_t} + \frac{G_t - T_t}{Y_t}. \tag{7.8}$$

[21] In a Ponzi scheme, an investor (or government) raises a return to the old investors (or creditors) by raising revenue from new investors (or creditors) that are inconsistent. When the inflow of new investors stops, the scheme falls apart. A prominent historical example of such a pyramid scheme occurred in Albania in 1997 which lead to the bankruptcy of some 25 firms; the liabilities of the scheme amounted to approximately \$1.2 billion (which was equal to half the annual GDP of Albania in 1997), and resulted in political unrests. As a consequence, the former economic advisor of Prime Minister Fatos Nano was arrested and imprisoned (see also Jarvis 1999). The no-Ponzi condition (7.6) excludes this possibility.

Equation (7.8) can be interpreted as a first-order difference equation in the debt-output ratio B/Y. Its stability depends on the value of its autoregressive coefficient $(1 + r_t^B)/(1 + \gamma_{t+1})$. Thus, for example, if the government follows a particular government deficit rule, e.g., that the primary deficit $G - T$ should not exceed 3% of GDP, $(G_t - T_t)/Y_t = 0.03$, government debt explodes if the real interest rate is larger than the growth rate, $(1 + r^B)/(1 + \gamma) > 1$, equivalently $r^B > \gamma$, and converges to

$$\frac{B}{Y} = \frac{1 + r^B}{\gamma - r^B} \frac{G - T}{Y},$$

if $r^B < \gamma$.

We have to issue a warning at this point. The real interest rate on government bonds r^B and the economic growth rate γ may not be independent of the deficit $G - T$ and, in particular, government debt B/Y. In the previous section, you learned about the empirical study of Reinhard and Rogoff, who argue there is a threshold level of B/Y beyond which economic growth falls. In Sect. 7.5, we will compare two different fiscal policies and their real impact on the US economy during the demographic transition between the years 2010 and 2150. The first holds the debt level constant at the 2010 level, meaning that the debt-output B/Y reaches 73.2% in 2150, while the second uses debt financing to compensate transitory generations for welfare losses, thereby effectively increasing the debt-output ratio B/Y to 214% by 2150. We show that these two policies imply a large difference in the real interest rate r^B which increases to 4.05% (low debt) and 6.32% (high debt) in the year 2150. Therefore, the arithmetic of stable debt and deficit policies critically depends on the effect of higher debt on economic growth γ and the interest rate r^B.

In this vein, the Stability and Growth Pact of the European Union for the members of the EMU sets two upper limits for public finances in Eurozone countries. The debt-output level B/Y and the fiscal deficit, $(G + i^B B - T)/Y = D/Y$, should not exceed 60% and 3%, respectively.[22] Let us counterfactually assume that the countries adhere to this policy and that, in addition, the economic growth rate γ and the real interest rate r^B are both equal to 3%, while inflation π amounts to 2%. To interpret the effects of this rule, let us use a different representation of the government budget (7.2) by assuming that the government bonds are no longer zero-coupon bonds but pay a nominal interest rate i_t^B (which adjusts such that the price of the bonds P_t^B is equal to one)

$$\tilde{B}_{t+1} - \tilde{B}_t = i_t^B \tilde{B}_t + P_t G_t - P_t T_t. \tag{7.9}$$

[22]Clearly, the Stability and Growth Pact has been violated dozens of times since its creation. Frequently during the period 2001–2015, all EMU countries presented in Figs. 7.4 and 7.5 – France, Germany, Italy, and Spain – ran deficits in excess of 3% in some years.

For this type of nominal debt instrument, we derive

$$\frac{\tilde{B}_{t+1} - \tilde{B}_t}{P_t Y_t} = \frac{P_t D_t}{P_t Y_t},$$

$$(1 + \gamma_{t+1})(1 + \pi_{t+1}) \frac{B_{t+1}}{Y_{t+1}} - \frac{B_t}{Y_t} = \frac{P_t D_t}{P_t Y_t}.$$

If the debt-output level is stationary, $\frac{B_{t+1}}{Y_{t+1}} = \frac{B_t}{Y_t} = \frac{B}{Y}$, and all other economic variables are constant, the above equation simplifies to:

$$[(1 + \gamma)(1 + \pi) - 1] \frac{B}{Y} = \frac{D}{Y}. \tag{7.10}$$

For a debt-output level of 60%, $B/Y = 0.6$, and a nominal economic growth rate of 5%, $(1 + \gamma)(1 + \pi) - 1 = 0.05$, the implied deficit-output ratio D/Y is equal to 3%.

A nominal growth rate of 5%, however, seems overly optimistic given the initial experience of the EMU during the period 2001–2015. During this period, when the euro became the official currency of the EMU, nominal growth was well below 3% on average. In this case, the implied deficit ratio for a target of B/Y equal to 60% (with $\pi + \gamma = 0.03$) would only amount to 1.8%. Examining the relationship in (7.10) in a different way by starting at a given constant fiscal deficit D/Y, we can also derive the implied long-run debt-output level. If we take the case of Italy, where the real GDP growth rate, inflation rate, and government deficit during the period 2001–2015 amounted to 0.28% (see Fig. 7.9), 1.85%, and 3.3%, respectively, we

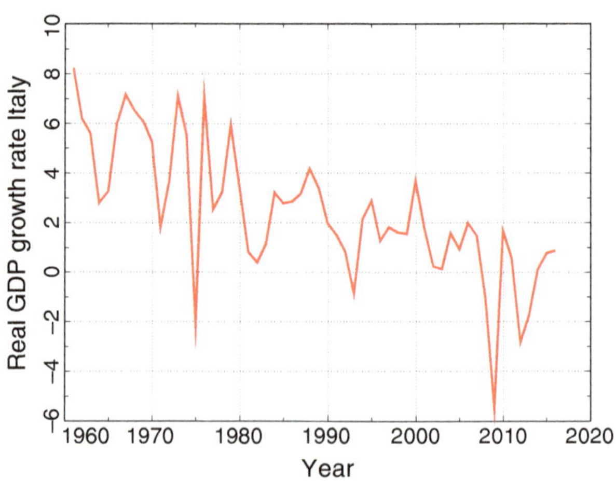

Fig. 7.9 Real GDP growth in Italy, 1961–2016

derive an implied long-run debt-output ratio B/Y of 155%. For comparison, the actual debt-GDP ratio amounted to 132% in Italy in 2015.

7.4 Ricardian Equivalence

Does it matter whether the government finances its expenditures with debt or (non-distortionary) lump-sum taxes? In his seminal article "Are Government Bonds Net Wealth?" in the Journal of Political Economy, Barro (1974) showed that the means of financing public expenditures does not matter for the real allocation of the economy if the following holds[23]:

1. Families act as infinitely lived dynasties because of intergenerational altruism.
2. Capital markets are perfect (i.e., all can borrow and lend at a single rate). In particular, private and public debt are perfect substitutes.
3. The path of government expenditures is fixed. Therefore, the government can precommit itself and its successors to a specific fiscal policy.

Under conditions (1)–(3), it does not matter whether the government finances deficits by issuing bonds rather than with lump-sum taxes. In the case of debt financing, the households completely offset the government policy and save the additional income (from lower lump-sum taxes) for the future periods when the government bonds have to be repaid, and the government raises (lump-sum) taxes. If the repayment of the debt will take place after the death of the currently living cohorts, the families simply accumulate bequests for their children such that they will be able to pay the higher taxes. This result is called the *Ricardian equivalence proposition* and is also known as the *Barro-Ricardo equivalence theorem.*[24]

In this section, we proceed as follows. We first show that Ricardian equivalence holds in the Ramsey model. Second, we show, both theoretically and by means of a numerical example, that Ricardian equivalence fails in the simple two-period OLG model without bequest motives. The crowding-out effect of public debt is shown to be quantitatively significant. Third, we demonstrate that altruism helps to restore Ricardian equivalence in the two-period OLG model.[25]

[23]The title from Barro's article derives from the fact that, in his model, an individual's consumption is proportional to net wealth. For this reason, if higher government debt increases net wealth, consumption rises, while savings decline.

[24]In 1974, Robert J. Barro provided the theoretical foundation for hesitant speculation by Ricardo (1817) that it "is only by saving from income, and retrenching in expenditures, that the national capital can be increased."

[25]In the next section, we analyze the effects of debt financing of the pension reform during the demographic transition in a large-scale OLG model and show that the effects of debt financing are dramatic and result in substantial long-run output and welfare losses.

7.4.1 The Ramsey Model with Government Debt

In the following, we introduce real government debt B_t into the standard Ramsey model from Chap. 2. For a given exogenous path of government consumption $\{G_t\}_{t=0}^{\infty}$, we will demonstrate that it does not matter whether G_t is financed by means of debt or (non-distortionary) lump-sum taxes in period t. The reason is straightforward. If the government uses debt rather than taxes today, the household knows that taxes have to increase in the future and saves the additional income for this date.

7.4.1.1 Demographics

Let us consider an economy in which the population N_t grows at the constant rate n:

$$N_t = (1+n)N_{t-1}. \tag{7.11}$$

Since each member of the household supplies one unit of labor, aggregate labor is also equal to N_t.

7.4.1.2 Production

Production is characterized by a Cobb-Douglas function:

$$F(K_t, L_t) = K_t^{\alpha}(A_t L_t)^{1-\alpha}, \tag{7.12}$$

where aggregate labor L_t is equal to population size N_t, $L_t = N_t$.[26] Labor-augmenting technology A_t grows at the exogenous rate γ.

Assuming perfect competition in the goods and factor markets, efficient labor $A_t L_t$ and capital K_t are rewarded by their marginal products:

$$w_t = (1-\alpha)\left(\frac{K_t}{A_t L_t}\right)^{\alpha} = (1-\alpha)k_t^{\alpha}, \tag{7.13a}$$

$$r_t = \alpha k_t^{\alpha-1} - \delta, \tag{7.13b}$$

where δ denotes the rate of depreciation.

7.4.1.3 Government

We assume that the government issues one-period bonds and pays the nominal interest i_t^B on outstanding debt. Therefore, we use the more convenient form of the government budget constraint (7.9) in the following.[27] The government finances

[26]We distinguish between aggregate labor L_t and population size N_t to allow us to use the same specification of the production sector in the next section on the OLG model.

[27]Both representations of the government budget are used in the literature. For example, Trabandt and Uhlig (2011) use the specification (7.9), while Heer and Scharrer (2018) use (7.3). The representation (7.9) has the advantage that we can use individual wealth $\omega_t = k_t + b_t$, which is

its consumption G_t, transfers Tr_t, and the real interest payment on outstanding debt $r_t^B B_t$ by means of debt. Since we also introduce economic growth into this model, we divide (7.9) by the product of the price level P_t, population size N_t, and labor efficiency level A_t to derive the real government budget constraint in per capita (and efficiency) units[28]

$$(1+n)(1+\gamma)b_{t+1} = (1+r_t^B)b_t + g_t + tr_t, \tag{7.14}$$

where we have replaced taxes T_t with (negative) lump-sum transfers Tr_t and defined $b_t \equiv \tilde{B}_t/(P_{t-1}A_tN_t)$.[29] Similarly, $g_t \equiv G_t/(A_tN_t)$ and $tr_t \equiv Tr_t/(A_tN_t)$ denote government consumption and transfers per capita (in efficiency units).

Again, we impose a no-Ponzi condition in a form similar to that of (7.6) to rule out Ponzi schemes:

$$B_0 + \sum_{t=0}^{\infty} \frac{1}{\prod_{s=0}^{t}(1+r_s^B)} G_t = -\sum_{t=0}^{\infty} \frac{1}{\prod_{s=0}^{t}(1+r_s^B)} Tr_t, \tag{7.15}$$

using the definition $B_t \equiv \tilde{B}_t/P_{t-1}$. According to (7.15), the discounted stream of taxes (or negative transfers) must equal the current value of outstanding government debt plus the present value of government consumption.

7.4.1.4 Households

In the Ramsey model with government debt, households may hold two types of assets, physical capital K_t and government bonds B_t. Total wealth is denoted by Ω_t:

$$\Omega_t = K_t + B_t. \tag{7.16}$$

Government bonds B_t and physical capital K_t provide real returns r_t^B and r_t, respectively. As a second source of income, the household receives wage income $w_tA_tN_t$. In addition, the household (of size N_t) receives lump-sum transfers Tr_t (or pays lump-sum taxes when $Tr_t < 0$) in period t, and its real budget constraint is

---

equal to the sum of the two assets, as the individual state variable, while Heer and Scharrer (2018) have to define $\tilde{\omega}_t = k_t + b_t/(1+r_{t-1}^B)$ as the individual state variable to solve their model. Since the former is easier to interpret and more convenient to handle, we will use it in the following.

[28] We commit a small notational sin here (since we have exhausted all arabic letters for the notation of variables in this book). In Sect. 4.5.2, we used the variable g_t to denote the mark-up. Here, it stands for real government consumption G_t divided by A_tN_t.

[29] Since we define real debt per capita b_t with respect to the price level in period $t-1$, P_{t-1}, the nominal interest rate is related to the real interest rate according to

$$1 + i_t^B = (1+r_t^B)(1+\pi_t)$$

rather than by (7.4).

represented by

$$(1 + r_t^B)B_t + (1 + r_t)K_t + w_t A_t N_t + Tr_t = C_t + K_{t+1} + B_{t+1}, \tag{7.17}$$

where C_t denotes household consumption. In per capita (efficiency) terms, the budget constraint (7.17) can be re-written as (after dividing by $A_t N_t$):

$$(1 + r_t^B)b_t + (1 + r_t)k_t + w_t + tr_t = c_t + (1 + n)(1 + \gamma)(k_{t+1} + b_{t+1}), \tag{7.18}$$

with $c_t \equiv C_t/(A_t N_t)$.

The household maximizes its intertemporal utility

$$U = \sum_{t=0}^{\infty} \beta^t u(c_t) \tag{7.19}$$

subject to (7.18), implying the first-order condition in the form of the *Euler-equation*:

$$u'(c_t) = \beta \frac{u'(c_{t+1})}{(1 + n)(1 + \gamma)} (1 + r_{t+1}), \tag{7.20}$$

and the equality of the two asset returns (since we do not consider uncertainty in the model):

$$r_t = r_t^B. \tag{7.21}$$

7.4.1.5 Equilibrium

If we insert (7.14) into (7.18), we derive

$$(1 + r_t)k_t + w_t = c_t + g_t + (1 + n)(1 + \gamma)k_{t+1}. \tag{7.22}$$

The equilibrium conditions for the household, (7.20) and (7.22), and for the firm, (7.13a) and (7.13b), together with the initial capital stock k_0, describe the dynamics of the model and are the same as in the Ramsey model without government debt ($b_t \equiv 0$). None of the four equilibrium equations depend on b_t or tr_t. As a consequence, only the time path of government consumption $\{G_t\}_{t=0}^{\infty}$ affects the dynamics of the capital stock k_t and, hence, output $y_t = k_t^{\alpha}$ and consumption c_t but not its financing. This result is called *Ricardian equivalence*.

7.4.2 The Two-Period OLG Model with Government Debt

Ricardian equivalence holds in the Ramsey model because households have infinite lifetimes. If the government chooses to finance its expenditures by higher debt today and, therefore, higher future taxes, the household saves the additional income for this repayment period. In the OLG model, by contrast, households have finite lifetimes. Therefore, if future generations have to pay higher taxes, current generations do not accordingly increase savings in the absence of altruism and a bequest motive. We will both qualitatively and quantitatively illustrate the effects of debt in the standard OLG model from Chap. 3.

As in the Ramsey model with government debt above, population grows at rate n. Let N_t denote the number of young households such that (7.11) holds. In addition, the production and government sectors are also described as above.

7.4.2.1 Households

Households live for two periods. Lifetime utility is additive in instantaneous utility $u(c)$ from consumption in young and old age c_t^1 and c_{t+1}^2[30]:

$$U_t = \frac{\left(c_t^1\right)^{1-\sigma} - 1}{1-\sigma} + \beta \frac{\left(c_{t+1}^2\right)^{1-\sigma} - 1}{1-\sigma}, \tag{7.23}$$

where β denotes the discount factor, and $1/\sigma$ is equal to the intertemporal elasticity of substitution. The young household inelastically supplies one unit of labor and does not work in old age.

The young household is born without any assets, $b_t^0 = k_t^0 = 0$, and thus, its savings are given by:

$$S_t = w_t A_t N_t + Tr_t - C_t^1, \tag{7.24}$$

or, after dividing by $A_t N_t$

$$s_t = w_t + tr_t - c_t^1. \tag{7.25}$$

We assume that the government only transfers income ($tr_t > 0$) to or levies lump-sum taxes ($tr_t < 0$) on the young household.[31]

The household finances its old-age consumption with the help of his savings and interest earnings[32]:

$$c_{t+1}^2 = (1 + r_{t+1}) s_t, \tag{7.26}$$

[30]In particular, we define $c_t^1 \equiv C_t^1/(A_t N_t)$ and $c_{t+1}^2 \equiv C_{t+1}^2/(A_t N_t)$ where C_t^1 and C_{t+1}^2 denote household consumptions in period t and $t+1$, respectively. See also Appendix 3.2 for a discussion of this assumption with respect to preferences and alternative specifications of the lifetime utility in the 2-period OLG model in the presence of economic growth.

[31]In Problem 7.1, you are asked to consider the case in which the government transfers tr_t equally to both the young and the old generation.

[32]In the following, we have already incorporated the result from the previous section that the two form of assets K_t and B_t need to generate the same rate of real return r_t.

implying the household's intertemporal budget constraint

$$c_t^1 + \frac{c_{t+1}^2}{1 + r_{t+1}} = w_t + tr_t. \tag{7.27}$$

Maximizing lifetime utility (7.23) subject to the intertemporal budget constraint (7.27) implies the first-order conditions:

$$\lambda_t = \left(c_t^1\right)^{-\sigma}, \tag{7.28a}$$

$$\lambda_t = \beta \left(c_{t+1}^2\right)^{-\sigma} [1 + r_{t+1}], \tag{7.28b}$$

and hence,

$$c_t^1 = \frac{w_t + tr_t}{1 + \beta^{\frac{1}{\sigma}} (1 + r_{t+1})^{\frac{1}{\sigma}-1}}. \tag{7.29}$$

Therefore, savings are equal to

$$s_t = (w_t + tr_t) \left(1 - \frac{1}{1 + \beta^{\frac{1}{\sigma}} (1 + r_{t+1})^{\frac{1}{\sigma}-1}}\right). \tag{7.30}$$

7.4.2.2 Equilibrium

In capital market equilibrium, aggregate savings S_t are equal to aggregate wealth, $\Omega_{t+1} = B_{t+1} + K_{t+1}$:

$$\begin{aligned} \Omega_{t+1} &= S_t \\ &= (w_t A_t N_t + Tr_t) \left(1 - \frac{1}{1 + \beta^{\frac{1}{\sigma}} (1 + r_{t+1})^{\frac{1}{\sigma}-1}}\right), \end{aligned}$$

and therefore,

$$(1+n)(1+\gamma)(b_{t+1}+k_{t+1}) = (w_t+tr_t) \left(1 - \frac{1}{1 + \beta^{\frac{1}{\sigma}} (1 + r_{t+1})^{\frac{1}{\sigma}-1}}\right), \tag{7.31}$$

with the factor prices given by (7.13a) and (7.13b).

Let us consider a steady state in which the capital stock and the debt per efficiency unit of labor, $k = K/(AN)$ and $b = B/(AN)$, are constant, implying

$$(1 + n)(1 + \gamma)(b + k) = (w + tr) \left(1 - \frac{1}{1 + \beta^{\frac{1}{\sigma}} (1 + r)^{\frac{1}{\sigma}-1}}\right). \tag{7.32}$$

From the government budget (7.14), equilibrium transfers (or lump-sum taxes) are given by

$$g + tr = (n + \gamma + n\gamma - r)\, b. \tag{7.33}$$

Accordingly, a government that finances its expenditures by debt rather than taxes in the short run accumulates higher debt in the long run and needs to impose higher lump-sum taxes on the household in the steady state (for the case $r > n + \gamma + n\gamma$). If we substitute (7.33) into (7.32), we notice that the financing decision of the government (as reflected in a lower or higher level of government transfers tr) affects the steady-state capital stock k:

$$(1+n)(1+\gamma)\left(\frac{g+tr}{n+\gamma+\gamma n-r}+k\right) = (w+tr)\left(1-\frac{1}{1+\beta^{\frac{1}{\sigma}}(1+r)^{\frac{1}{\sigma}-1}}\right) \tag{7.34}$$

With $r = \alpha k^{\alpha-1} - \delta$ and $w = (1-\alpha)k^{\alpha}$, the implicit solution of this equation for the steady-state capital stock k depends on government transfers tr or, equally, debt b. In other words, Ricardian equivalence fails.

We will illustrate the dependency of the capital stock on the government's financing policy by means of a numerical example. For this reason, we calibrate the model as in Chap. 3 and use a period length of 30 years. We choose an annual discount rate of 4%, implying $\beta = 0.96^{30} = 0.294$, and the production elasticity of capital $\alpha = 0.36$; capital depreciates by $\delta = 100\%$. Annual growth amounts to 2%, implying $\gamma = 0.81$. The intertemporal elasticity is set equal to $1/\sigma = 1/2$, and population grows at the rate $n = 20\%$. In our example, we ignore government consumption $G = 0$ and set the debt-output ratio equal to zero in the benchmark. The solution is computed with the help of the Gauss program `Ch7_debt1.g`.

Figure 7.10 presents the effects of the steady-state debt-output ratio B/Y on the capital stock.[33] The debt-output level B/Y is expressed in percentage points. Notice two observations: (1) The debt level has a significant effect on the capital stock and, hence, output. If the debt-output level B/Y increases from 0% to 10%, the capital stock k falls by 27%, from 0.0090 to 0.0066. Similarly, output y falls by 16.4%, from 0.184 to 0.164. Clearly, the quantitative steady-state effects of debt are tremendous in the simple OLG model.[34] (2) Government debt is able to completely crowd out physical capital, meaning that the economy collapses to the no-production case.

[33] We annualized the debt-output level by multiplying the model's 30-year value B/Y by 30.

[34] In Theorem 20.1 on page 321, Azariadis (1993) proves that, in any asymptotically stable stationary equilibrium, capital intensity and per capita saving are decreasing functions of per capita national debt under the following assumptions: (1) Government purchases are zero, (2) consumption goods in young and old age are normal and gross substitutes, (3) a constant stock of per capita public debt is serviced by lump-sum taxes on old individuals, (4) labor supply is exogenous, and (5) production is characterized by constant returns to scale, and goods and factor markets are competitive.

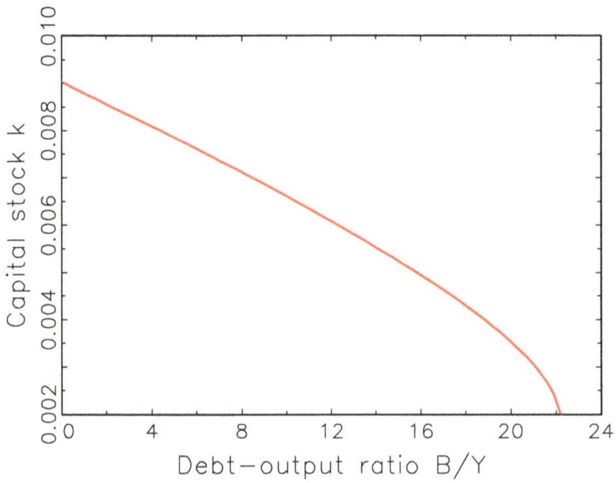

Fig. 7.10 Capital stock and debt in the OLG model

In our simple example, this takes place at the modest debt level of approximately 22% of GDP. For high debt and low capital stock, interest rates on government debt increase without bound for a capital stock that approaches zero. As a consequence, the household has to pay lump-sum taxes that exceed its income.[35]

Of course, our simple OLG model lacks realism. We do not observe that a country's economy collapses if the level of debt relative to GDP exceeds 20%. In Sect. 7.5, we will specify an OLG model that is more realistic and features multiple cohorts in each period. Therefore, we will be better able to match the number of retirees to workers and the amount of life-cycle savings. In this large-scale OLG model, we will find that the economy is able to sustain a much higher threshold value of public debt than in our simple two-period OLG model. In fact, the large-scale OLG model will behave somewhere in between the Ramsey model and the two-period OLG model, and the quantitative effects of higher debt on real variables, such as the capital stock, will be smaller. Nevertheless, we will identify a threshold value for government debt in Sect. 7.5 beyond which the economy collapses.

[35]Diamond (1965) shows in a competitive OLG model that, in steady state, higher debt reduces (increases) utility when the economy is efficient (in the case of over-accumulation of capital). However, as we noted in Chap. 3 on the OLG model and Chap. 6 on social security, it is rather unlikely that we would observe over-accumulation of capital in industrialized countries in face of the low population growth and the large unfunded public-pay-as-you pension systems. Therefore, debt is likely to decrease utility in steady state, as we will also find in the subsequent section for the case of the US economy.

7.4.3 Altruism and Ricardian Equivalence

In the standard OLG model, parents do not care for their offspring. In the following, we extend the model of the previous section along the lines of Sect. 3.3.3. The size of the young generation is again denoted by N_t and grows at rate n. In addition, let the household's lifetime utility V_t in period t be given by (3.29), which we restate for the reader's convenience:

$$V_t = u(c_t^1) + \beta u(c_{t+1}^2) + \frac{1}{1+R} V_{t+1}. \tag{7.35}$$

Parents discount lifetime utility of their children V_{t+1} at rate $R > 0$. In this specification, V_t represents the lifetime utility of a representative member of the household, and c_t^s, $s = 1, 2$ denotes per capita consumption in efficiency units with $c_t^1 \equiv C_t^1/(A_t N_t)$ and $c_{t+1}^2 \equiv C_{t+1}^2/(A_t N_t)$. Accordingly, the number of children n does not change lifetime utility, ceteris paribus.

Postponing the time index t in (7.35) and recursive substitution into (7.35) results in

$$V_t = \sum_{i=0}^{\infty} \frac{1}{(1+R)^i} \left[u(c_{t+i}^1) + \beta u(c_{t+i+1}^2) \right].$$

Parents who are born in period t can leave (per capita) bequests $beq_{t+1} \geq 0$ to their children at the end of period $t+1$. Similarly, they receive bequests $beq_t \geq 0$ from their parents at the end of period t. Accordingly, the budget constraints of generation t are represented by:

$$c_t^1 + s_t = w_t + beq_t + tr_t, \tag{7.36a}$$

$$c_{t+1}^2 + (1+n)(1+\gamma) beq_{t+1} = (1 + r_{t+1}) s_t, \tag{7.36b}$$

implying the intertemporal budget constraint

$$c_t^1 + \frac{c_{t+1}^2 + (1+n)(1+\gamma) beq_{t+1}}{1 + r_{t+1}} = w_t + beq_t + tr_t.$$

For simplicity, we assume that only young households receive government transfers.

Assuming an operative bequest motive, $beq_{t+1} > 0$, the first-order conditions of the household are represented by[36]:

$$\frac{\partial}{\partial c_t^1} : \lambda_t = u'(c_t^1), \tag{7.37a}$$

$$\frac{\partial}{\partial c_{t+1}^2} : \frac{\lambda_t}{1+r_{t+1}} = \beta u'(c_{t+1}^2), \tag{7.37b}$$

$$\frac{\partial}{\partial beq_{t+1}} : \lambda_t \frac{(1+n)(1+\gamma)}{1+r_{t+1}} = \frac{\lambda_{t+1}}{1+R}, \quad \text{if } beq_{t+1} > 0, \tag{7.37c}$$

where λ_t denotes the Lagrange multiplier of the generation-t budget constraint. We can simplify the first-order conditions by eliminating λ_t. Therefore,

$$u'(c_t^1) = \beta u'(c_{t+1}^2)\,(1+r_{t+1})\,, \tag{7.38a}$$

$$u'(c_t^1)\frac{(1+n)(1+\gamma)}{1+r_{t+1}} = \frac{u'(c_{t+1}^1)}{1+R}, \quad \text{if } beq_{t+1} > 0. \tag{7.38b}$$

The first equation is the familiar Euler condition that equals the marginal utility from consuming an additional unit today and from saving it. In the second equation, the marginal utility from the consumption today is compared with the marginal utility of saving an additional unit and leaving it as a bequest. If the household consumes the additional unit today, it derives the marginal utility $u'(c_t^1)$. If it saves the extra unit, it will be able to leave $(1 + r_{t+1})$ units of the consumption good to the next generation. Since the next generation has the size $(1 + n)$ relative to the present generation and also applies the higher productivity level A_{t+1} to the normalization of the consumption good, $c_{t+1}^1 \equiv C_{t+1}^1/(A_{t+1}N_{t+1})$, the additional per capita consumption in efficiency units decreases by the divisor $(1+n)(1+\gamma)$. In addition, the present households discounts the utility of the next generation at the rate R.

The production sector is described as in the previous section such that production is given by the Cobb-Douglas function (7.12), and profit maximization implies that the factor prices of labor and capital, w and r, are equal to their marginal products as represented by (7.13a) and (7.13b).

For ease of exposition, we abstract from government consumption. Consequently, government expenditures only consist of transfers that are financed by debt B_t, $Tr_t = B_{t+1} - (1 + r_t)B_t$, or in per capita terms:

$$tr_t = ((1+n)(1+\gamma))\, b_{t+1} - (1+r_t)b_t. \tag{7.39}$$

[36] See Appendix 3.1 for the computation of this Kuhn-Tucker problem.

Again, we have already incorporated the no-arbitrage condition from the capital market, $r_t^B = r_t$.

In equilibrium, savings are equal to next-period assets, $\Omega_{t+1} = N_t A_t s_t$, or after dividing by $N_t A_t$,

$$(1+n)(1+\gamma)(k_{t+1} + b_{t+1}) = s_t. \tag{7.40}$$

For this economy, Barro (1974) demonstrates that an increase in transfers does not change the equilibrium capital stock. We discussed the idea behind this result above. If the present generation receives a higher transfer tr, it leaves the complete amount to the next generation, which will use it to repay the debt. As a consequence, the additional savings (for the next generations) in the capital market just offset the additional debt.

We will demonstrate this by means of an example for the steady state. Therefore, assume the following function for instantaneous utility:

$$u(c) = \frac{c^{1-\sigma} - 1}{1-\sigma},$$

which is characterized by the constant intertemporal elasticity of substitution $1/\sigma$. The equilibrium conditions for the steady state with constant consumption (c^1, c^2), bequests beq, and capital stock k are given by eight equations in the eight unknowns $w, r, tr, k, beq, b, c^1, c^2$ and the exogenous variable, the debt-output ratio b/y:

$$w = (1-\alpha)k^{\alpha}, \tag{7.41a}$$

$$r = \alpha k^{\alpha-1} - \delta, \tag{7.41b}$$

$$tr = (n + \gamma + n\gamma - r)\, b, \tag{7.41c}$$

$$b = \frac{b}{y} k^{\alpha}, \tag{7.41d}$$

$$c^2 = [\beta(1+r)]^{\frac{1}{\sigma}} c^1, \tag{7.41e}$$

$$\frac{1+r}{(1+n)(1+\gamma)} = 1 + R, \tag{7.41f}$$

$$c^1 + \frac{c^2}{1+r} = w + beq\left(1 - \frac{(1+n)(1+\gamma)}{1+r}\right) + tr \tag{7.41g}$$

$$(1+n)(1+\gamma)(k+b) = w + beq + tr - c^1. \tag{7.41h}$$

The equilibrium values of k, c^1, c^2, and y are independent of the debt-output ratio b/y. To see this, observe that (7.41f) determines r for given R, γ, and n irrespective of the debt-output ratio b/y. Then, (7.41b) determines k, and hence, (7.41a) implies the wage w. For this reason, production, $y = k^{\alpha}$, is also determined by

k. Accordingly, none of the variables $\{r, w, k, y\}$ depend on b/y, and Ricardian equivalence holds with respect to capital and total production. Furthermore, c^1/c^2 depends only on r according to (7.41e) and, therefore, depends on k but not on b/y.

In the goods market equilibrium of the economy,[37]

$$(1+n)(1+\gamma)k_{t+1} + c_t^1 + \frac{c_t^2}{(1+n)(1+\gamma)} = k_t^\alpha + (1-\delta)k_t. \tag{7.42}$$

Therefore, in steady state, c^1 and c^2 depend only on k and n but not on b/y:

$$(n+\gamma+n\gamma+\delta)k + c^1 + \frac{c^2}{(1+n)(1+\gamma)} = k^\alpha.$$

Finally, (7.41d) determines public debt per capita b, and (7.41c) determines tr. Bequests beq adjust such that (7.41h) holds.

In conclusion, we have demonstrated that Ricardian equivalence holds in the OLG model with altruistic bequests. In particular, the debt-output level only determines transfers and bequests but does not change savings, consumption, or output. We have also demonstrated that this result holds irrespective of the discount factor R. Higher debt does not affect k, c^1, c^2, or y for a given R. Therefore, if parents discount their children's utility by a lower (or higher) discount rate than they apply to their old-age consumption, $R < 1/\beta - 1$ ($R > 1/\beta - 1$), Ricardian equivalence continues to hold.[38]

7.5 Putting It All Together: Demographic Transition, Pensions, and Debt

The debt situation of modern industrialized countries is likely to deteriorate in the coming decades due to the demographic transition. As you learned in Sect. 6.6 on the fiscal space, public expenditures are likely to rise while tax revenue will shrink in an aging economy. The fiscal space narrows, and thus, government might resort to higher debt financing, in particular if it reforms the public pay-as-you-go pension system and needs to compensate the losing generations during the transition.

In the following, we consider the effects of the demographic transition in an OLG model that is able to closely match the US economy with respect to its generational structure, pension system, and tax revenue. The model is based upon

[37] In Problem 7.2, you are asked to derive the goods market equilibrium condition (7.42). Therefore, notice that consumption in both young and in old age, c_t^1 and c_{t+1}^2, is normalized (made stationary) by dividing by A_t. When you derive the goods market equilibrium, take care to sum total consumption according to $C_t = N_t A_t c_t^1 + N_{t-1} A_{t-1} c_t^2$.

[38] If the discount factor R changes, of course, equilibrium values of the variables k, y c^1, and c^2 will adjust.

Heer, Polito, and Wickens (2017). We will first describe the model before we analyze the steady-state effects of higher debt. We confirm the results of the previous section that financing through higher government debt has real effects and crowds out investment in both the short and long run. Next, we derive the transition dynamics for the US economy for the projected population dynamics during the period 2010–2100 and study the effects of different forms of public financing. Since Ricardian equivalence fails in our model, the form of financing has real effects on output and welfare. Therefore, we distinguish four fiscal policies for financing higher government expenditures during the adjustment dynamics: (1) labor income taxes, (2) lump-sum taxes (or, equally, lower lump-sum transfers), (3) labor income taxes combined with transitory debt financing, and (4) lump-sum taxes combined with transitory debt financing. As we will find, the four different forms of financing imply different effects on generational welfare and long-run income.

7.5.1 Model Assumptions

The model is similar to that in Sect. 6.4; however, we simplify the model with respect to two assumptions. We do not consider cohort heterogeneity, and thus, all members of a given cohort are equal. In addition, we choose a longer time period of 5 years, which greatly facilitates the numerical problem without sacrificing any important model elements such as population demographics or life-cycle savings.

7.5.1.1 Demographics and Timing

A period, t, corresponds to 5 years. Newborns have a real-life age of 20–24, denoted by $s = 1$. All generations retire at the beginning of age $R = 10$ (corresponding to real-life age 65) and live up to a maximum age of $J = 15$ (real-life age 94). The number of periods during retirement is equal to $J - R + 1 = 6$. We denote total population in t by N_t and the number of s-year old cohort by $N_t(s)$ so that the cohort share in total population amounts to $\mu_t^s \equiv N_t(s)/N_t$. The survival probability of the s-year-old household in period t to survive until age $s+1$, $\phi_{t,s}$, is nonzero, except in the last period of its life. In addition, $\phi_{t,0} = 1.0$ and $\phi_{t,J} = 0$.

7.5.1.2 Households

Each household comprises one (possibly retired) individual. Households maximize the expected intertemporal lifetime utility

$$U_t = \sum_{s=1}^{J} \beta^{s-1} \left(\prod_{j=1}^{s} \phi_{t+j-2,j-1} \right) \left(u(c_{t+s-1}^s, l_{t+s-1}^s) + v(g_{t+s-1}) \right), \tag{7.43}$$

where the instantaneous utility is specified as a function of consumption and labor, as in Trabandt and Uhlig (2011)[39]:

$$u(c_{t+s-1}^s, l_{t+s-1}^s) = \frac{1}{1-\sigma}\left(\left(c_{t+s-1}^s\right)^{1-\sigma}\left[1 - \nu_0(1-\sigma)\left(l_{t+s-1}^s\right)^{1+1/\nu_1}\right]^{\sigma} - 1\right). \tag{7.44}$$

In this function, ν_1 and $1/\sigma$ denote the Frisch labor supply elasticity and the intertemporal elasticity of substitution. During the working life, the labor supply of the s-year-old amounts to $l_t^s \geq 0$, $s = 1, \ldots, R-1$, while it is set to $l_t^s \equiv 0$ during retirement, for $s = R, R+1, \ldots, J$. Utility from government consumption $v(g_t)$ is additive, meaning that government consumption per capita g_t does not have any direct effect on household behavior (it only indirectly affects behavior through its effects on transfers and taxes). We, therefore, can drop it in the consideration of the utility optimization problem in the following.

Let $\bar{y}^s$ denote the productivity of the s-year-old household, where the age-productivity profile $\{\bar{y}^s\}_{s=1}^{R-1}$ is a hump-shaped function, as estimated by Hansen (1993). Accordingly, total gross labor income $w_t \bar{y}^s A_t l_t^s$ is the product of the wage rate per efficiency unit w_t, the age-efficiency factor $\bar{y}^s$, aggregate labor productivity A_t, and working hours l_t^s. The retired household receives a lump-sum pension pen_t such that net non-capital income x_t^s is represented by

$$x_t^s = \begin{cases} (1-\tau_t^w - \tau_t^p) w_t \bar{y}^s A_t l_t^s & s = 1, \ldots, R-1, \\ pen_t & s = R, \ldots, J. \end{cases} \tag{7.45}$$

Wage income is taxed at the rate τ_t^w. In addition, the household also pays contributions to the pension system at rate τ_t^p.

The budget constraint of the household at age $s = 1, \ldots, R-1$ is given by

$$(1+\tau_t^c)c_t^s = x_t^s + \left(1 + (1-\tau_t^K)r_t\right)\omega_t^s + tr_t - \omega_{t+1}^{s+1}, \tag{7.46}$$

where assets of the s-year-old household at the beginning of period t, ω_t^s, consist of capital stock k_t^s and government bonds b_t^s, $\omega_t^s = k_t^s + b_t^s$. The household is born without assets and leaves no bequests at the end of its life, implying $k_t^1 = k_t^{J+1} = 0$ and $b_t^1 = b_t^{J+1} = 0$. The household receives interest income r_t on capital and government bonds and pays income taxes on capital income at rate τ_t^K. Consumption is taxed at the constant rate $\tau_t^c = \tau^c$. In addition, the household receives government transfers in the amount of tr_t in period t.

[39]Trabandt and Uhlig (2011) show that the utility function (7.44) has the two properties: (1) It is consistent with long-run growth, and (2) it features a constant Frisch elasticity of labor supply ν_1.

Maximizing lifetime utility (7.43) subject to budget constraint (7.46) results in the following first-order conditions for the s-year-old household[40]:

$$\lambda_t^s(1+\tau^c) = (c_t^s)^{-\sigma}\left[1-\nu_0(1-\sigma)(l_t^s)^{1+1/\nu_1}\right]^{\sigma}, \quad s=1,\ldots,J \tag{7.47a}$$

$$\lambda_t^s(1-\tau_t^w-\tau_t^p)\bar{y}^s A_t w_t = \nu_0\sigma\left(1+\frac{1}{\nu_1}\right)(c_t^s)^{1-\sigma}\left[1-\nu_0(1-\sigma)(l_t^s)^{1+1/\nu_1}\right]^{\sigma-1} \cdot (l_t^s)^{1/\nu_1}, \quad s=1,\ldots,R-1 \tag{7.47b}$$

$$\lambda_t^s = \beta\phi_{t,s}\,\lambda_{t+1}^{s+1}\left[1+(1-\tau_{t+1}^K)r_{t+1}\right], \quad s=1,\ldots,J-1. \tag{7.47c}$$

In Eq. (7.47a), $l_t^s \equiv 0$ for $s \geq R$.

The solution to the household maximization problem implies that the household is indifferent between holding assets in the form of physical capital and government debt, as both yield the same (certain) after-tax return. If we only had one household living for two periods, this would pose no problem because the proportion of asset holdings would be the same at the individual and aggregate levels. However, with many periods, the portfolio allocation is indeterminate. Therefore, we assume, without loss of generality, that each household holds the two assets in the same proportion, which is determined as the share of capital in total aggregate assets and equal to $k_t/(k_t+b_t)$.

7.5.1.3 Production

The production technology is described by a Cobb-Douglas function:

$$Y_t = K_t^{\alpha}(A_t L_t)^{1-\alpha}. \tag{7.48}$$

Capital depreciates at rate δ and A_t grows at exogenous rate γ:

$$A_{t+1} = (1+\gamma)A_t. \tag{7.49}$$

Firms operate in competitive goods and factor markets. They maximize profits

$$\Pi_t = Y_t - r_t K_t - w_t A_t L_t - \delta K_t, \tag{7.50}$$

[40] See Appendix 7.2 for the derivation.

resulting in the first-order conditions

$$r_t = \alpha K_t^{\alpha-1}(A_t L_t)^{1-\alpha} - \delta, \tag{7.51a}$$

$$w_t = (1-\alpha) K_t^{\alpha}(A_t L_t)^{-\alpha}. \tag{7.51b}$$

Notice that w_t is the wage rate per efficiency unit (and not per working hour).

7.5.1.4 Government

The government expenditures consist of public consumption G_t, transfers Tr_t, and interest on public debt B_t. Government expenditures are financed by taxes T_t, debt, $B_{t+1} - B_t$, and confiscated accidental bequests Beq_t according to:

$$G_t + Tr_t + (1 + r_t)B_t = B_{t+1} + T_t + Beq_t. \tag{7.52}$$

Government consumption per capita g_t grows at exogenous rate γ such that, for example, $\tilde{g}_t \equiv \frac{G_t}{A_t N_t}$ denotes stationary government expenditures per capita in efficiency units.

Accidental bequests are collected from the households that do not survive:

$$Beq_{t+1} = \sum_{s=1}^{J} (1 - \phi_{t,s}) N_t(s) \left(\left[1 + (1 - \tau_{t+1}^K) r_{t+1} \right] \omega_{t+1}^{s+1} \right). \tag{7.53}$$

Taxes are levied on consumption, interest income, and wage income:

$$T_t = \tau^c C_t + \tau_t^w w_t A_t L_t + \tau_t^K r_t \Omega_t, \tag{7.54}$$

with

$$C_t = \sum_{s=1}^{J} N_t(s)\, c_t^s, \tag{7.55a}$$

$$L_t = \sum_{s=1}^{R-1} N_t(s)\, \bar{y}^s l_t^s, \tag{7.55b}$$

$$\Omega_{t+1} = \sum_{s=1}^{J} N_t(s)\, \omega_{t+1}^{s+1}. \tag{7.55c}$$

According to (7.55a) and (7.55b), aggregate consumption and labor are simply equal to the sum of the individual variables. Equation (7.55c) represents the condition for the capital market equilibrium according to which the beginning of next-period aggregate wealth is equal to the sum of individual savings at the end of period t.

Aggregate wealth Ω_t is equal to the sum of aggregate capital K_t and government bonds B_t.

7.5.1.5 Social Security

The social security authority runs a balanced budget:

$$\tau_t^p w_t A_t L_t = \sum_{s=R}^{J} N_t(s)\ pen_t. \tag{7.56}$$

7.5.1.6 Equilibrium Conditions

The goods market equilibrium is given by

$$Y_t = C_t + G_t + K_{t+1} - (1-\delta)K_t. \tag{7.57}$$

The results for the stationary equilibrium and the transition are computed with the help of the Gauss programs `Ch7_US_debt.g` and `Ch7_US_transition.g`. The stationary equilibrium and the computation of the model are described in greater detail in Appendix 7.2.

7.5.2 Calibration

In parameterizing the model, we follow Trabandt and Uhlig (2011) as closely as possible. Table 7.4 summarizes the model calibration.

7.5.2.1 Demographics

In the initial steady state in 2010, the annual population growth rate is set equal to 0.95%, as estimated by UN (2015) for the average population growth rate prevailing during the years 1995–2010.[41] The 5-year survival probabilities for the 15 different age groups are taken from UN (2015). These data show that survival probabilities have increased over time and are subject to a higher rate of growth for the older age groups.[42] For our benchmark simulation, we use the average survival probabilities during the period 1990–2010.[43] Since we consider the stationary age distribution in the steady state, the dependency ratio of the stationary population implied by these demographic variables, in the amount of 30.6%, is higher than the empirical value of 21.6% (24.7%) that is projected by UN (2015) for the US economy in the year 2010 (2015).

[41] Trabandt and Uhlig (2011) use the calibration period 1995–2007.

[42] Please see also Fig. 6.8 in Chap. 6.

[43] For the projection of future survival probabilities that serve as inputs into our subsequent quantitative analysis, we continue to use moving averages of four periods.

Table 7.4 Calibration of the OLG model with pensions and debt

Parameter	Value	Description
n	0.95%	Population growth rate (annual)
$OADR2$	30.6%	Old-age dependency ratio (20–65)/65+
β	1.0361	Discount factor (annual)
ν_0	21.5	Preference parameter: weight of labor
σ	2.0	(Inverse of) Intertemporal elasticity of substitution
ν_1	0.30	Frisch labor elasticity
α	0.35	Production elasticity of capital
δ	8.3%	Depreciation (annual)
γ	2.0%	Growth rate (annual)
G/Y	18.0%	Government consumption/GDP
Tr/Y	12.3%	Government transfers/GDP
B/Y	63%	Government debt/GDP (annual)
$pen/(wA\bar{l})$	35.2%	Gross replacement rate of pensions
τ^p	10.8%	Social security tax rate
τ^c	5.0%	Consumption tax rate
τ^K	36.0%	Capital tax rate
$\tau^w + \tau^p$	28.0%	Labor tax rate

7.5.2.2 Production

Following Trabandt and Uhlig (2011), the production elasticity of capital is set equal to $\alpha = 0.35$. The annual depreciation rate of $\delta = 8.3\%$ implies a 5-year period value of depreciation equal to 35.2%, $\delta = 1-(1-0.083)^5$. Also in accordance with these authors, we choose an annual economic growth rate of 2%.

7.5.2.3 Preferences

The value of the utility parameter, $\nu_0 = 21.5$, is chosen such that the implied steady-state average working hours are equal to 0.30. Following our discussion of the empirical evidence on the Frisch labor supply elasticity in Sect. 4.4.5, we set $\nu_1 = 0.3$, while we also retain the calibration of the intertemporal elasticity of substitution, $1/\sigma = 1/2$, in accordance with that in Chaps. 2–6. The annual value for the discount factor, $\beta = 1.0361$, is chosen to imply an annual real interest rate of 4%, or 21.9% over a period of 5 years.

7.5.2.4 Government

The average annual debt-GDP ratio of 63% during the period 1995–2007 corresponds to a 5-year value of 12.6%. Fiscal policy is described by the tax rates of 28%, 36%, and 5% on labor income, capital income, and consumption, respectively. Notice that the labor income tax rate in Trabandt and Uhlig (2011) is estimated following the methods of Mendoza, Razin, and Tesar (1994), meaning that our social security contribution tax rate τ^p is already a component of the labor tax rate of 28.0%. Therefore, $\tau^w + \tau^p$ is calibrated at 28.0%, and $\tau^w = 17.2\%$ is computed as a residual with the help of τ^p. Government consumption amounts to 18.0% of GDP, while government transfers, being determined endogenously to satisfy the government budget constraint, amount to 12.2% of GDP. We fix the gross replacement rate of 35.2% in accordance with the data provided in OECD (2017) for men with average earnings (percentage of pre-retirement income) and compute endogenously the social security tax rate, $\tau^P = 10.8\%$, that balances the social security budget in the benchmark calibration.[44]

7.5.3 Stationary Allocation

For the benchmark calibration of the stationary equilibrium in 2010, the life-cycle profiles of individual assets, labor supply, and the consumption and savings rates are presented in Fig. 7.11. Wealth increases over the working life and peaks prior to retirement at real lifetime age 60–64. Working hours are hump-shaped over the working life and peak during the age period 30–34. The labor supply mirrors the efficiency-age profile but peaks earlier due to the effect of increasing wealth. Consumption increases throughout the working life because the discount rate $1/\beta - 1$ is smaller than the interest rate after taxes $(1 - \tau^K)r$. At the age of retirement, consumption falls because leisure increases to 100%. During retirement, consumption is hump-shaped as the survival probability $\phi_{t,s}$ declines, and hence, the inverse of the product $\phi_{t,s} \cdot \beta$ rises above the interest rate $1 + (1 - \tau^K)r$ after age 80 (so that the factor $\phi_{t,s}\beta(1 + (1 - \tau^K)r)/(1 + \gamma)^\sigma$ falls below one – see the Euler equation (7.64c) in Appendix 7.2). The savings rate is defined as the ratio of savings $y - (1+\tau^c)c$ to disposable income y, with income y being equal to net labor and capital income plus transfers. Savings are positive during the working life and fall below zero during retirement. The maximum savings rate is observed during the period with the highest labor supply at age range 30–34 and amounts to 40% of net income.

7.5.4 Steady-State Results on Debt and Crowding-Out of Capital

Figure 7.12 presents the effects of higher debt B/Y on the capital stock.[45] Ricardian equivalence fails just as in the simple two-period OLG model, and higher debt

[44]In 2016, the US social security contribution rates amounted to 6.2% for each the employer and the employee. Our endogenous value of τ^p falls somewhat short of this value for two main reasons. (1) In the US, social security also encompasses disability insurance, which we do not model. (2) In addition, there is a maximum threshold level of wage income for which the individual pays social security contributions. In 2016, this limit amounted to \$118,500. Accordingly, the effective average social security contribution rate on wage income is lower than 12.4%.

[45]The results that are computed with the Gauss program *Ch7_US_debt.g* fluctuate around the line presented in Fig. 7.12 due to numerical inaccuracies. We, therefore, interpolated the results by a

Fig. 7.11 Age profiles of individual variables in the benchmark large-scale OLG model

crowds out capital. An increase in the debt-output ratio B/Y from 50% to 70% decreases the capital stock k by 0.14%, from 0.11680 to 0.11664. We notice that higher debt crowds out capital to a much smaller quantitative extent than in Fig. 7.10 that we computed in the stylized two-period OLG model in Sect. 7.4.2. Since we consider a much more realistic demographic structure in the present model, the household accumulates larger savings in an economy with higher public debt. In this regard, a large-scale OLG model behaves much more akin to the Ramsey model than to the simplified two-period model. Therefore, although our lifetime is finite, it is sufficiently long to induce a relatively small failure of Ricardian equivalence, ceteris paribus.

Furthermore, we have another countervailing effect of higher debt on savings. In the standard two-period model presented in Sect. 7.4.2, labor supply is inelastic. In the present model, labor supply is elastic. Since higher public debt necessitates lower lump-sum transfers (or, equivalently, higher lump-sum taxes), we observe a positive income effect on labor supply in the sense that labor supply increases.[46] Therefore, aggregate production and, hence, income and savings increase, ceteris paribus. The magnitude of this income effect critically depends on the Frisch labor supply elasticity. If we even choose a Frisch labor supply elasticity $\nu_1 = 1.0$, we

smooth cubic function. The vertical distance between the computed values and the line in Fig. 7.12 is on the order of 10^{-5}.

[46]Of course, the total effect on labor supply would be much different if higher debt were instead financed by distortionary labor income taxes.

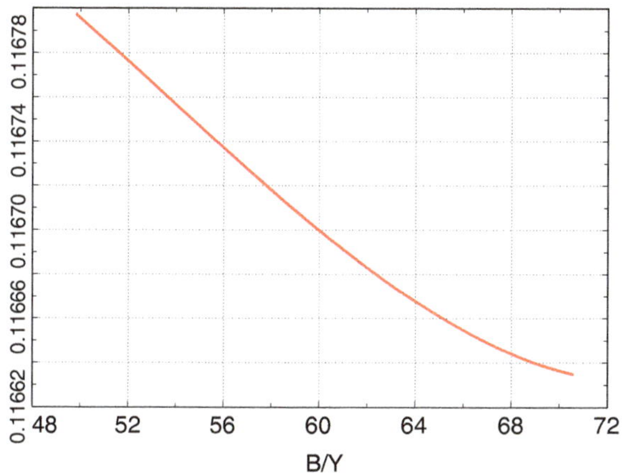

Fig. 7.12 Capital stock and debt in the large-scale OLG model with pensions

observe that debt no longer crowds out capital and k actually increases slightly for higher debt B/Y (not presented).[47]

7.5.5 Transition Results Regarding Fiscal Policy and Demographics

In this section, we analyze the effects of fiscal policy on the dynamics of aggregate wealth, employment, and welfare. To do so, we assume that the economy is in the steady state in 2010 that is implied by the stationary population for the 2010 survival probabilities and population growth rate. During the period 2010–2100, the population evolves according to the medium forecast of the United Nations, as described in UN (2015). After 2100, the survival probabilities and population growth rates remain constant at their 2100 levels, meaning that the population is approximately stationary circa 2150. As a consequence, the share of the labor force in the total population shrinks from 76.6% in 2010 to 65.8% in 2150, as depicted in the bottom-right panel of Fig. 7.13. In addition, we assume that the net pension replacement rate remains constant at 35.2% throughout the transition period. To balance the budget of the social security authority, the pension contribution rate τ^p has to adjust and increases from 10.8% to 18.3%.

We distinguish four different fiscal policies (1)–(4). Under the benchmark policy (1) (the red solid line in Fig. 7.13), real debt B and transfers Tr (both variables relative to population N_t and technology A_t) remain constant at their 2010 levels, while the wage income tax rate τ^w adjusts to balance the government budget. As a consequence, τ^w has to increase from 17.2% to 27.0%, and thus, the new long-run

[47]The reader is asked to verify this by adjusting the parameter value in the program *Ch7_US_debt.g*.

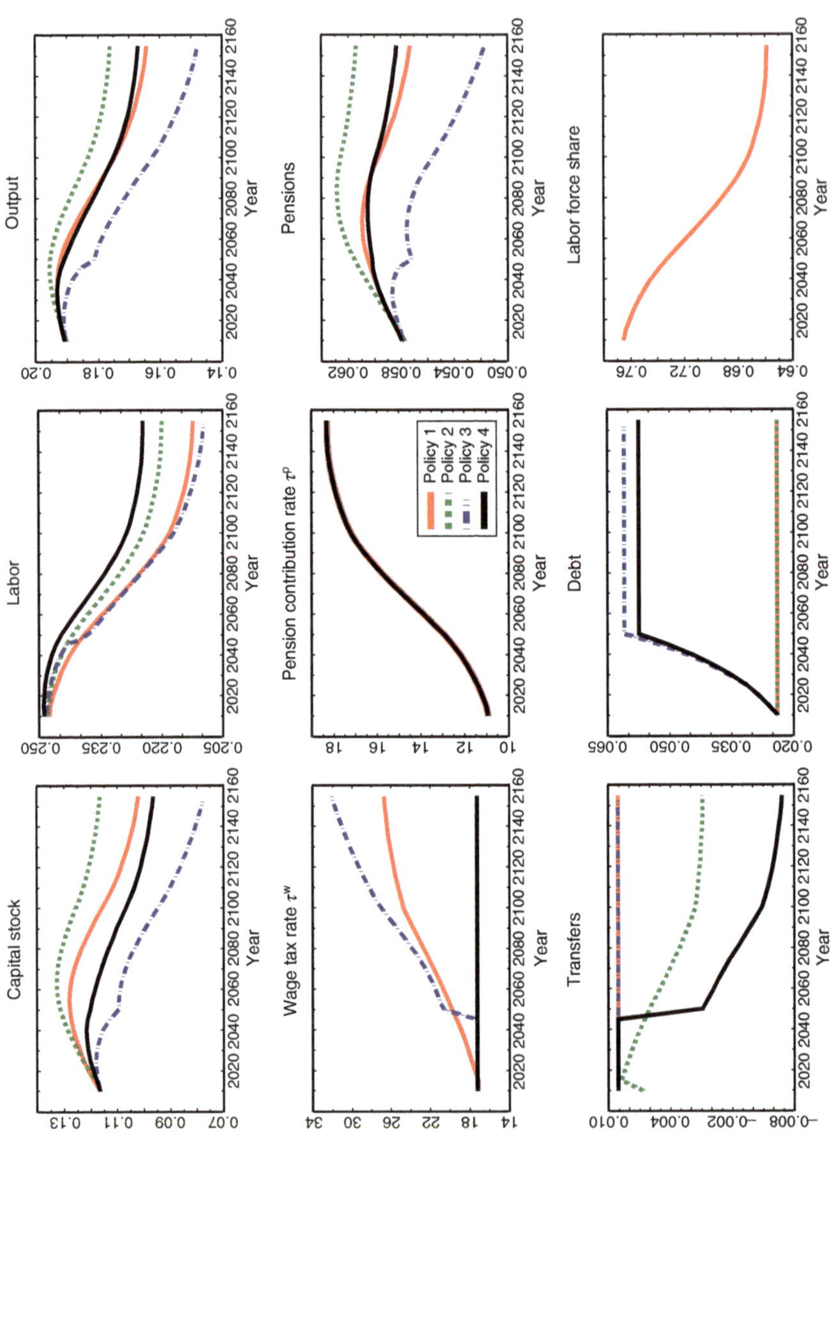

Fig. 7.13 Demographic transition and fiscal policy

value for the tax on wage income (wage tax τ^w plus pension contribution rate τ^p) amounts to 45.3%. In Fig. 7.13, the dynamics of τ^w and τ^p are illustrated in the middle-left and middle panels for the years 2010–2160.

Under policy scenario 2, government transfers Tr rather than the wage tax τ^w adjust to balance the government budget. This scenario is presented by the broken green line in Fig. 7.13. Policies (3) and (4), as presented by the broken blue and solid black lines, consider the cases in which, during an initial phase of 40 years, both the labor income tax rates and transfers are held constant at their 2010 levels while the fiscal deficit is financed by means of debt. From 2055 onward, the debt level remains constant, and either the wage tax rate (policy 3) or transfers (policy 4) adjust to balance the government budget.

The dynamics for the benchmark case under policy 1 are similar to those studied in Sect. 6.4. Due to the decrease in aggregate labor, income and savings fall in the long run. As a consequence, the capital stock declines in the medium and long run. The time profile of the capital stock, however, is hump-shaped, as aggregate savings increase initially because young agents save a higher proportion of their income. Compared with the young households in the stationary population of 2010, the young households during the transition face a higher survival probability and, therefore, increase their precautionary savings for old age.[48] After the initial phase of the transition, however, the decline in the share of the labor force in the total population and the simultaneous rise in labor income taxes reduce aggregate disposable income to such an extent that aggregate savings fall. In the case of debt financing during the years 2015–2055, the decline in the capital stock is even stronger. Under policy 3, higher debt increases threefold, and the (annual) debt-output ratio increases to 215% in the final steady state (while it remains at 73.2% in the case of policy 1). The high debt level crowds out aggregate capital so strongly that capital falls by approximately one-third.[49] As a consequence, the (annualized) real interest rate r is also considerably higher in the case of debt financing and amounts to 6.32% in the long run (compared with 4.05% in the case of policy 1). Therefore, government interest payments $r_t B_t$ under policy 3 increase not only because of higher debt B but also because of a simultaneous increase in the interest rate.[50]

The dynamics of aggregate labor L, as depicted in the top-middle panel of Fig. 7.13 are also sensitive to the fiscal policy considered. Labor supply declines

[48]There are multiple other effects on savings; for example, the age composition of the labor force changes and average productivity increases in the older labor force ceteris paribus. Moreover, the savings rate of the 20–24-year old workers is below the average savings rate of the workers (as presented in Fig. 7.11).

[49]If we increased the number of periods during which the government resorts to debt financing of additional expenditures to 50 years, our economy would collapse. In this case, labor income taxes would be insufficient to finance government expenditures in the long run and the fiscal space would shrink to zero. In other words, the maximum tax revenues at the peak of the Laffer curve are insufficient to balance the government budget in this case. See also Sect. 6.6 for the concept of the fiscal space.

[50]Recall our discussion of the intertemporal government budget constraint in Sect. 7.3.

more strongly if the government uses labor income taxes rather than transfers to finance expenditures. In addition, higher debt also affects the labor supply in the short to medium rum. First, during the initial phase until the year 2045, higher debt helps to keep wage tax rates low so that the labor supply is higher under policy 3 than under policy 1. In the case of policies 2 and 4, the direction of the effect is reversed. Higher debt results in higher transfers under policy 4 compared with that under policy 2 so that the workers reduce their labor supply. Second, debt crowds out capital, meaning that the marginal product of labor and, hence, wages fall. The latter effect is also present in the long run, meaning that labor supply is approximately 2% lower under policies 3 compared with that under policies 1. In the case of a constant wage tax rate under policies 2 and 4, higher debt results in lower transfers so that labor supply is higher under policy 4 than that under policy 2 (due to the income effects). Since both labor and capital shrink in the long run, per capita output (relative to technology A_t) also declines. Under fiscal policies 1–4, the decline in long-run output amounts to 14.4%, 7.6%, 25.4%, and 12.6%. Notice the devastating effect of higher debt on output.

Unsurprisingly, the fiscal policies have a significant effect on generational welfare. Figure 7.14 presents the lifetime utility of the generations alive during the transition, 2010–2160. The oldest generation that is affected by the policy changes in 2015 is the generation born in 1940, for whom the last period of life is the 5-year period starting in 2015.[51] Since we assume the economy to be in steady state prior to 2015, the effect on lifetime utility is negligible. To express welfare effects in an interpretable way, we compare lifetime utilities for the individual generations born in the same year to those under policy 1 and express the change in consumption equivalent changes. As a consequence, the welfare change under policy 1 as presented by the red solid line is zero. For the case of policy 2, all generations born after 2070 benefit, while the generations prior to 2070 lose from transfers rather than tax financing. The reason is straightforward. During the transition, higher transfers benefit retired households, while workers suffer from higher taxes. In the long run, the distortionary effects of higher labor income dominate.

Furthermore, older cohorts benefit from the temporary use of debt financing. In particular, welfare effects of 4% of total consumption or less accrue to the households born prior to 2120 if policy 3 rather than policy 1 is implemented. In the case of a constant wage tax, the benefitial welfare effects of debt financing are reduced. In this case, generations born prior to 1985 benefit if the government increases debt during the transition. In the (very) long run, however, the welfare effects of transitory debt financing are detrimental. Under policies 3 and 4, welfare eventually falls by 1.5–4.5% if debt is used to smooth the transition. We, therefore, conclude that the government should be careful using debt during the demographic transition to ease the burden of a shrinking population. The effects of debt on both output and welfare are dramatic.

[51] Remember that we assume that households are born at age 20.

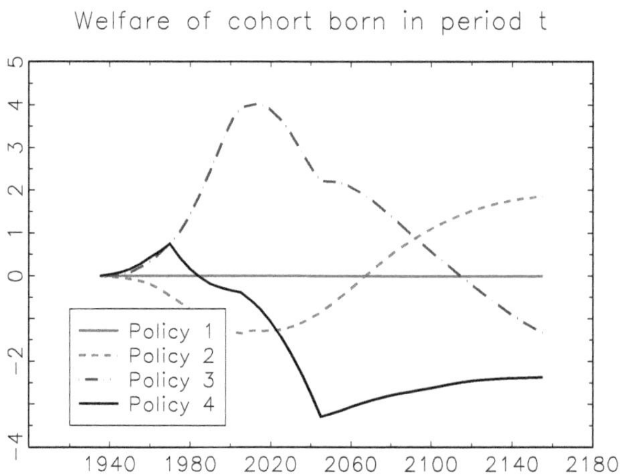

Fig. 7.14 Generational welfare and fiscal policy

Also notice the consequences of our results for the political feasibility of debt financing. If the government asked the voters if it should use debt financing to ease the transition, it obtained a clear majority. All generations alive in 2015 would prefer policy 3 (policy 4) to policy 1 (policy 2). The dotted blue (solid black) line and the solid red (dotted green) line only intersect in the year 2115 (2025). In Sect. 6.4, we also observed that the voter would reject a reform of the pension system (a cut in the pension benefits). Therefore, both our results emphasize that the implementation of policies to improve fiscal sustainability in modern democracies such as the US will be confronted with obstacles. Voters prefer higher debt, lower taxes, and more generous public pensions.

Our model is related to the results of D'Erasmo, Mendoza, and Zhang (2016), who study the sustainability of debt in a two-country model calibrated to the US economy and the EU15. In contrast to our model, they assume an infinitely lived representative household in each country and do not consider the consequences of the demographic transition or the burden of higher pension payments. Ricardian equivalence fails in their model because the government has to finance public expenditures with the help of distortionary taxes. Therefore, as in our case, the fiscal space depends on the debt level via the dynamic Laffer curves of capital and labor income taxation.[52] The results of their structural model casts doubt "that the high

[52]In addition to the factors considered in our model, D'Erasmo, Mendoza, and Zhang (2016) also include endogenous utilization of capital and limited tax deprecation allowances of capital. As a consequence, they are able to more accurately model the (stronger) capital response to higher capital income tax rates. The two-country setup also allows them to model the externality of higher domestic capital taxes on the foreign country. Since capital is mobile, the revenue from higher capital income taxation is reduced. As we do not consider the use of capital income taxes to finance higher debt in our analysis, we refrained from implementing these important model components.

debt ratios of some countries reached by many advanced economies in the years since 2008 will be fully repaid" (p. 2557). They also assert that there is fundamental difference in the abilities of the US and Europe to cope with the recent increase in debt. While in the US, only a small increase in labor income taxation makes debt sustainable, Europe's tax system has nearly exhausted all its financing abilities.

İmrohoroğlu, Kitao, and Yamada (2016) study the sustainability of fiscal policy in Japan. Their model is somewhat similar to ours. In some respects, they considerably improve upon the model above. For example, they much more accurately consider heterogeneity among consumers, e.g., by explicitly considering gender and different forms of employment by distinguishing among regular jobs, contingent jobs, and self-employment. In addition, the Japanese pension system is modeled in much greater detail. However, some simplifying assumptions not made in our model are imposed. For example, the consumption age profile is not endogenous but assumed to be time-invariant. Furthermore, both labor supply and interest rates are exogenous. Therefore, the costs of higher debt, which results in a significant increase in interest rates in our model, is likely to be underestimated in their model.[53] They find that government debt (relative to GDP) reaches levels of 210% in 2030 and 370% in 2050 under current policies and no further reforms. As mentioned in the previous chapter, Braun and Joines (2015) also conduct an analysis with respect to fiscal sustainability and debt in Japan that is similar to ours for the US economy. To identify sustainable fiscal policies, they also consider reductions in government purchases and/or an increase of consumption taxes. For example, an immediate, but very dramatic, increase in the consumption tax rate from 5% to 36% restores sustainability. Braun and Joines (2015) find that Japan will face a severe fiscal crisis if reforms are not implemented soon. Somewhat surprisingly, they also show that a higher fertility rate might even exacerbate Japan's fiscal problems in the short and intermediate run.

7.6 Epilogue

Thus far, we have assumed perfect foresight. The government is able to repay its debt; otherwise, the government does not secure credit from private investors. However, public debt has not always been honored in (recent) history. Reinhart and Rogoff (2011) document the debt default histories of 64 countries over the period 1800–2010 and identify 250 cases of external debt default and 68 cases of overt default where, in the latter case, even domestic debt holders were punished, in the form of lower coupon rates, a suspension of payments, a unilateral reduction in principal or a forcible conversion, e.g., after a currency reform. While most studies have addressed the case of external defaults, few studies have analyzed the case

[53] İmrohoroğlu, Kitao, and Yamada (2016) perform a sensitivity analysis for the returns on government debt and find that "if the interest rate on government debt is higher than the 1% assumed for the benchmark case, the resulting impact on fiscal balance can be disastrous".

of a domestic default. We will briefly review the two types of studies. For more comprehensive surveys of debt default studies, we recommend the two articles by Aguiar, Chatterjee, Cole, and Stangebye (2016) and Stähler (2013) and Chapter 13 in Schmitt-Grohé and Uribe (2017) on external sovereign default and D'Erasmo, Mendoza, and Zhang (2016) on domestic debt crises.

7.6.1 Sovereign External Default

During the 1980s and 1990s, we observed many episodes of external default in Latin American and Asian countries, including the crises in Argentina in 1982 and 2001, Mexico in 1983, and Thailand in 1993. Debt crises during this period, however, were not confined to Latin America and Asia but also emerged in Europe, e.g., in Turkey, and in Africa, e.g., in Nigeria. The circumstances differed across the individual countries. In Mexico in 1983, for example, the onset of the crisis was initiated by a collapse of commodity prices and a steep rise in interest rates in the early 1980s, while the crises in South Korea, Thailand, and Indonesia were associated with the (often implicit) currency peg against the dollar and a loss of investor confidence.

A large literature on sovereign default grounded in the pioneering work of Eaton and Gersovitz (1981) evolved in the wake of government crises over the two final decades of the last century. Its focus differs from that studied in this chapter. In the model in Sect. 7.5, for example, we analyzed debt policies that are sustainable and identified the level of debt beyond which fiscal policy becomes unsustainable in the US economy. In the literature on sovereign debt, however, the government is assumed to be able to service its debt but optimally decides either against or in favor of doing so. These studies base their analysis of the sovereign's optimal default decision on the assumption that debt is financed by investors from abroad. A government decision to default on its debt results in punitive action in the form of exclusion from international capital markets. In addition, a default cost is usually imposed exogenously, taking the form of output loss.[54] Production in these models is assumed to be both exogenous and stochastic.[55] In particular, the interest

[54]In the early work on sovereign default, the default costs are usually assumed to be linear in output, while in latter work such as Arellano (2008) and Aguiar, Chatterjee, Cole, and Stangebye (2016), the focus has shifted to non-linear default costs that increase in output. Non-linear default costs help to improve the performance of these types of models with respect to the replication of empirical facts, particularly with regard to the volatility of interest spreads, which often serve as a measure of default. For example, Aguiar, Chatterjee, Cole, and Stangebye (2016) identify a "crisis" episode with a period that features an interest rate spread of government debt in the top 5% of the distribution of quarterly changes. Mendoza and Yue (2012) provide a model with a micro-foundation of non-linear costs in which domestic producers cannot import foreign intermediate inputs during periods of default.

[55]The results from this literature are sensitive to the stochastic nature of the shock. Most of these models assume a deterministic trend, which, at a minimum, seems questionable for a multitude of developing countries; alternatively, a stochastic growth rate is considered. With the assumption of deterministic growth, the economy is more responsive to a negative output shock because it

rate that is charged by foreign investors does not affect output. The sovereign is usually identified with the representative infinitely-lived household in the economy and maximizes intertemporal utility as in the Ramsey model.[56] The government faces a budget constraint. Either it honors its debt obligations and retains access to international capital markets, or it does not repay, which increases its present consumption in the amount of the forgone debt but decreases consumption by the imposed output costs, while it also reduces the possibility of consumption smoothing over future periods because of its exclusion from capital markets.[57]

The timing in these models of sovereign default is as follows. The government observes the state of the economy, which is described by the production (income) and all other exogenous shocks, for example the wealth of the investors and, hence, the external demand for government debt. There is usually a bond auction only once per period. In the following step, the literature considers two different sequences. Either the government first decides about default and, if it does not default, auctions its bonds or, vice versa, first auctions its bonds and decides about default thereafter, meaning that newly auctioned bonds also face a within-period default risk. Next, the government consumes its exogenous income (subject to possible default costs) and moves to the next period, where it either has access to or is excluded from international capital markets. In the event of default, it is often assumed that the government re-enters international credit markets without debt with a certain probability in each period.[58] Once it re-enters, it retains access until it defaults again. Reputation effects are usually absent from these models, and the probability of re-entry (or, equally, average length of its absence) does not depend on the previous number and extent of defaults.

The literature has expanded in many directions including, among others, risk-averse investors, different maturities of government debt, news shocks, fiscal policy rules, exchange rates, ideas from political economy (myopia and political turnover in multi-party systems), and reputation effects.[59] Important results of this literature include the following: (1) Bond prices fall and, hence, interest rates rise when

---

implies a recovery to trend and, therefore, a higher future level of output. As a consequence, the government issues more debt to intertemporally smooth consumption.

[56]As one exception to the assumption of one sovereign, Cuadra and Sapriza (2008) incorporate the presence of two parties, e.g., a left-wing versus a conservative party, into a model of a political economy. Both parties give preferential treatment to their voters, which consist of two types of households, each preferring a different public good. The incumbent party also has an incentive to finance its expenditures by borrowing and impose potential default risk on its successor.

[57]In general, the possibility of smoothing consumption with the help of domestic credit markets is also not considered in these types of models.

[58]In the benchmark model of Eaton and Gersovitz (1981), the government remains excluded from international capital markets after a default.

[59]See, for example, Arellano (2008) on risk-averse investors, Alfaro and Kanczuk (2007) and Aguiar and Amador (2016) on different maturities of government debt, Durdu, Nunes, and Sapriza (2013) on news shocks, Gosh, Kim, Mendoza, Ostry, and Qureshi (2013) on fiscal policy rules, Asonuma (2016) on exchange rates, Cuadra and Sapriza (2008) on political economy, and D'Erasmo (2012) on reputation.

debt increases and/or productivity is expected to be low in the future. Accordingly, the sovereign default models are in line with the empirical observation that highly indebted countries have a greater incentive to default and that interest spreads move countercyclically. (2) Shocks to long-run growth have a more substantial impact on default rates than do transitory shocks. (3) Politically unstable economies display higher default rates and greater volatility of sovereign interest rate spreads. (4) For emerging countries, interest rates on the long-term debt are, on average, higher than those on short-term debt. (5) The issuance of new debt may decrease the value of existing debt (debt dilution). (6) Empirically, fiscal policy in developing economies acts pro-cyclically by increasing public spending and cutting taxes in good times. This can be shown to also prevail in models in which the government also considers the probability of default and the effects of its debt policy on interest rates. (7) Negative shocks to productivity can lead to a decline in the real exchange rate and a higher likelihood of a default on sovereign debt.

7.6.2 Domestic Default

Reinhart and Rogoff (2011) note that domestic debt represents the lion's share of total public debt, amounting to approximately two-thirds during the period that extends from 1800 until the present. Large domestic debt helps to explain why many countries default on their external debt at low levels. The literature on quantitative models of debt default, however, has mostly ignored domestic debt, with only a few exceptions.

As one prominent study on domestic default, D'Erasmo and Mendoza (2016) examine a government which has sufficient revenue to repay its debt but may nevertheless optimally decide to default. Therefore, the government weights the respective changes in the utilities of the bondholders and non-bondholders. As a consequence of a default, wealth is redistributed domestically, from bondholders to non-bondholders. In addition, an exogenous default cost is imposed, which is non-increasing in (stochastic) government expenditures and proportional to total (exogenous) income. Both D'Erasmo, Mendoza, and Zhang (2016) and D'Erasmo and Mendoza (2016) calibrate the model with respect to European data[60] to show that there is a reasonable parameter space that supports an equilibrium with sustainable debt subject to default risk. As an implication of their model, d'Erasmo and Mendoza show that public debt is more sustainable if the government payoff function is biased in favor of bondholders. The latter assumption may be justifiable with the help of arguments from the political economy literature. For

[60]For example, D'Erasmo, Mendoza, and Zhang (2016) assume that the (annual) logarithmic government expenditure-GDP ratio follows an AR(1) process with an autoregressive coefficient of 0.8802 and a standard deviation of 1.7%, while the output costs of a default amount to a minimum of 2% and increase linearly with government expenditures. Default occurs with a probability of 1%.

example, the electoral participation of older households, which may represent a larger share of the bondholders, is higher than that of the younger households, which have just begun to accumulate wealth and invest a larger share of their wealth in stocks than in government bonds.[61] Notice also that a default may not especially harm the rich or benefit the poor. For the rich households, government bonds usually constitute a smaller proportion of total wealth since the portfolio share of (supposedly) more risky assets in the form of stocks increases with wealth.[62] In addition, government debt may provide utility-enhancing services such as liquidity to credit-constrained individuals or a self-insurance mechanism as emphasized by Aiyagari and McGrattan (1998).[63]

RBC and dynamic stochastic general equilibrium (DSGE) models are the standard tools to analyze the effects of fiscal policy. At present, it is rather difficult to include the quantitative models of sovereign and debt default in the standard growth and business cycle models and, thereby, endogenize the accumulation of physical capital.[64] A laudable exception is Gordon and Guerron-Quintana (2018), who study a standard business cycle model with endogenous capital accumulation that simultaneously accounts for the empirical features of sovereign default episodes and business cycle properties of small open economies. As one major component of their analysis, they integrate capital adjustment costs. When the economy is hit by a positive productivity shock, firms gradually increase investment. The sovereign also increases international borrowing during these times. When the economy slides into a recession as a consequence of adverse productivity shocks, the sovereign mitigates their impact on consumption by rolling over debt and reducing investment. Consequently, debt grows relative to output. If the negative shocks persist, the sovereign defaults, and output, consumption, and investment all decline substantially. Capital acts as a smoothing channel against shocks, while it also

[61]As a different interpretation of their model, D'Erasmo, Mendoza, and Zhang (2016) argue, "in the European debt crisis, a Greek default can be viewed as redistributing from German tax payers to Greek households" (on p. 2559).

[62]D'Erasmo and Mendoza (2016) also show that default is more costly in the presence of physical capital and, hence, an endogenous portfolio choice by the household.

[63]Aiyagari and McGrattan (1998) find that the optimal debt-GDP ratio amounts to approximately 2/3 in the US, which is close to the post-war average level. In Problem 7.4, you are asked to compute the insurance mechanism that is provided by the public debt system in a simple three-period OLG model with income uncertainty. In essence, the provision of public debt in a non-Ricardian economy increases the interest rate, meaning that the return on savings increases, and a (stochastic) drop in income can be more easily compensated.

[64]One reason for the difficulties in integrating endogenous default into the standard DSGE model (that can be overcome) is the inherent non-linear nature of the sovereign's optimization problem. The choice is binary (default versus honoring debt payments), and the equilibrium bond price is a highly non-linear function of the state of the economy. In addition, Gordon and Guerron-Quintana (2018) note that policy functions (and the value functions of the sovereign) might become non-monotone in the presence of capital accumulation and long-term government debt; the latter is necessary for matching empirical spread and default statistics. As a consequence, the linearization method for solving the DSGE model that we introduced in Chap. 2 is no longer applicable.

makes external default more attractive, as foreign creditors cannot seize domestic capital. The net effect of physical capital on the interest rates of external debt is positive, meaning that the risk premium on government debt decreases. In addition, the model is also able to match the empirical countercyclical behavior of spreads and net exports in emerging economies.

Appendix 7.1: Government Budget with Money Finance

In Sect. 7.3, we assumed that the government only finances its expenditures by means of taxes and public debt. In this appendix, we also consider money financing of the government budget. Therefore, we include high-powered money or central bank money $\tilde{H}_t$ in the government budget constraint (7.2):

$$\tilde{H}_{t+1} - \tilde{H}_t + P_t^B \tilde{B}_{t+1} - \tilde{B}_t = P_t G_t - P_t T_t. \tag{7.58}$$

The revenue from money creation, $\tilde{H}_{t+1} - \tilde{H}_t$, is called *seignorage* and can be written in real terms as[65]

$$S_t = \frac{\tilde{H}_{t+1} - \tilde{H}_t}{P_t} = (1 + \pi_{t+1}) H_{t+1} - H_t = \pi_{t+1} H_{t+1} + \Delta H_{t+1},$$

where $\Delta H_{t+1} = H_{t+1} - H_t$ denotes the change in real central bank money with $H_t \equiv \tilde{H}_t / P_t$. Notice that only central bank money (inside money) is included in the definition of H_t, while outside money that is created in the banking sector with the help of credit does not provide revenues for the government.

With this approximation, the real government budget (7.3) is therefore given by

$$\frac{1}{1 + r_t^B} B_{t+1} + \pi_{t+1} H_{t+1} + \Delta H_{t+1} = B_t + G_t - T_t. \tag{7.59}$$

Similarly, one can divide (7.59) by real GDP Y_t to derive

$$\frac{1 + \gamma_{t+1}}{1 + r_t^B} \frac{B_{t+1}}{Y_{t+1}} + (\gamma_{t+1} + \pi_{t+1}) \frac{H_{t+1}}{Y_{t+1}} + \Delta \frac{H_{t+1}}{Y_{t+1}} = \frac{B_t}{Y_t} + \frac{G_t - T_t}{Y_t}, \tag{7.60}$$

where we have used the approximation $\pi \gamma \approx 0$. In steady state with constant B/Y and H/Y, revenues from seignorage (relative to GDP) become

$$\frac{S}{Y} = (\gamma + \pi) \frac{H}{Y}.$$

[65] There exist many definitions of seignorage in the literature; for example one different definition also encompasses interest-bearing government bonds held by the central bank, which we included in the variable B_t. For an overview, see Section 4.2 in Walsh (2010).

Let us consider a rough approximation of real seignorage revenues for the US in 2016. For simplification, let us assume that the US is in steady state in 2016. Nominal GDP in 2016 amounted to \$18,621 billion, while high-powered money was equal to \$3,744 billion as of July 6, 2016, according to data from the Federal Reserve Bank of St. Louis.[66] The inflation rate (for consumer prices) and real GDP growth amounted to 2.18% and 0.81%, respectively, in the US economy during the period 2001–2016. Therefore, as a very crude approximation,

$$\frac{S}{Y} = (\gamma + \pi)\frac{H}{Y} \approx (0.0201 + 0.0081) \times 0.201 = 0.60\%.$$

This amount of seignorage is a very high value in the context of the post-World War II monetary history of the United States. Prior to the financial crisis of 2007–2008, high-powered money relative to GDP was much smaller. In the wake of this crisis and the start of the "Quantitative Easing" program of the Federal Reserve, however, high-powered money increased almost fivefold, from \$848 billion on January 8, 2008, to \$4,150 billion on September 17, 2014. Therefore, seignorage (relative to GDP) used to be much smaller in the US. In particular, King and Plosser (1985) report that seignorage only equalled 0.3% of GDP during the period 1952–1982.

Appendix 7.2: Computation of the Large-Scale OLG Model in Sect. 7.5

In this Appendix, we first define the stationary equilibrium for the model in Sect. 7.5 before we describe the computation of the transition.

Stationary Equilibrium

To describe the model in stationary variables, we define the following individual stationary variables:

$$\tilde{c}_t^s \equiv \frac{c_t^s}{A_t}, \quad \tilde{\omega}_t^s \equiv \frac{\omega_t^s}{A_t}, \quad \tilde{k}_t^s \equiv \frac{k_t^s}{A_t}, \quad \tilde{b}_t^s \equiv \frac{b_t^s}{A_t}, \quad \tilde{tr}_t \equiv \frac{tr_t}{A_t}, \quad \tilde{pen}_t \equiv \frac{pen_t}{A_t},$$

$$\tilde{\lambda}_t^s \equiv \frac{\lambda_t^s}{A_t^{-\sigma}},$$

[66]The data were downloaded from https://fred.stlouisfed.org/series/GDPA and https://fred.stlouisfed.org/series/BASE and are also included in the Gauss program *Ch7_data.g*.

and aggregate stationary variables:

$$\tilde{k}_t \equiv \frac{K_t}{A_t N_t}, \quad \tilde{y}_t \equiv \frac{Y_t}{A_t N_t}, \quad \tilde{b}_t \equiv \frac{B_t}{A_t N_t}, \quad \widetilde{beq}_t \equiv \frac{Beq_t}{A_t N_t}, \quad \widetilde{tax}_t \equiv \frac{T_t}{A_t N_t},$$
$$\tilde{L}_t \equiv \frac{L_t}{N_t},$$

implying the factor prices

$$r_t = \alpha \tilde{k}_t^{\alpha-1} \tilde{L}_t^{1-\alpha} - \delta, \tag{7.61a}$$

$$w_t = (1-\alpha) \tilde{k}_t^{\alpha} \tilde{L}_t^{-\alpha}. \tag{7.61b}$$

Stationary non-capital income $\tilde{x}_t^s = x_t^s / A_t$ of the s-year-old household in period t is represented by:

$$\tilde{x}_t^s = \begin{cases} (1-\tau_t^w - \tau_t^p) w_t \bar{y}^s l_t^s & s = 1, \ldots, R-1, \\ \widetilde{pen}_t & s = R, \ldots, J. \end{cases} \tag{7.62}$$

The stationary budget constraint of the household at age $s = 1, \ldots, R-1$ is given by

$$(1+\tau_t^c)\tilde{c}_t^s = \tilde{x}_t^s + \left[1 + (1-\tau_t^K) r_t\right] \tilde{\omega}_t^s + \tilde{tr}_t - (1+\gamma)\tilde{\omega}_{t+1}^{s+1}, \tag{7.63}$$

where individual wealth $\tilde{\omega}_t^s$ is equal to the sum of the two assets: capital $\tilde{k}_t^s$ and government bonds $\tilde{b}_t^s$, $\tilde{\omega}_t^s = \tilde{k}_t^s + \tilde{b}_t^s$.

To derive the first-order conditions, the household maximizes the Lagrange function

$$\begin{aligned}\mathscr{L} = \sum_{s=1}^{J} \beta^{s-1} \left(\prod_{j=1}^{s} \phi_{t+j-2,j-1} \right) \Bigg[\frac{1}{1-\sigma} \left((c_{t+s-1}^s)^{1-\sigma} \left[1 - \nu_0 (1-\sigma) \left(l_{t+s-1}^s \right)^{1+1/\nu_1} \right]^{\sigma} - 1 \right) \\ + \lambda_{t+s-1}^s \bigg(x_{t+s-1}^s + \left[1 + (1-\tau_{t+s-1}^K) r_{t+s-1} \right] \omega_{t+s-1}^s + tr_{t+s-1} - \omega_{t+s}^{s+1} \\ - (1+\tau_t^c) c_{t+s-1}^s \bigg) \Bigg]\end{aligned}$$

with respect to c^s_{t+s-1}, l^s_{t+s-1}, and ω^{s+1}_{t+s}. The first-order conditions of the s-year-old household are represented by (7.47).[67] In terms of stationary variables, (7.47) can be expressed as follows:

$$\tilde{\lambda}^s_t(1+\tau^c) = (\tilde{c}^s_t)^{-\sigma}\left[1-\nu_0(1-\sigma)(l^s_t)^{1+1/\nu_1}\right]^{\sigma}, \quad s=1,\ldots,J \tag{7.64a}$$

$$\tilde{\lambda}^s_t(1-\tau^w_t-\tau^p_t)\bar{y}^s w_t = \nu_0\sigma\left(1+\frac{1}{\nu_1}\right)(\tilde{c}^s_t)^{1-\sigma}\left[1-\nu_0(1-\sigma)(l^s_t)^{1+1/\nu_1}\right]^{\sigma-1} \cdot (l^s_t)^{1/\nu_1}, \quad s=1,\ldots,R-1 \tag{7.64b}$$

$$(1+\gamma)^{\sigma}\tilde{\lambda}^s_t = \beta\phi_{t,s}\tilde{\lambda}^{s+1}_{t+1}\left[1+(1-\tau^K_{t+1})r_{t+1}\right], \quad s=1,\ldots,J-1. \tag{7.64c}$$

The stationary budget constraint of the government in per capita terms is represented by:

$$\tilde{g}_t + \tilde{tr}_t + (1+r_t)\tilde{b}_t = (1+\gamma)(1+n)\tilde{b}_{t+1} + \tilde{tax}_t + \tilde{beq}_t. \tag{7.65}$$

The resource constraint of the economy in stationary equilibrium is given by:

$$\tilde{y}_t = \tilde{c}_t + \tilde{g}_t + (1+\gamma)(1+n)\tilde{k}_{t+1} - (1-\delta)\tilde{k}_t. \tag{7.66}$$

Steady State Computation

To compute the steady state, we solve a non-linear equations problem in 28 variables consisting of the 14 individual asset levels, $\tilde{\omega}^s = \tilde{k}^s + \tilde{b}^s$, $s = 1,\ldots,15$, (with $\tilde{\omega}^1 \equiv 0$), the 9 individual labor supplies, l^s, $s = 1,\ldots,9$, and the aggregate variables, $\tilde{k}$, $\tilde{L}$, $\tilde{\omega}$, τ^p, and $\tilde{tr}$.

The system of non-linear equations consists of the 23 first-order conditions of the household (the 14 Euler conditions and the 9 first-order conditions of the household with respect to the labor supply) as presented in (7.64b) and (7.64c)

[67] More specifically, Eq. (7.47) represent the first-order conditions of the s-year-old household who was born in period $t-s+1$. Therefore, we are able to present the first-order conditions of all s-year old households, $s = 1,\ldots,J$, who are alive in period t.

(after the substitution of $\tilde{\lambda}_t^s$ from (7.64a)) and the following 5 aggregate equilibrium conditions[68]:

$$(1+n)\tilde{\omega} = \sum_{s=1}^{J} \mu_s \tilde{\omega}^{s+1}, \tag{7.67a}$$

$$\tilde{L} = \sum_{s=1}^{R-1} \mu^s \bar{y}^s l^s, \tag{7.67b}$$

$$\tilde{k} = \tilde{\omega} - \tilde{b}, \tag{7.67c}$$

$$\tilde{tr} = \tilde{tax} + \tilde{beq} + (n + \gamma + n\gamma - r)\tilde{b} - \tilde{g}, \tag{7.67d}$$

$$\tau^p = \frac{\sum_{s=R}^{J} \mu_s \tilde{pen}}{w\tilde{L}}, \tag{7.67e}$$

where μ^s represents the stationary share of the population $N_t(s)/N_t$ and $\tilde{tax} = \tau^w w\tilde{L} + \tau^K r\tilde{\omega} + \tau^c \tilde{c}$ with

$$\tilde{c} = \sum_{s=1}^{J} \mu_s \tilde{c}^s, \quad (1+n)\tilde{b} = \sum_{s=1}^{J} \mu_s \tilde{b}^{s+1}. \tag{7.68}$$

Accidental bequests in steady state (with $\phi_{t,s} = \phi_s$) amount to[69]

$$(1+n) \cdot \tilde{beq} = \sum_{s=1}^{J} \mu_s (1 - \phi_s) \left[1 + (1 - \tau^K) r\right] \tilde{\omega}^{s+1}.$$

To compute $\tilde{b}^s$, we used the condition that all agents hold the two assets $\tilde{k}^s$ and $\tilde{b}^s$ in the same proportion.

All other variables, e.g., individual consumption, factor prices, and aggregate bequests and taxes, can be computed with the help of the 28 endogenous variables.

---

[68]Notice that (7.67a) states the capital market equilibrium. In particular, we use the stationarity condition $\tilde{\omega}' = \tilde{\omega}$, so that (7.67a) is the analog to the capital market equilibrium condition $(1 + n)k_{t+1} = s_t$. In order to make this correspondence even more evident, consider the economy with just two periods, $J = 2$. In this case, the individual savings of the young generation at age 1 at the end of period t amount to $\tilde{\omega}_{t+1}^1$. The savings of the old households at age 2 at the end of period t are equal to zero so that total savings are equal to $N_t(1)\tilde{\omega}_{t+1}^1$. Therefore, in capital market equilibrium

$$\tilde{\Omega}_{t+1} = N_t(1)\tilde{\omega}_{t+1}^1.$$

After division by N_{t+1}, (7.67a) follows noticing that $N_t(1)/N_{t+1} = \mu^1/(1+n)$ and $\tilde{\Omega}_{t+1}/N_{t+1} = \tilde{\omega}' = \tilde{\omega}$ in steady state. The same reasoning applies to the computation of $\tilde{b}$ in (7.68) below.

[69]The set-up of the equation for bequests is motivated in Chap. 6 and, in particular, Appendix 6.1. Notice that accidental bequests also include net interest payments at the end of period $t + 1$.

For example, for the computation of individual consumption levels $\tilde{c}^s$, we can use the individual budget constraint. For the computation of the factor prices w and r, we use the first-order conditions of the firms.

We solve this non-linear equations problem with a modified Newton-Rhapson algorithm as described in Section 11.5.2 and applied to a large-scale OLG model in Section 9.1.2 of Heer and Maußner (2009). The main challenge for the solution is to determine good initial values for the individual and aggregate state variables.

Therefore, we start from a simple nine-period OLG model with exogenous labor in which all cohorts are workers. The exogenous labor supply is set equal to 0.3, and the initial value for the aggregate capital stock is set equal to the corresponding value in the Ramsey model. Next, we add one additional cohort of retirees in each step and use the solution of the model in the previous step as an input for the initial value of the next step. Finally, we introduce endogenous labor into the model. During these initial computations, we compute the solution for the individual optimization problem in an inner loop and update the aggregate capital variables in an outer loop with a dampening iterative scheme as described in Section 3.9 of Judd (1998) that helps to ensure convergence. For the final calibration and the computation of the steady states for different tax rates, we apply the modified Newton-Rhapson algorithm to the complete set of the 28 individual and aggregate equilibrium conditions. The algorithm is implemented in the Gauss programs *Ch7_US_debt.g*.

The transition dynamics are computed as described in Appendix 6.2. Different from this case, however, the final steady state for policies 3 and 4 is not known until we have computed the transition because we do not know the accumulated government debt in 2215 in these cases. We begin with an initial guess that debt remains constant. We update the final steady state using the solution from the transition dynamics in each iteration.

The transition is computed much faster than in the case of the large-scale OLG model in Sect. 6.4.1 and only amounts to approximately 2 min (rather than several hours). The computation is executed by the Gauss program `Ch7_US_transition.g`. In the absence of individual income uncertainty, we can compute the solution of the individual optimization problem with the help of the Newton-Rhapson algorithm and do not have to apply a time-consuming algorithm that is based upon value function iteration. In addition, we use a period length of 5 years rather than 1 year.

Appendix 7.3: Data Sources

In addition to the macroeconomic data presented in Appendices 2.4 and 4.6 and the population data described in Appendix 6.3, we introduce the following variables in our empirical analysis:

- **Debt-GDP ratios** The gross and net debt-ratios presented in Table 7.1 are retrieved from the IMF database (Accessed on 15 December 2017). https://www.imf.org/external/pubs/ft/weo/2017/02/weodata/index.aspx.

The series for gross debt and net debt as percent of GDP are denoted with the identifier 'GGXWDG_NGDP' and 'GGXWDN_NGDP', respectively. The series will be read by the Gauss program *Ch7_data.g* from the Excel files *IMF_grossdebt.xls* and *IMF_netdebt.xls*.
The time series data for the US gross debt-GDP ratio are taken from the series 'GFDEGDQ188S' from the Federal Reserve Bank of St. Louis (Accessed on 15 December 2017).
https://fred.stlouisfed.org/series/GFDEGDQ188S.

- **Budget deficits** Data on general government deficit are retrieved from the OECD, Government at a Glance, 2017 (Accessed on 15 December 2017). The government deficit is defined as the fiscal position of government after accounting for capital expenditures. The series can be downloaded from the OECD at
https://data.oecd.org/gga/general-government-deficit.htm.
- **Government revenue** Data for the 1920s are retrieved from the 'Historical Statistics of the United States 1789–1945' provided by the US Bureau of the Census with the cooperation of the Social Science Research Council (Accessed on 15 December 2017). The document can be downloaded at
https://www.census.gov/library/publications/1949/compendia/hist_stats_1789--1945.html.
- **Government bond yields** The data displayed in Fig. 7.6 are taken from the FRED data base of the Federal Reserve Bank of St. Louis (Accessed on 15 January 2018). For France, for example, the series name of the 10-year nominal government bond yield is 'IRLTLT01FRM156N'.
- **Implicit debt-GDP ratios** European Commission, Eurostat. Calculations: Research Centre for Generational Contracts.
- **Real GDP growth** The Data for Italy in Fig. 7.9 are taken from the World Bank (Accessed on 20 February 2018) and can be downloaded at https://data.worldbank.org/indicator/NY.GDP.DEFL.KD.ZG,
and
https://data.worldbank.org/indicator/NY.GDP.MKTP.KD.ZG.
The data are included in the download files on my homepage that serve as an input into the Gauss computer program *Ch7_data.g*.

Problems

7.1. Recompute the dynamics of the OLG model with government debt in Sect. 7.4.2 under the assumption that the government applies the same lump-sum transfers $\tilde{tr}_t$ to both the young and the old generation such that aggregate transfers are the same in both cases:

$$\tilde{tr}_t = \frac{1+n}{2+n} tr_t.$$

How does this policy affect equilibrium capital stock and output in the numerical example presented in Fig. 7.10?

7.2. Derive the goods market equilibrium (7.42).

7.3. Barro (1979) argues that even in the presence of Ricardian equivalence, debt financing of government expenditures may be optimal if lump-sum taxes are not available and only distortionary taxes such as income taxes can be used. For this reason, reconsider the numerical example of a temporary increase in government consumption from Sect. 4.3, where the transition dynamics are illustrated by the solid green line in Fig. 4.11. Instead, assume now that only distortionary labor income taxation and bond financing of government expenditures are available. The initial and final levels of public debt in periods 0 and 40 are equal to zero. What is the optimal fiscal policy? Is it characterized by tax smoothing?

7.4. Optimal Debt in the Three-Period OLG Model with Income Uncertainty Assume that an agent lives for three periods. Each period length is equal to 20 years. In the first two periods, the agent is working; in the third period, he receives a pension. Each generation has mass $1/3$. We will only consider the steady state.

Lifetime utility is given by

$$U = \mathbb{E}\left\{\sum_{s=1}^{3} \beta^{s-1} u(c^s, 1-l^s)\right\}. \tag{7.69}$$

Instantaneous utility is represented by

$$u(c, 1-l) = u(c, 1-l) = \frac{(c(1-l)^{\iota})^{1-\sigma}}{1-\sigma},$$

with $\iota = 2.0$ and $\sigma = 2.0$. Assume that $\beta = 0.50$. Time is allocated to either work or leisure.

During the first two periods, households work; in the third period, they retire ($l^3 \equiv 0$). Agents are born without assets, $a^1 = 0$. Furthermore, households cannot borrow, $a^s \geq 0$ for $s \in \{2, 3\}$. In addition, the workers pay contributions to the pension system equal to $\tau = 10\%$ of their gross labor income. Gross labor income depends on individual labor productivity, which amounts to e^s at ages $s \in \{1, 2\}$. Therefore, the budget constraint at age $s = 1, 2$ is given by

$$(1-\tau)we^s l^s + (1+r)a^s = c^s + a^{s+1}.$$

In the first period of life, $e^1 = 1.0$. In the second period of life, the individual faces income uncertainty, and insurance markets are missing. Due to health problems, 10% of the agents in the second cohort experience a decline in their labor productivity to $e_l^2 = 0.1$, while the remaining 90% of the cohort maintain constant productivity with $e_h^2 = 1.0$. Therefore, average productivity drops to $\bar{e}^2 = 0.91$ for the two-period-old agent.

During retirement, agents receive pensions *pen* that do not depend on the individual contribution history but are provided lump-sum. The budget of the social security authority is balanced:

$$\frac{pen}{3} = \tau w \frac{1}{3}\left(e^1 l^1 + 0.9 e_h^2 l_h^2 + 0.1 e_l^2 l_l^2\right),$$

where l_j^2, $j \in \{h, l\}$ denotes the labor supply of the agent with high and low productivity at age 2. The budget constraint of the retired worker is given by:

$$pen + (1+r)a^3 = c^3.$$

Production is described by a Cobb-Douglas function:

$$Y = K^\alpha L^{1-\alpha},$$

with

$$L = \frac{e^1 l^1 + 0.9 e_h^2 l_h^2 + 0.1 e_l^2 l_l^2}{3},$$

and $\alpha = 0.36$.

Factors are rewarded by their marginal products:

$$w_t = (1-\alpha)\left(\frac{K_t}{L_t}\right)^\alpha,$$

$$r_t = \alpha\left(\frac{K_t}{L_t}\right)^{\alpha-1} - \delta.$$

The depreciation rate is set equal to $\delta = 0.5$.

1. Consider the equilibrium without government debt in which total assets are equal to the capital stock. Solve the problem with the help of direct computation (solving a system of non-linear equations).
2. Consider the economy with a government sector that is subject to the public budget constraint

$$Tr_t + rB_t = B_{t+1} - B_t.$$

Aggregate transfers are equal to individual transfers, $Tr_t = tr_t$. In equilibrium, aggregate assets Ω_t are equal to $K_t + B_t$. Higher government debt helps to insure the one-period-old agent against negative income shocks because it increases interest rates and, hence, savings. However, it crowds out capital and, hence, lowers per capita consumption. In addition, steady-state transfers decrease with higher debt. Compute the optimal level of steady-state public debt, $B \geq 0$, that maximizes the expected lifetime utility of the households.

3. Compute the optimal level of debt B that maximizes the ex post lifetime utility of the individuals facing a decline in labor productivity (according to the maximin criterion).
4. Is your result for the optimal level of public debt robust with respect to the assumption that labor supply is exogenous, $l^1 = l^2 = 0.3$, where households cannot insure themselves against income uncertainty by adjusting their labor supply?

References

Aguiar, M., & Amador, M. (2016). *Take the short route: Equilibrium default and debt maturity* (NBER working paper, 23552).

Aguiar, M., Chatterjee, S., Cole, H., & Stangebye, Z. R. (2016). Quantitative models of sovereign debt crises. In J. B. Taylor & H. Uhlig (Eds.), *Handbook of the macroeconomics* (pp. 1697–1755). Amsterdam: Elsevier.

Aiyagari, R. S., & McGrattan, E. R. (1998). The optimum quantity of debt. *Journal of Monetary Economics, 42*, 447–469.

Alfaro, L., & Kanczuk, F. (2007). Sovereign debt as a contingent claim: A quantitative approach. *Journal of International Economics, 65*, 890–905.

Arellano, C. (2008). Default risk and income fluctuations in emerging economies. *American Economic Review, 89*(3), 690–712.

Asonuma, T. (2016). *Sovereign defaults, external debt, and real exchange rate dynamics* (IMF working papers, 16/37).

Azariadis, C. (1993). *Intertemporal macroeconomics*. Oxford: Blackwell.

Barro, R. J. (1974). Are government bonds net wealth? *The Journal of Political Economy, 82*(6), 1095–1117.

Barro, R. J. (1979). On the determination of public debt. *The Journal of Political Economy, 87*(5), 940–971.

Braun, R. A., & Joines, D. H. (2015). The implications of a graying Japan for government policy. *Journal of Economic Dynamics and Control, 57*, 1–23.

Cuadra, G., & Sapriza, H. (2008). Sovereign defaults, interest rates and political uncertainty in emerging economies. *Journal of International Economics, 76*, 77–88.

D'Erasmo, P. (2012). *Government reputation and debt repayment in emerging economies*. Mimeo.

D'Erasmo, P., & Mendoza, E. G. (2016). Distributional incentives in an equilibrium model of domestic sovereign default. *Journal of the European Economic Association, 14*(1), 7–44.

D'Erasmo, P., Mendoza, E. G., & Zhang, J. (2016). What is a sustainable public debt? In J. B. Taylor & H. Uhlig (Eds.), *Handbook of the macroeconomics* (pp. 2493–2597). Amsterdam: Elsevier.

Diamond, P. (1965). National debt in a neoclassical growth model. *American Economic Review, 55*, 1126–1150.

Durdu, C. B., Nunes, R., & Sapriza, H. (2013). News and default risk in small open economies. *Journal of International Economics, 91*(1), 1–17.

Eaton, J., & Gersovitz, M. (1981). Debt with potential repudiation: Theoretical and empirical analysis. *Review of Economic Studies, 48*(2), 289–309.

Gordon, G., & Guerron-Quintana, P. A. (2018). Dynamics of investment, debt, and default. *Review of Economic Dynamics, 28*, 71–95.

Gosh, A. R., Kim, J. I., Mendoza, E. G., Ostry, J. D., & Qureshi, M. S. (2013). Fiscal fatigue, fiscal space and debt sustainability in advanced economies. *Economic Journal, 123*, 4–30.

Hagist, C., Moog, S., Raffelhüschen, B., & Vatter, J. (2009). Public debt and demography – An international comparison using generational accounting. *CESifo Dice Report, 4*, 29–36.

Hansen, G. (1993). The cyclical and secular behavior of the labor input: Comparing efficiency units and hours worked. *Journal of Applied Econometrics, 8*, 71–80.

Heer, B., & Maußner, A. (2009). *Dynamic general equilibrium modeling: Computational methods and applications* (2nd ed.). Heidelberg: Springer.

Heer, B., & Scharrer, C. (2018). The age-specific burdens of short-run fluctuations in government spending. *Journal of Economic Dynamics and Control, 90*, 45–75.

Heer, B., Polito, V., & Wickens, M. R. (2017). *Population aging, social security and fiscal limits* (CEPR Discussion Paper Series, DP11978).

Herndon, T., Ash, M., & Pollin, R. (2014). Does high public debt consistently stifle economic growth? A critique of Reinhart and Rogoff. *Cambridge Journal of Economics, 38*, 257–279.

İmrohoroğlu, S., Kitao, S., & Yamada, T. (2016). Achieving fiscal balance in Japan. *International Economic Review, 57*, 117–154.

Jarvis, C. (1999). *The rise and fall of Albania's pyramid schemes* (IMF working paper, WP 99/98).

Judd, K. L. (1998). *Numerical methods in economics*. Cambridge: MIT Press.

King, R. G., & Plosser, C. I. (1985). Money, deficits, and inflation. *Carnegie-Rochester Conference Series on Public Policy, 22*, 147–196.

Mendoza, E. G., & Yue, V. Z. (2012). A general equilibrium model of sovereign default and business cycles. *Quarterly Journal of Economics, 127*(2), 889–946.

Mendoza, E. G., Razin, A., & Tesar, L. L. (1994). Effective tax rates in macroeconomics: Cross country estimates of tax rates on factor incomes and consumption. *Journal of Monetary Economics, 34*, 297–323.

OECD. (2017). *Pensions at a Glance 2017: OECD and G20 indicators*. Paris: OECD Publishing.

Polito, V., & Wickens, M. R. (2015). Sovereign credit ratings in the European Union: A model-based fiscal analysis. *European Economic Review, 78*, 220–247.

Reinhart, C. M., & Rogoff, K. S. (2009). *This time is different*. Princeton: Princeton University Press.

Reinhart, C. M., & Rogoff, K. S. (2010). Growth in a time of debt. *American Economic Review, 100*, 573–78.

Reinhart, C. M., & Rogoff, K. S. (2011). The forgotten history of domestic debt. *Economic Journal, 121*, 319–350.

Reinhart, C. M., & Rogoff, K. S. (2012). Public debt overhangs: Advanced-economy episodes since 1800. *Journal of Economic Perspectives, 26*, 69–86.

Ricardo, D. (1817). *On the principles of political economy and taxation*. Cambridge: Cambridge University Press.

Schmitt-Grohé, S., & Uribe, M. (2017). *Open economy macroeconomics*. Princeton: Princeton University Press.

Stähler, N. (2013). Recent developments in quantitative models of sovereign default. *Journal of Economic Surveys, 27*(4), 605–633.

Trabandt, M., & Uhlig, H. (2011). The Laffer curve revisited. *Journal of Monetary Economics, 58*, 305–327.

UN (2015). *World population prospects: The 2015 revision, methodology of the United Nations population estimates and projections* (ESA/P/WP. 242).

Walsh, C. E. (2010). *Monetary theory and policy*. Cambridge: MIT Press.

Author Index

B. Heer, *Public Economics*, Springer Texts in Business and Economics,
https://doi.org/10.1007/978-3-030-00989-2

Subject Index

B. Heer, *Public Economics*, Springer Texts in Business and Economics,
https://doi.org/10.1007/978-3-030-00989-2

GPSR Compliance
The European Union's (EU) General Product Safety Regulation (GPSR) is a set of rules that requires consumer products to be safe and our obligations to ensure this.

If you have any concerns about our products, you can contact us on

ProductSafety@springernature.com

In case Publisher is established outside the EU, the EU authorized representative is:

Springer Nature Customer Service Center GmbH
Europaplatz 3
69115 Heidelberg, Germany

www.ingramcontent.com/pod-product-compliance
Ingram Content Group UK Ltd.
Pitfield, Milton Keynes, MK11 3LW, UK
UKHW021410091026
15624UKWH00009B/182
9783030009878